Dynamics of Crystal Surfaces and Interfaces

FUNDAMENTAL MATERIALS RESEARCH

Series Editor: M. F. Thorpe, *Michigan State University*
East Lansing, Michigan

ACCESS IN NANOPOROUS MATERIALS
Edited by Thomas J. Pinnavaia and M. F. Thorpe

DYNAMICS OF CRYSTAL SURFACES AND INTERFACES
Edited by P. M. Duxbury and T. J. Pence

ELECTRONIC PROPERTIES OF SOLIDS USING CLUSTER METHODS
Edited by T. A. Kaplan and S. D. Mahanti

Dynamics of Crystal Surfaces and Interfaces

Edited by

P. M. Duxbury and T. J. Pence
Michigan State University
East Lansing, Michigan

Plenum Press • New York and London

Library of Congress Cataloging in Publication Data

Dynamics of crystal surfaces and interfaces / edited by P. M. Duxbury and T. J. Pence.
p. cm.—(Fundamental materials research)
Proceedings of a workshop held in August 1996.
Includes bibliographical references and index.
ISBN 0-306-45619-2
1. Crystalline interfaces—Congresses. 2. Crystals—Surfaces—Congresses. I. Duxbury, P. M. II. Pence, T. J. III. Series.
QC173.458.C78D96 1997
530.4'17—dc21 97-15447
CIP

Proceedings of a workshop on Dynamics of Crystal Surfaces and Interfaces, held August 4–8, 1996, at the Park Place Hotel, Traverse City, Michigan

ISBN 0-306-45619-2

A Division of Plenum Publishing Corporation
233 Spring Street, New York, N. Y. 10013

http://www.plenum.com

10 9 8 7 6 5 4 3 2 1

Printed in the United States of America

SERIES PREFACE

This series of books, which is published at the rate of about one per year, addresses fundamental problems in materials science. The contents cover a broad range of topics from small clusters of atoms to engineering materials and involve chemistry, physics, and engineering, with length scales ranging from Ångstroms up to millimeters. The emphasis is on basic science rather than on applications. Each book focuses on a single area of current interest and brings together leading experts to give an up-to-date discussion of their work and the work of others. Each article contains enough references that the interested reader can access the relevant literature. Thanks are given to the Center for Fundamental Materials Research at Michigan State University for supporting this series.

M.F. Thorpe, Series Editor
E-mail: thorpe@pa.msu.edu

PREFACE

During the period 4th-8th August 1996, a conference with the same title as this book was held in Traverse City, Michigan. That conference was organized as a sequel to an interesting and successful WEM workshop in a similar area run by Profs. Hans Bonzel and Bill Mullins in May 1995. This book contains papers presented at the Traverse City conference. The book focuses on: atomic processes, step structure and dynamics; and their effect on surface and interface structures and on the relaxation kinetics of larger length-scale nonequilibrium morphologies. A second theme is nucleation and growth in the presence of strain. Although the main application of these ideas has so far been to surface processes, their extension to grain boundaries, dislocations, and other bulk defects is natural and likely to be important in the near future. Some work along these directions is presented in this proceedings. Needless to say, these subjects have enormous importance in a variety of surface and interface materials problems, both scientific and technological. As can be seen from the participant list, the conference drew researchers from several research communities. To service this audience, many of the papers contain brief overviews as well as new results. We hope that non-experts can find an introduction to the literature by starting with this book.

The conference was supported by the Condensed Matter Theory Conference Fund at Michigan State University, administered by Prof. Mike Thorpe. We gratefully acknowledge that support and the financial support of the Center for Fundamental Materials Research at Michigan State University, who underwrote the publication of the proceedings. We also thank the staff at Michigan State University; especially Lorie Newman, Jeanette Dubendorf, and Janet King for their assistance; and the staff at the Park Place Hotel in Traverse City for use of their wonderful venue and for their cordial efficiency.

The cover illustration is Figure 4 from the paper by H.P. Bonzel and S. Surnev in these proceedings. We thank them for allowing us to use their intriguing STM figure.

Phillip M. Duxbury
Thomas J. Pence

CONTENTS

Dynamics of Crystal Surfaces and Interfaces

FROM ATOMIC DIFFUSION TO STEP DYNAMICS

Branislav Blagojevic and Phillip M. Duxbury

Department of Physics/Astronomy and
Center for Fundamental Materials Research,
Michigan State University,
East Lansing, MI 48824-1116

1. INTRODUCTION

The equilibrium structure [1-6] and dynamics [7-17] of steps and step trains are now routinely probed and modelled with atomic precision. The dynamics of steps is often mediated by atomic diffusion, either along the step edge or across the terraces which lie between steps. Step motion in turn can lead to the relaxation of larger-scale non-equilibrium surface structures, such as gratings, STM tips, quantum dots and interconnect structures(many of the articles in this proceeding describe step mediated relaxation of non-equilibrium surface structures). Traditionally, the relaxation of these structures is treated using linear relaxation theory [18-20]. However as realised even in those early seminal papers, the linear theory usually needs significant modification below roughening [21-30] where steps are well defined. Actually, when the steps are well defined, their (capillary) fluctuations are usually treated linearly, and the non-linear evolution is though the center-of-mass motion of the steps. A nice review of the capillary theory for the dynamics of step fluctuations, including comparison with experiment, is given by Einstein and Khare in this proceedings. Here we present a complimentary formalism based on atomic diffusion, and we show explicitly the atomic diffusion processes which lead to the various limiting behaviors treated by previous workers [8-16]. Our method naturally relates the continuum sticking coefficients to the lattice ones, which are in turn related to step-edge energy parameters. We also show when "non-universal" behavior can occur, and provide a formalism which describes the crossover between the various limits discussed by Pimpinelli et al. [13].

Usually, experimentalists quantify step fluctuations by averaging the data to find the correlation function $G(t) = 0.5 < (h(x,t) - h(x,0))^2 >$, where $h(x,t)$ specifies the step position at time t and the average is over many sample points, x. $G(t)$ measures how far a position on a step wanders with time. If that position were completely free to wander, it would obey a diffusive law $G(t) \sim t$. However, its motion is restricted by the fact that it is connected to the other parts of the step. For that reason $G(t)$ is sub-diffusive. The detailed law which $G(t)$ obeys is dependent on the atomic processes which mediate step motion. For example, if the step edge is able to freely exchange atoms with a heat bath (non-conserved dynamics), then $G(t) \sim t^{1/2}$ this behavior has been observed on Au(110) [10] and on Si(111) [12] surfaces. In contrast, if step-edge diffusion dominates (conserved dynamics), $G(t) \sim t^{1/4}$ and this result has been observed on $Cu(1,1,19)$ and $Cu(1,1,79)$ surfaces [11]. It should be noted however, that in later work on the same surface Giesen at al.[15] found that the time exponent *increases* as the temperature *decreases*, probably due to the small values of roughness found in those experiments. At early times, before $G(t)$ is larger than an atomic spacing, the motion is closer to "free" diffusion (see Sections 2 and 3). Although the asymptotic time exponents are interesting, perhaps the main reason for discriminating between them is to extract the correct prefactors from the expressions for $G(t)$. These prefactors are related to step-edge kinetic barriers, and the study of $G(t)$ provides an important technique for estimating these critical parameters.

Somewhat surprisingly, the simplest solid-on-solid model (with step-edge sticking coefficient one, and step-edge (Schwoebel) barrier zero) of the dynamics of an isolated step leads to the behavior $G(t) \sim t^{1/3}$ [16], and a "non-local" continuum Langevin equation was developed to model this behavior. Bartelt et al. [16] also suggested that if the step-edge sticking coefficient is small, the behavior $G(t) \sim t^{1/2}$ is recovered. Note however that at long enough times (and in the thermodynamic limit) one always has $t^{1/3}$ behavior for any finite sticking coefficients (see later). However, in the experiments, the $t^{1/2}$ law arises due to the presence of other steps [13]. (they actually calculated a relaxation time $\tau(L)$ for a bump or mode of linear size L, this is related to $G(t)$ - see below).

The paper is arranged as follows. In Section 2, we discuss the Langevin(rate) theory for the dynamics of an isolated step. The different atomic mechanisms are mathematically described by different diffusion kernels. For example, this allows us to smoothly interpolate between the step-edge diffusion limit and the attachment-detachment limit. We show that in the perfect sticking limit, the diffusion kernel, P(l), for an isolated step decays with distance l as $1/l^2$. In Section 3, we extend the method to the study of a step train, but ignoring step-step interaction terms in the step free energy (This will be treated elsewhere[31]). There are then diffusion kernels between steps as well as a diffusion kernel to the same step. Section 4 contains a conclusion. There are three Appendices. Appendix A summarizes the calculation of $G(t)$ for a linear Langevin equation. Appendix B relates continuum sticking coefficients to lattice sticking coefficients and hence to step-edge energy barriers. In Appendix C, we calculate the diffusion kernels used in the text.

2 AN ISOLATED STEP

Often, the energy of a step configuration is modelled by,

$$E = J \sum_{i=1}^{L} (h_i - h_{i-1})^p \tag{1}$$

where h_i are in discrete units of $a_\perp$, the lattice constant perpendicular to the step edge. $p = 1$ is the solid-on-solid (SOS) model, $p = 2$ is the discrete Gaussian model, while $p = \infty$ is the restricted solid-on-solid model (RSOS).

In analytic work it is more convenient to use a model which takes on continuum values of the variable h_i. Moreover, it can be shown that for an isolated step, the free energy derived from the energy (1) is often well approximated(at small slopes) by,

$$F \sim \frac{\tilde{\Sigma}}{2} \int_0^L (\nabla h)^2 dx \tag{2}$$

where the "stiffness" $\tilde{\Sigma} = \Sigma(0) + \Sigma''(0)$ depends on the value of p in (1). $\Sigma(\theta)$ is the angle-dependent interface free energy derived from (1). $h(x)$ is a continuous height variable. Three cases in which the stiffness can be calculated exactly[32-34] are,

$$\tilde{\Sigma} = 2k_BT sinh^2(J/2k_BT) \quad for\ the\ SOS\ model, \tag{3}$$

$$\tilde{\Sigma} = k_BT\frac{1+2e^{-J/k_BT}}{2e^{-J/k_BT}} \quad for\ the\ RSOS\ model, \tag{4}$$

$$\tilde{\Sigma} = k_BT sinh(J/k_BT) \quad for\ the\ Ising\ model. \tag{5}$$

In the Ising case, J used in Eq. (5) is *twice* the nearest neighour coupling strength. J is the kink energy, and in the low temperature and large L limit, Eqs. (3-5) all reduce to

$$\tilde{\Sigma} = \frac{k_BT}{2}e^{J/k_BT} \quad for \quad k_BT << J \tag{6}$$

From the Hamiltonian (2), it is straightforward to find the equilibrium width, $w_\infty^2 =< h^2(t) > - < h(t) >^2$ of the step edge, (we take $< h(t) >= 0$ henceforth, as we are interested in fluctuations about the mean),

$$w_\infty = (\frac{k_BTL}{6\tilde{\Sigma}})^{1/2} = (\frac{L}{6\tilde{s}})^{1/2}, \quad for\ pinned\ ends \tag{7}$$

where $\tilde{s} = \tilde{\Sigma}/k_BT$ (there was a factor of 2 error in some of the earlier work in this area - see paragraph after Eq. (1) in ref. [13] for a discussion). Note that Eq. (7) is in dimensionless units, and should be multiplied by $a_\perp$ to return to an expression with units. In fact, we shall work henceforth with dimensionless L and $\tilde{s}$, and make the substitutions

$$L \rightarrow La_{||}, \quad and \quad \tilde{s} \rightarrow \frac{a_{||}}{a_\perp^2}\tilde{s}$$

in the final expressions for $G(t)$, where $a_{||}$ is the lattice constant parallel to the step edge. The "small-slope approximation" corresponds to $w/L << 1$. We assume that this holds. The result (7) is for periodic boundary conditions and for a string pinned, at $h = 0$, at both ends. If the ends are free, w_∞ is larger by a factor of $\sqrt{2}$. In experiment, the step fluctuations are measured over a portion of length L, and the ends are effectively pinned by the unmeasured portions of the string.

The chemical potential, $\mu(x)$ (from Eq. (2)) is,

$$\mu(x) = -\frac{\delta F}{\delta h} = -\tilde{\Sigma}\nabla^2 h \tag{8}$$

Within non-equilibrium thermodynamics, the driving force for relaxation is provided by deviations in the local chemical potential from it's equilibrium value. The rate at which such deviations relax is determined by the dominant *kinetics* in the physical system of interest. In addition, the thermal noise in the system randomly generates fluctuations. We thus describe the dynamics of a step edge by the equation,

$$\frac{\partial h(x,t)}{\partial t} = \frac{\Gamma_h}{k_BT}J(x,t) + \eta(x,t). \tag{9}$$

where:

$$J(x,t) = \int_0^{L/2} P(l)\{\mu(x+l,t) - 2\mu(x,t) + \mu(x-l,t)\}dl \tag{10}$$

L is the length of the step along the x-direction, $a_{||}$ is the lattice constant along the step and Γ_h is the rate; (following [8])

$$\Gamma_h = a_\perp^2 a_{||}/\tau_h,$$

where τ_h is the time between detachment events from a site on the step edge, and is related to the step-edge energies(see Appendix B). $\eta(x,t)$ is the noise term, which must reproduce the equilibrium fluctuations of the step at long times(see Appendix A). Eq. (9) states that the relaxation rate of a chemical potential difference between two step positions (separated by distance l) is a product of the chemical potential difference between those sites *and* the flux of atoms which is exhanged between the two positions.

A Fourier transform of (9) and (10) yields,

$$\frac{\partial h_q(t)}{\partial t} = -2\tilde{s}\Gamma_h q^2 I(q) h_q(t) + \eta_q(t) \tag{11}$$

where

$$I(q) = \int_0^{L/2} P(l)(1 - \cos ql)dl, \tag{12}$$

and,

$$h(x,t) = \sum_{n=1}^{L} h_q(t) sin(qx), \quad with \quad q = \pi n/L$$

Note that if the noise is removed from the right hand side of Eq. (11), we have

$$h_q(t) = h_q(0)e^{-t/\tau(q)} \quad with \quad no \quad noise \tag{13}$$

with the relaxation time, $\tau(q)$, of a mode of wavelength $\lambda = 2\pi/q$ given by,

$$\tau^{-1}(q) = \omega(q) = 2\tilde{s}\Gamma_h q^2 I(q). \tag{14}$$

A general expression for $\tau^{-1}(q)$ was found by Bales and Zangwill [14](corrected by Pimpinelli et al.[13]). We rederive their expressions from a different physical perspective in Section 3 and Appendix C.

In the presence of noise, the solution to the linear Langevin equation (11) is (see Appendix A, and also Eq. (3-5) of the article by Einstein in this proceedings),

$$G_q(t) = < h_q^2(t) - h_q^2(0) > = < h_q^2(t) > = \frac{2}{L\tilde{s}q^2}(1 - e^{-2\tilde{s}\Gamma_h q^2 I(q)t}). \sim \frac{4}{L}\Gamma_h I(q)t \tag{15}$$

where the last expression on the RHS is valid for $t << \tau_{eq}$, with τ_{eq} defined below. An inverse transform of $G_q(t)$ then gives (see Appendix A),

$$G(t) = \sum_{n=1}^{L} \frac{1}{L\tilde{s}q^2}(1 - e^{-2\tilde{s}\Gamma_h q^2 I(q)t}) \tag{16}$$

We also have $w^2(t) = G(2t)$, where $w^2(t) =< h^2(x,t) >$ is the rate at which an *initially flat* step develops its thermal fluctuations. $w(t)$ is a *non-equilibrium* correlation function, while $G(t)$ is a time dependent *equilibrium* correlation function. Within the linear Langevin Eq. (11) they are essentially equivalent. At long times, the exponential part in Eq. (16) is dominated by the $n = 1$ term in the sum. We thus define the equilibration time τ_{eq}, by the time it takes for the $(n = 1)$ exponential term to decay to e^{-2} so that,

$$\tau_{eq} = \frac{1}{\tilde{s}\Gamma_h(\frac{\pi}{L})^2 I(\frac{\pi}{L})}. \tag{17}$$

This gives a simple and accurate estimate of the equilibration time. Thus the "natural" scaled variables are $G(t)/w_\infty^2$ and t/τ_{eq}.

In several cases, we find that

$$q^2 I(q) \sim A q^\gamma, \tag{18}$$

in which case, the continuum limit of Eq. (16) yields(see Eq. (83) of Appendix A),

$$w(t) = G(2t) \sim \frac{1}{\pi \tilde{s}}(4\Gamma_h \tilde{s} A t)^{\frac{1}{\gamma}} \Gamma(1 - \frac{1}{\gamma}), \tag{19}$$

where Γ is the Gamma integral as defined for example in [31].

2.1 Step-edge diffusion

Step-edge diffusion is described by (see Fig. 1a),

$$P(l) = \frac{1}{2}(\delta(l - a_{\|}) + \delta(l + a_{\|})), \tag{20}$$

which restricts motion to the two adjacent neighbours along the step edge.

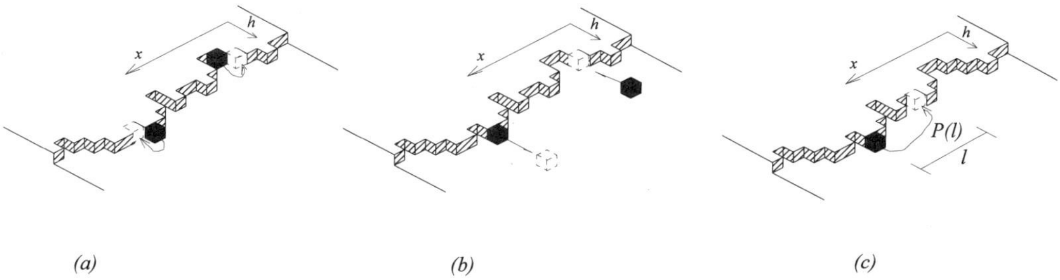

Figure 1. Three limiting mechanisms for atomic processes which mediate step fluctuations. **a)** Step-edge diffusion; **b)** Evaporation-recondensation; **c)** Terrace diffusion with diffusion kernel $P(l)$. By appropriate choice of $P(l)$, this case can reduce to cases a) and b) (see text).

Substituting this into Eq. (12) and expanding for small q gives,

$$I(q) = \frac{1}{2}(1 - cos(qa_{\|})) \approx \frac{q^2 a_{\|}^2}{4}, \tag{21}$$

Substitution of this into Eq. (15) gives $G_q(t) \sim q^2$ for this case, and using $\gamma = 4$ and, $A = \frac{a_{\|}^2}{4}$ in Eq. (19) yields,

$$w^2(t) = G(2t) = \frac{\Gamma(3/4)}{\pi}(\frac{a_{\|}^2 \Gamma_{se} t}{\tilde{s}^3})^{1/4} \to 0.39 a_{\perp}^2 e^{\frac{-E_{se}}{4k_B T}} \frac{1}{\tilde{s}^{3/4}}(\nu_{se} t)^{1/4}. \tag{22}$$

($\Gamma(3/4) = 1.22541..$). In that last expression we have used the substitutions after Eqs. (7) and (10). E_{se} is the activation energy for an atom (or other dominant diffusing species, e.g. a dimer) to jump to a nearest neighbour along the step edge, and ν_{se} is the attempt frequency associated with that process. These quantities can also be found from an analysis of $G_q(t)$ from Eq. (15) with Eq. (21). Note that Eq. (22) disagrees with ref. [8] by a factor of $2^{-1/4}$ (In their Eq. (22), $a_{\|}^2$ is absorbed into Γ_{he}), and is case F of [13]. The origin of the discrepancy with ref. [8] is that Eq. (10) with (20) leads to $0.5\partial^4 h/\partial x^4$ which is smaller by a factor of 2 when compared to the continuum model assumed in ref. [8].

A behavior similar to Eq. (22) occurs for any short range diffusion kernel. For example, taking

$$P(l) = \frac{1}{2\xi} e^{-\frac{l}{\xi}} \quad with \quad a_{\|} << \xi << L \tag{23}$$

yields

$$G(t) = const.(\frac{\xi^2\Gamma_h t}{\tilde{s}^3})^{1/4}. \tag{24}$$

A little later in this section, we also show that if $P(l) = B/l^\alpha$, with $\alpha \geq 3$, then a similar behavior pertains.

2.2 Attachment-detachment model

The attachment-detachment model (see Fig. 1b) can be described by using the probability function Eq. (23) but now assuming that the correlation length, $\xi \gg L$. Physically this means that infinite range conserved-order-parameter dynamics is the same as non-conserved order parameter dynamics. In that case the normalized probability $P(l)$ is,

$$P(l) = const. = \frac{1}{L}. \tag{25}$$

The integral (12) then yields, $I(q) = 1/2$. Then $A = 1/2$ and $\gamma = 2$, and $\Gamma(1/2) = \sqrt{\pi}$ in Eq. (19) yields,

$$w^2(t) = G(2t) = (\frac{2\Gamma_{ec}t}{\pi\tilde{s}})^{1/2} \rightarrow 0.798 a_\perp^2 e^{\frac{-E_{ec}}{2k_BT}} \frac{1}{\tilde{s}^{1/2}}(\nu_{ec}t)^{1/2}. \tag{26}$$

In obtaining the last expression we have used the substitutions after Eqs. (7) and (10). E_{ec} is the activation energy for detachment from a step edge into the "vapor", and ν_{ec} is the attempt frequency for that process. (see also [8] and case A of [13]). As can be seen from Eq. (15), in this case, $G_q(t)$ is independent of q at short times.

2.3 Terrace diffusion

Over broad temperature regimes, terrace diffusion is the dominant kinetic mechanism. In this process, an atom detaches onto the lower terrace with probability p_L, and onto the upper terrace with probability p_U. It then diffuses on the terrace and reattaches to the step edge(see Fig. 1c). We define the "sticking coefficient" on approach to the step edge from the lower terrace to be α_L, while the sticking coefficient on approach to the step edge from the upper terrace is α_U (here we use continuum sticking coefficients, and relate them to lattice parameters in Appendix B). In Appendix C we calculate $P(l)$ for terrace diffusion (see Eq. (106) for $P_1(l)$ with $d \rightarrow \infty$), from which we find,

$$P(l) = p_U \frac{\alpha_U}{\pi} \int_0^\infty \frac{cos(kl)e^{-ka_\perp}dk}{k+\alpha_U} + p_L \frac{\alpha_L}{\pi} \int_0^\infty \frac{cos(kl)e^{-ka_\perp}dk}{k+\alpha_L} \tag{27}$$

Using this expression in Eq. (12) leads to,

$$I(q) = 1 - \alpha_U p_U \frac{e^{-|q|a_\perp}}{\alpha_U + |q|} - \alpha_L p_L \frac{e^{-|q|a_\perp}}{\alpha_L + |q|}. \tag{28}$$

In the limit of perfect sticking $\alpha_U = \alpha_L \rightarrow \infty$, we find the simple result

$$I(q) = |q|a_\perp \tag{29}$$

and hence from Eq. (19), we have

$$w^2(t) = G(2t) = \frac{\Gamma(2/3)}{\pi}(4\frac{a_\perp \Gamma_h t}{\tilde{s}^2})^{1/3} \rightarrow 0.862 a_\perp^2 (\frac{a_\perp}{a_{||}})^{1/3} e^{\frac{-E_0}{3k_BT}} \frac{1}{\tilde{s}^{2/3}}(\nu_t t)^{1/3} \tag{30}$$

where $\Gamma(2/3) = 1.3541$, and we have used the substitutions after Eqs. (7) and (10) along with the expression (93) of Appendix B (in the limit $E = E_B = 0$). E_0 is the activation energy for detachment of an atom (or other dominant diffusing species) onto a terrace, and ν_t is the attempt frequency for that process. Eq. (30) was derived using a different perspective by Bartelt et al.[16], and it corresponds to case B of [13]. It has the same qualitative behavior as the case $f(q) = 4D_{ter}c_{ter}\Omega^2|q|$ in the article by Einstein and Khare in these proceedings. The physical differences between their perspective and ours will be discussed in the conclusion (Section 4).

It is evident from Eq. (28), that for q sufficiently small, the behavior of Eq. (29) is always recovered. That is for long enough times, Eq. (30) should hold. However, if the sticking coefficients are small, there is a significant regime in which a different behavior holds. This is evident from Eq. (28) if one defines $\alpha_{av} = \alpha_U p_U + \alpha_L p_L$ and $q >> \alpha_L$, then

$$I(q) \sim 1 - \alpha_{av}/|q| \quad \alpha_{av} << |q| \tag{31}$$

The leading term in this expression is a constant and leads to a behavior equivalent to that found for the "attachment-detachment" mechanism (see e.g. Eq. (26)). This has been suggested before[16] and is illustrated in Fig. 2.

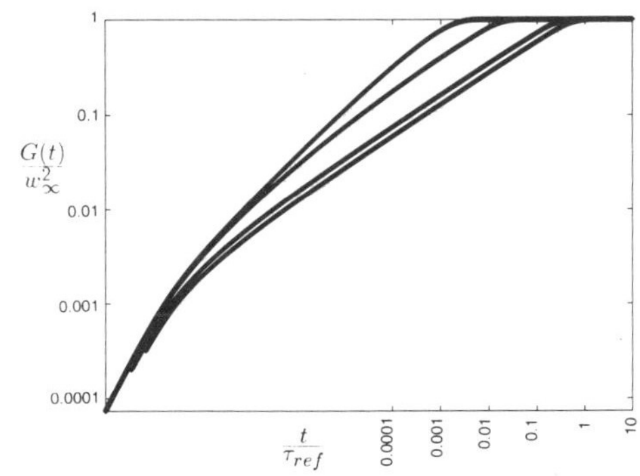

Figure 2. The correlation function $G(t)$, which measures the rate at which a step position "anomalously diffuses" away from a starting position. This diffusion is limited to the equilibrium width $G(t \to \infty) = w_\infty^2$. The time, t is scaled by τ_{ref}, which is the equilibration time (τ_{eq} defined in Eq. (17)) for the case $\alpha_U = \alpha_L = \infty$ (perfect sticking). The curves are for (from the top): $\alpha_L = \alpha_U = 0.001$; $\alpha_L = \alpha_U = 0.03$; $\alpha_L = \alpha_U = 1.0$; $\alpha_L = \alpha_U = 1000.0$. In all cases $L = 1000$, and the other parameters are absorbed into $w_\infty = 17.5$ and τ_{ref}. In particular Γ_h is the same for all curves. In general p_U and p_L are related to α_U and α_L, as seen in Appendix B, Eqs. (88-89). In the particular cases presented here, $p_U = p_L = 1/2$.

To find the data for this figure, we have used the full expression for $I(q)$ in Eq. (28) and have numerically performed the sum in Eq. (16). The data are scaled by the equilibrium width w_∞ (from Eq. (7)) , and the equilibrium time in the perfect sticking limit (defined in Eq. (17) with Eq. (29) for perfect sticking). The linear behavior $G(t) \sim t$ holds for $G(t) < 1$, when the "free diffusion" of each point on the step edge is essentially unrestrained by its neighbours. The $t^{1/3}$ behavior holds in the regime $1/(6\tilde{s}\alpha_{max}) << G(t) << w_\infty^2$, where α_{max} is the *largest* of α_U and α_L. It is at first surprising that the small sticking coefficient case relaxes more rapidly than the large sticking coefficient case. However small sticking coefficients lead to long range $P(l)$ and hence faster relaxation on long length scales.

In experiment, the step-edge, terrace diffusion and evaporation-recondensation mechanisms all occur, albeit with different activation energies. It is straightforward to combine the diffusion kernels (20), (25) and (27), with arbitrary weights and hence find,

$$I(q) = a\frac{q^2 a_{||}^2}{4} + \frac{b}{2} + c(1 - \alpha_L p_L \frac{e^{-|q|a_\perp}}{\alpha_L + |q|} - \alpha_U p_U \frac{e^{-|q|a_\perp}}{\alpha_U + |q|}) \qquad (32)$$

with $a + b + c = 1$ to preserve the normalisation of the kernel $P(l)$. The weights a, b, c are expected to have an activated form, dependent on the various step-edge and kink energies, so the crossover between the various mechanisms as a function of temperature is quite sharp. At present atomistic simulations are not sufficiently precise to predict in detail the relative importance of the various terms in Eq. (32).

2.4 Anomalous diffusion

It is interesting to note that there are cases in which the step dynamics is non-universal. We illustrate this point through use of the probability distribution

$$P(l) = \frac{B}{l^\alpha} \quad for \quad l \geq a_{||}$$

where $\alpha = 0$ recovers the attachment-detachment limit, while $\alpha > 3$ (see below) corresponds to "short-range" hopping and recovers the step-edge diffusion behavior. $B^{-1} = 2((L/2)^{1-\alpha} - a_{||}^{1-\alpha})/(1-\alpha)$ normalises the probability on the interval $(a_{||}, \infty)$. By evaluating the integral Eq. (12), we find that:
For $0 \leq \alpha < 1.0$

$$I(q) \sim \frac{1}{2} + O((\frac{2q}{L})^{2-2\alpha});$$

For $1 < \alpha < 3$,

$$I(q) \approx \frac{(1-\alpha)}{2}\Gamma(1-\alpha)cos(\frac{(1-\alpha)\pi}{2})(qa_{||})^{\alpha-1} + \frac{1-\alpha}{4(3-\alpha)}(qa_{||})^2$$

(Note that for integer α, $I(q)$ is finite despite the divergence in $\Gamma(1-\alpha)$);
While for $\alpha > 3$

$$I(q) \sim \frac{\alpha-1}{4(\alpha-3)}(qa_{||})^2 - \frac{\alpha-1}{48(\alpha-5)}(qa_{||})^4.$$

Using Eq. (19), we then find

$$G(t) \sim t^{1/2} \quad for \quad 0 < \alpha < 1,$$

$$G(t) \sim t^{\frac{1}{\alpha+1}} \quad for \quad 1 \leq \alpha \leq 3,$$

and

$$G(t) \sim t^{1/4} \quad for \quad \alpha \geq 3.$$

Thus, there is a continuous variation in the dynamical exponent for $1 < \alpha < 3$, while the attachment-detachment universality class holds for $\alpha < 1$ and the step-edge universality class holds for $\alpha \geq 3$.

3 AN N-STEP BUNCH

Direct step-step interaction terms in the step energy ("direct" interactions are: entropic repulsion, strain terms, electronic structure effects etc.) do influence the step fluctuations, and they also drive the spreading of step trains, wires and bumps. Nevertheless, it is instructive to first ignore these "direct" step-step repulsion, as is done in this section and in much of the prior work on step fluctuations. Step-step repulsion effects can be straightforwardly included in the treatment given below and this extension will be discussed elsewhere [35].

Ignoring direct interactions, neighbouring steps do not influence each other if the dynamics is dominated by evaporation-recondensation or by step-edge diffusion. In either of these cases, the single step results derived in Section 2 (i.e. Eq. (22) and (26)) then hold. However, if the dynamics is mediated by terrace diffusion, neighbouring steps influence each other through the diffusion field on the terraces, and a coupled set of Langevin equations must be solved, as shown below (see also [13-17]).

The geometry is depicted in Fig. 3.

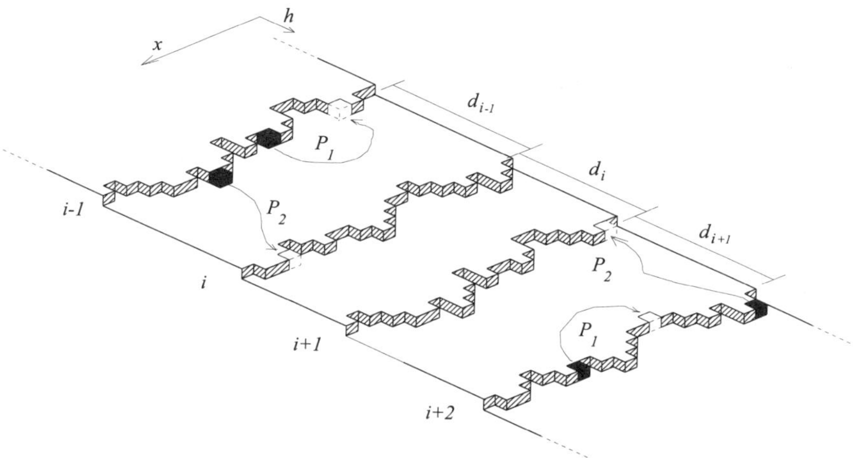

Figure 3. The diffusion processes contributing to the fluctuations of a step edge in a step bunch.

We number the steps with $i = 1, 2, ..., N$, with $i = 1$ and $i = N$ being steps at the beginning and end of the step train. Let $h_i(x,t)$ describe the random motion of the ith *step* in the train *about its center of mass, which is assumed to be fixed - "direct" interaction terms are needed to produce center of mass dynamics*; $\mu_i(x,t)$ is the local chemical potential and d_i, with $d_0 = \infty = d_N$ is the average distance between the centers of mass of adjacent steps (see Fig. 3). d_i are time-independent in this analysis. In terms of these variables, the Langevin Eq. for the ith step is;

$$\frac{\partial h_i(x,t)}{\partial t} = \frac{\Gamma_h}{k_B T}\{J_{i,i} + J_{i-1,i} + J_{i+1,i}\} + \eta_i(x,t), \tag{33}$$

where

$$J_{i,i} = \int_0^{L/2} P^i(l)\{\mu_i(x+l,t) - 2\mu_i(x,t) + \mu_i(x-l,t)\}dl, \tag{34}$$

with,

$$P^i(l) = p_U P_1(\alpha_U, \alpha_L, d_{i-1}, l) + p_L P_1(\alpha_L, \alpha_U, d_i, l), \tag{35}$$

where $P^i(l)$ is the probability that an atom detaches from the ith step, diffuses on either the upper or lower terrace and reattaches to the same step. The probability P_1 is calculated in Appendix C (see Eq. (106)). We also have,

$$J_{i-1,i} = \int_0^{L/2} P_n^{i-1}(l)\{\mu_{i-1}(x+l,t) - 2\mu_i(x,t) + \mu_{i-1}(x-l,t)\}dl, \tag{36}$$

and

$$J_{i+1,i} = \int_0^{L/2} P_n^{i}(l)\{\mu_{i+1}(x+l,t) - 2\mu_i(x,t) + \mu_{i+1}(x-l,t)\}dl, \tag{37}$$

with

$$P_n^{i-1} = (p_L P_2(\alpha_L, \alpha_U, d_{i-1}, l) + p_U P_2(\alpha_U, \alpha_L, d_{i-1}, l))/2, \tag{38}$$

where P_n^{i-1} is the probability that an atom is exchanged between the step i and the neighbouring step $i-1$. The probability P_2 for diffusing from one step to the other is calculated in Appendix C (see Eq. (106)).

A Fourier transform of Eq. (33) with (34,36,37) yields the N-dimensional set of equations,

$$\frac{\partial \vec{h_q}}{\partial t} = -Cq^2 \tilde{M} \vec{h_q} + \vec{\eta_q} \tag{39}$$

where

$$C = 2\tilde{s}\Gamma_h \tag{40}$$

with $\tilde{M}$ given by,

$$\tilde{M} = \begin{pmatrix} I_a^1(q) & I_b^1(q) & 0 & \cdots & 0 \\ I_b^1(q) & I_a^2(q) & I_b^2(q) & \cdots & 0 \\ 0 & I_b^2(q) & \ddots & \ddots & \vdots \\ \vdots & \cdots & \ddots & I_a^{N-1}(q) & I_b^{N-1}(q) \\ 0 & \cdots & \cdots & I_b^{N-1}(q) & I_a^N(q) \end{pmatrix} \tag{41}$$

where

$$I_a^i(q) = \int_0^{L/2} \{P^i(l)(1 - cosql) + P_n^{i-1} + P_n^i\} dl,$$

$$I_b^i(q) = -\int_0^{L/2} P_n^i cosqldl \tag{42}$$

Substituting the results (106) and (113) of Appendix C into Eqs. (35) and (38) and doing the integrals in Eq. (42) yields,

$$\begin{aligned} I_a^i(q) &= p_U\alpha_U\{r(0, d_{i-1}, b_{i-1}, \alpha_L) - r(q, d_{i-1}, b_{i-1}, \alpha_L)\} \\ &\quad + p_L\alpha_L\{r(0, d_i, b_i, \alpha_U) - r(q, d_i, b_i, \alpha_U)\} \\ &\quad + p_U\alpha_L r(0, d_i, a_\perp, \alpha_U) + p_L\alpha_U r(0, d_{i-1}, a_\perp, \alpha_L) \end{aligned} \tag{43}$$

and

$$I_b^i(q) = -(p_U\alpha_L r(q, d_i, a_\perp, \alpha_U) + p_L\alpha_U r(q, d_i, a_\perp, \alpha_L))/2, \tag{44}$$

where,

$$r(q, d, x, \alpha) = \frac{qcosh(qx) + \alpha sinh(qx)}{q^2 sinh(qd) + q(\alpha_U + \alpha_L)cosh(qd) + \alpha_U\alpha_L sinh(qd)}, \tag{45}$$

and in the limit $q \to 0$, we get,

$$r(0, d, x, \alpha) = \frac{1 + \alpha x}{\alpha_U + \alpha_L + d\alpha_U\alpha_L}. \tag{46}$$

(Note $r(0, d, x, \alpha)$ solves the one dimensional diffusion problem) We find the eigenvalues, $\tilde{\epsilon}$ of the symmetric matrix $\tilde{M}$, using the orthogonal transformation $\tilde{\epsilon} = \tilde{U}^T \tilde{M} \tilde{U}$ (the lth column of $\tilde{U}$ contains the lth eigenvector of $\tilde{M}$). Then the quantities $\vec{v}_q(t) = \tilde{U}^T \vec{h}_q(t)$ are independent and obey the Langevin equations

$$\frac{\partial v_{iq}(t)}{\partial t} = -Cq^2 \epsilon_i(q) v_{iq}(t) + \eta_{iq}(t), \tag{47}$$

whose solution is presented in Appendix A. The statistics of the noise is unaltered by the othogonal transformation, $\tilde{U}$ provided it is normalized. The correlation functions are then given by,

$$< h_k(x,t)h_{k'}(x',t') >=< \sum_{qq'} h_{kq}(t)sin(qx)h_{k'q'}(t')sin(q'x') >, \tag{48}$$

or (after assuming translational invariance along the steps and hence averaging over $x + x'$),

$$< h_k(0,t)h_{k'}(X,t') >= \sum_{l=1}^{N}\sum_{q} U_{kl}U_{k'l} < v_{lq}(t)v_{lq}(t') > \frac{1}{2}cosqX, \tag{49}$$

where $X = x' - x$. From Eq. (79) of Appendix A, we have,

$$< v_{lq}(t)v_{lq}(t') >= \frac{2}{Lq^2\tilde{s}}(e^{-Cq^2\epsilon_l(q)(t'-t)} - e^{-Cq^2\epsilon_l(q)(t'+t)}). \tag{50}$$

As for the one-step case, we have $G_k(2t) = w_k^2(t)$, for the kth step in the step train. Since $G_k(t) =< h_k^2(x,t_1) > - < h_k(x,t_1)h_k(x,t_2) >$ with $t = t_2 - t_1$, and (49) and (50) (setting $t = t'$, $X = 0$) gives,

$$G_k(t) = \sum_{ql}(U_{kl})^2\frac{1}{L\tilde{s}q^2}(1 - e^{-C\epsilon_l(q)q^2t}), \tag{51}$$

which reduces to Eq. (16) as $N \to 1$. For general d_i, it is not possible to solve for the eigenvalues and eigenvectors of $\tilde{M}$ analytically. However in special cases of experimental interest, an analytic analysis is possible, as discussed below.

3.1 Two steps

In this case the matrix $\tilde{M}$ is given by,

$$\tilde{M} = \begin{pmatrix} I_a^1(q) & I_b^1(q) \\ I_b^1(q) & I_a^2(q) \end{pmatrix}, \tag{52}$$

and we use $d_0 = \infty$, $d_1 = d$, $d_2 = \infty$, where d is the distance between the two steps whose fluctuations are described by $h_1(x,t)$ and $h_2(x,t)$. For example, we have,

$$I_a^1(q) = p_U(1 - \alpha_U\frac{e^{-|q|a_\perp}}{\alpha_U + |q|}) + p_L\alpha_L\{r(0,d,b,\alpha_U) - r(q,d,b,\alpha_U)\} + p_U\alpha_L r(0,d,a_\perp,\alpha_U), \tag{53}$$

Expanding to order $|q|$ yields (with $a_\perp > 1/\alpha$),

$$\tilde{M} = \begin{pmatrix} m + p_U|q|a_\perp & -m \\ -m & m + p_L|q|a_\perp \end{pmatrix}, \tag{54}$$

where $m = p_U\alpha_L r(0,d,a_\perp,\alpha_U) = p_L\alpha_U r(0,d,a_\perp,\alpha_L)$, with r given in Eq. (46). Thus m is given by,

$$m = \frac{p_U\alpha_L(1 + \alpha_U a_\perp)}{\alpha_L + \alpha_U + d\alpha_U\alpha_L} = \frac{p_L\alpha_U(1 + \alpha_L a_\perp)}{\alpha_L + \alpha_U + d\alpha_U\alpha_L}, \tag{55}$$

m gives the probability, in one dimension, that an atom detaches from a step edge, diffuses across a terrace of width d and attaches to a second step edge. A more physically meaningful way of writing this is,

$$m = \frac{p_U}{\alpha_U^l}\frac{a_\perp/d_s}{(1 + d/d_s)} = \frac{p_L}{\alpha_L^l}\frac{a_\perp/d_s}{(1 + d/d_s)} \tag{56}$$

where

$$d_s = \frac{1}{\alpha_U} + \frac{1}{\alpha_L} = a_\perp(\frac{1-\alpha_U^l}{\alpha_U^l} + \frac{1-\alpha_L^l}{\alpha_L^l}) = a_\perp(e^{\frac{E_B}{k_B T}} + e^{\frac{E}{k_B T}} - 2), \tag{57}$$

and the dimensionless "lattice" sticking coefficients are related to the continuum ones by, (see Appendix B)

$$\alpha^l = \frac{\alpha a_\perp}{1 + \alpha a_\perp}. \tag{58}$$

Physically, d_s measures the distance over which the probability of finding a diffusing atom decays as one moves away from the step edge. In the small q limit, the eigenvalues of $\tilde{M}$ are given by,

$$\epsilon_1 = |q|a_\perp/2, \quad and \quad \epsilon_2 = 2m + |q|a_\perp/2, \tag{59}$$

From a physical perspective, the m term is due to the diffusion field *between* the two steps, and the $|q|$ term is due to atomic diffusion on the infinite terraces on the outer sides of the two steps. When the steps are close, and if the sticking coefficients are not too small ($m >> a_\perp \pi/L$), we can ignore the linear term in $|q|$, in which case, the eigenvector matrix $\tilde{U}$ is given by,

$$\tilde{U} = \frac{1}{\sqrt{2}}\begin{pmatrix} 1 & -1 \\ 1 & 1 \end{pmatrix}, \tag{60}$$

Since only ϵ_2 contributes, both steps exhibit the behavior,

$$w^2(t) = G(2t) \sim \frac{1}{2\tilde{s}L}\sum_q \frac{1 - e^{-4Cmq^2 t}}{q^2} \sim (\frac{2m\Gamma_h t}{\pi \tilde{s}})^{1/2} \tag{61}$$

$$\rightarrow 0.798(\frac{a_\perp/d_s}{1 + d/d_s})^{1/2} a_\perp^2 e^{\frac{-E_0}{2k_B T}} \frac{1}{\tilde{s}^{1/2}}(\nu_t t)^{1/2}$$

Which has the same time dependence as the evaporation-recondensation case for an isolated step [13], Eq. (26), but of course there are different prefactors in this case. Thus the presence of another step changes the behavior of $G(t)$ from the $t^{1/3}$ law characteristic of terrace-diffusion-mediated relaxation of an isolated step to the $t^{1/2}$ law characteristic of evaporation-recondensation at an isolated step. The steps provide a "heat bath" for each other [13] and so from the point of view of one step in a train, there is a "vapor" to exchange with. For $d << d_s$ Eq. (61) becomes, $G(t) \sim t^{1/2}$, while for $d >> d_s$, we have $G(t) \sim (t/d)^{1/2}$.

However, if $\alpha_U = 0$, or $\alpha_L = 0$ (infinite Schwoebel barrier - case E of [13]), then the analysis of the previous paragraph is invalid. The matrix $\tilde{M}$ is diagonal, as there is no exhange of atoms between the steps. Nevertheless even in this limit, the two steps do influence each other's fluctuations through the d dependence of the quantities I_a^i. Consider the case, $\alpha_U = 0$, $\alpha_L \rightarrow \infty$. From Eqs. (43) - (46), we find,

$$I_a^1(q) = 1 - \frac{cosh(qb)}{cosh(qd)},$$

$$I_a^2(q) = 1 - e^{-|q|a_\perp}. \tag{62}$$

Thus the second step (described by $h_2(x,t)$) fluctuates exactly as given by Eq. (26), while in the small q limit, we find,

$$I_a^1(q) = q^2 d a_\perp. \tag{63}$$

Comparing this with Eq. (21) demonstrates that the dynamics of the first step is in the step-edge universality class, and (from Eq. (19))

$$w^2(t) = G(2t) = \frac{1}{\pi}\Gamma(\frac{3}{4})(\frac{4\Gamma_h d a_\perp t}{\tilde{s}^3})^{1/4}. \quad (64)$$

$$\rightarrow 0.552 a_\perp^2 e^{\frac{-E_0}{4k_BT}} \frac{1}{\tilde{s}^{3/4}}(\frac{a_\perp d}{a_{||}^2})^{1/4}(\nu_t t)^{1/4}$$

. The physical origin of the short range diffusion is the reflection, by the large Schwoebel barrier, of diffusing particles back to their original step. This leads to an effectively shorter range diffusion kernel for a given step. This diffusion kernel is asymptotically exponential for any finite d, and so the process is in the step-edge universality class. The fluctuations grow more rapidly as the steps move apart, due to the fact that the diffusion kernel becomes *longer ranged* with increasing d. Note that the behavior (64) still assumes that the $|q|$ term is negligible, which is probably not true for two steps. Thus for two steps, the $t^{1/3}$ law may hold in the large Schwoebel barrier limit. However, as shown below, it is difficult to find experimental situations where two steps are truly isolated.

3.2 An infinite uniform step train $d_i = d$, $N \rightarrow \infty$

We need only treat the case $\alpha_U > 0$, $\alpha_L > 0$, as if there is an infinite Schwoebel barrier, all of the steps behave in the same manner as for step 1 of the two-step case treated in the previous paragraph (i.e. Eq. (64)).

Assuming $d_i = d$, and $|q| < \alpha_L, \alpha_U, 1/d$, expansion of Eqs. (43),(44) then yields,

$$I_a^i = 2m + O(q^2) \quad and \quad I_b^i = -m + O(q^2) \quad (65)$$

These quantities are now i *independent* for a uniform step train. One feature to note is that the $O(|q|)$ term is missing for an infinite step train. It is only present in the steps at the beginning and end of a step bunch, or equivalently, if $d_i \rightarrow \infty$. Thus the $t^{1/3}$ typical of the isolated step (see Eq. (30)) is absent in the infinite step train. It is difficult to imagine an experimental situation in which the behavior of Eq. (30) could be expected.

The eigenvalues ϵ_k of the matrix $\tilde{M}$, ignoring the q^2 terms in Eq. (65), lie in a band, with eigenvalues

$$\epsilon_l = 2m(1 - cos\frac{2\pi l}{N}), \quad with \quad l = 0, ..., N-1, \quad (66)$$

and the eigenvectors are two-fold degenerate for N odd (except for $l = 0$ which is unique),

$$(\vec{U}_l)^T = \frac{1}{\sqrt{N}}(1, cos\frac{2\pi l}{N}, ..., cos\frac{2\pi l(N-1)}{N}) \quad for \quad l = 0, ...(N-1)/2$$

$$(\vec{U}_l)^T = -\frac{1}{\sqrt{N}}(1, sin\frac{2\pi l}{N}, ..., sin\frac{2\pi l(N-1)}{N}) \quad for \quad l = 1, ...(N-1)/2 \quad (67)$$

Then for any one of the steps,

$$G_q(t) = < h_q^2(t) > = \frac{1}{N}\sum_{j=0}^{N-1}\frac{2}{\tilde{s}Lq^2}(1 - e^{-Cq^2\epsilon_j t}), \quad (68)$$

In the large t, continuum, limit evaluation of the correlation function (51) (again using Eq. (19)), gives

$$w^2(t) = G(2t) = A(N)\frac{1}{\pi\tilde{s}}(4mCt)^{1/2}\Gamma(1/2), \tag{69}$$

where,

$$A(N) = \frac{1}{N}\sum_{k=0}^{N-1}(1 - cos\frac{2\pi k}{N})^{1/2} \sim \frac{2\sqrt{2}}{\pi} \tag{70}$$

The last expression on the RHS is valid in the large N limit. Notice that the expression (69), with (70) is the same as the expression (61) for the two step case, provided we replace the numerical prefactor there (0.798) with (1.436). We illustrate the asymptotic results (69) and (64) in Fig. 4.

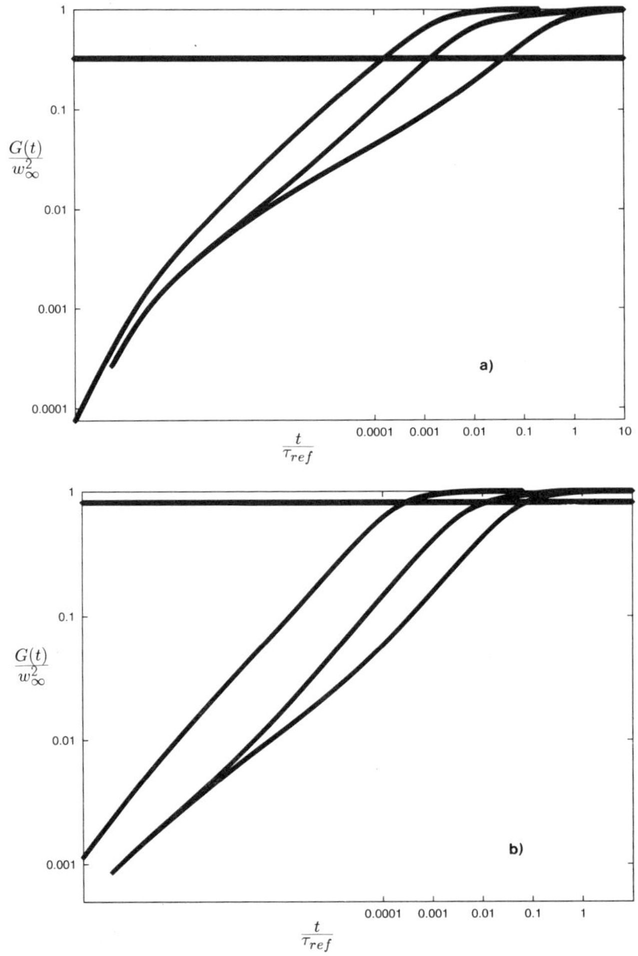

Figure 4. The fluctuations of the 5th. step in a N=10 step bunch. In both figures, the sticking coefficients are (from the top): $\alpha_U = \alpha_L = 0.001$; $\alpha_U = \alpha_L = 1000.0$ and the asymmetric, large Schwoebel barrier case; $\alpha_U = 0.001$, $\alpha_L = 1000.0$. In **a)** the steps are $d = 10$ atomic spacings apart and $L = 1000$, while in **b)** $d = 50$ and $L = 10000$. In both cases, the time is scaled by the same τ_{ref} used in Fig. 2. The flat line indicates d^2/w_∞^2, beyond which a theory which ignores step spreading is dubious.

The data in Fig. 4 was found by taking the full expressions for $I_a^i(q)$ and $I_b^i(q)$ and forming the 10×10 matrix $\tilde{M}$ of Eq. (41). A numerical eigenanalysis at each q yields $U_{kl}(q)$ and $\epsilon_l(q)$. These results are substituted into Eq. (51) and the sum numerically performed. From this analysis, we find that a step near the center of the step train has three limiting types of terrace-diffusion-mediated fluctuations: $G(t) \sim t$ for $G(t) < 1$; $G(t) \sim (mt^{1/2})$ for $t > \tau_b$ and; $G(t) \sim (dt)^{1/4}$ for $t < \tau_b$ (see Fig. 4). τ_b is a time-scale set by the relative size of the sticking coefficients. If $\alpha_U > \alpha_L$, then $\tau_b \sim (\alpha_U^l/\alpha_L^l)\tau_h$. Physically it measures the time required before the probability of surmounting the largest reflecting barrier is significant. These results are valid provided $G(t) < d^2$ the terrace width. In the other limit, step spreading is important, and hence d is not a constant. One might also wonder if the the $t^{1/3}$ behavior characteristic of an isolated step (see Fig. 2) ever occurs in a step train. However the $|q|$ behavior only arises in Eq. (45,46) in the limit $d > L$, which is unlikely to occur experimentally. A preliminary study of the effects of random d_i on step dynamics indicates that the results above are quite insensitive to weak disorder.

CONCLUSON

We have elucitated the link between atomic diffusion and step fluctuations on vicinal surfaces, including the relation between the continuum kinetic parameters and the step-edge energy barriers (see Appendix B). We have shown how atomic diffusion leads to the limiting results found by the "Maryland" Group [8,16], Pimpenelli et al. [13] (Use the following connections between their parameters and ours $\tau \to t$, $h^2 \to G(t)$. Also note that in their paper $L \sim h^2$) and Bales and Zangwill [14](set $x_s = 0$ in their equations and check [13] for a correction to [14]). In addition, we have shown that in some cases the time exponent in the step fluctuations may be non-universal, although we have not found an experiment in which to expect this. The key results for a step train whose kinetics are dominated by terrace diffusion are summarised in Fig. 4 and the discussion following it. The step-edge and evaporation-recondensation cases are given in Eqs. (22) and (26) respectively. To include all of these mechanisms in one formalism, the "a" and "b" terms in Eq. (32) need to be added to the diagonal terms of the matrix (41). It is also relatively straightforward to include different types of terrace diffusion, for example as occurs due to parallel and perpendicular dimerization on Si(001), by using different P_1 and P_2 on alternating terraces in Eq. (41). A more sophisticated, but straightforward, generalisation is necessary to include the possibility of transmission across a step edge, a processes which may be important at many step edges [36].

The key physics of our model (see Eqs. (9) and (10)) is contained in the "non-local" diffusion kernels which occur after integrating over the atomic processes which produce step fluctuations. We have calculated these kernels for a variety of physically interesting cases (see Appendix C) and have related the parameters in those kernels to atomic energy barriers (see Appendix B). The model used here is close in spirit to the work of Pimpinelli et al. [13], who developed a scaling analysis based on diffusion ideas. The theory of Einstein and co-workers and Bales and Zangwill is based on an "equilibrated gas" of atoms on each terrace. The concentration of this gas of atoms obeys Laplace's equation just as our probability P does. To make complete contact between the two methods however, we would need to treat the effect of a gas of atoms on the diffusion probabilities we have studied. Actually there are two effects that could be included. (1) The effect of step roughness on $P(l)$ - we checked this numerically and found it to be quite small and (2) The effect of atom interactions on the terrace - This leads to the "tracer" diffusion problem. It is known that in the presence of interactions, Laplace's equation still holds for the calculation of $P(l)$, but there is a concentration dependence in the prefactors. In the small "gas" concentration limit, our analysis and that of Einstein et al. should be equivalent.

Acknowledgements: This work was supported by the NSF under grant DMR-9312839.

Appendix A - Solution to the linear Langevin equation.

The linear Langevin equation is, (in q-space):

$$\frac{\partial h_q(t)}{\partial t} = -Cq^2 I(q) h_q(t) + \eta_q(t) \tag{71}$$

where $C = 2\tilde{s}\Gamma_h$ is a real constant. Note that by symmetry this sets $< h_q(t) >=< h(x,t) >= 0$ for $L \to \infty$. In order to reproduce the thermal fluctuations, the noise is assumed to satisfy the relation,

$$< \eta_{q_1}(t_1)\eta_{q_2}(t_2) >= f(q_1, q_2)\delta(q_1 + q_2)\delta(t_1 - t_2). \tag{72}$$

The general solution of (71) is,

$$h_q(t) = e^{-Cq^2 I(q)t} \int_0^t e^{Cq^2 I(q)t'} \eta_q(t')dt'. \tag{73}$$

The height-height correlation function in Fourier space is then,

$$\begin{aligned} < h_{q_1}(t_1)h_{q_2}(t_2) > &= e^{-Cq_1^2 I(q_1)t_1 - Cq_2^2 I(q_2)t_2} \times \\ & \int_0^{t_1}\int_0^{t_2} e^{Cq_1^2 I(q_1)t_1' + Cq_2^2 I(q_2)t_2'} < \eta_{q_1}(t_1')\eta_{q_2}(t_2') > dt_1' dt_2'. \end{aligned} \tag{74}$$

Using (A2), we find (assuming $t_2 \geq t_1$)

$$\begin{aligned} < h_{q_1}(t_1)h_{q_2}(t_2) > &= \frac{f(q_1, q_2)\delta(q_1 + q_2)}{C(q_1^2 I(q_1) + q_2^2 I(q_2))} \times \\ & (e^{-C(q_2^2 I(q_2)(t_2 - t_1)} - e^{-C(q_1^2 I(q_1)t_1 + q_2^2 I(q_2)t_2)}). \end{aligned} \tag{75}$$

The delta function imposes $q_1 = -q_2 = q$, and since all of the functions of q are symmetric in q, we have

$$< h_q(t_1)h_q(t_2) >= \frac{f(q)}{2Cq^2 I(q)}(e^{-Cq^2 I(q)(t_2 - t_1)} - e^{-Cq^2 I(q)(t_1 + t_2)}) \tag{76}$$

When $t_1 = t_2 = t$, we have

$$< h_q^2(t) >= \frac{f(q)}{2Cq^2 I(q)}(1 - e^{-2Cq^2 I(q)t}). \tag{77}$$

Since eq. (A6) must saturate (when $t \to \infty$) to the value of the equilibrium width [2]

$$< h_q^2(\infty) >= \frac{2}{\tilde{s}Lq^2} = \frac{f(q)}{2Cq^2 I(q)} \tag{78}$$

We require the result $< h_q^2(\infty) >= 2/\tilde{s}Lq^2$ to ensure that the equilibrium with $w_\infty^2 = L/6\tilde{s}$ is found at large times (see the correction in [13] after [25]). From Eq. (A8), then we must impose the condition $f(q) = 8\Gamma_h I(q)/L$. The key result which should be compared with experiment is then,

$$< h_q(t_1)h_q(t_2) >= \frac{2}{Lq^2\tilde{s}}(e^{-Cq^2 I(q)(t_2 - t_1)} - e^{-Cq^2 I(q)(t_1 + t_2)}) \tag{79}$$

or its inverse transform,

$$< h(x_1,t_1)h(x_2,t_2) >= \frac{2}{L\tilde{s}}\sum_q (e^{-Cq^2I(q)(t_2-t_1)} - e^{-Cq^2I(q)(t_1+t_2)})\frac{sin(qx_1)sin(qx_2)}{q^2}$$

$$\approx \frac{2}{L\tilde{s}}\frac{L}{\pi}\int_0^\infty (e^{-Cq^2I(q)(t_2-t_1)} - e^{-Cq^2I(q)(t_1+t_2)})sin(qx_1)sin(qx_2)\frac{dq}{q^2} \tag{80}$$

For a string pinned at both ends, $q = \pi n/L$, with n a positive integer on the interval [1,L], and the integral form is only valid in the intermediate time regime, before the lower limit of the q-sum dominates. If the lower limit dominates, the interface width saturates at its equilibrium value.

If a step is initially flat, we find that its square width $w^2(t) =< h^2(t) >$ grows according to (averaging over $sin^2(x)$ leads to a factor of 1/2),

$$< h^2(x,t) >= \frac{1}{2}\sum_q < h_q^2(t) >= \frac{1}{L\tilde{s}}\sum_q \frac{1}{q^2}(1-e^{-4\Gamma_h\tilde{s}q^2I(q)t}) \approx \frac{1}{\tilde{s}\pi}\int_0^\infty \frac{dq}{q^2}(1-e^{-4\Gamma_h\tilde{s}q^2I(q)t}), \tag{81}$$

We often find that $q^2I(q) \approx Aq^\gamma$, in which case,

$$w^2(t) =< h^2(x,t) >= \frac{1}{\pi\tilde{s}}(2C_1t)^{1/\gamma}\frac{1}{\gamma}\int_0^\infty x^{-(\gamma+1)/\gamma}(1-e^{-x})dx \tag{82}$$

where $C_1 = CA$. The integral in (82) is related to the standard gamma function (see e.g. [31]), Γ, and we find,

$$w^2(t) = \frac{1}{\pi\tilde{s}}(2C_1t)^{1/\gamma}\Gamma(1-\frac{1}{\gamma}) \quad for \quad t < \frac{1}{2C_1}(\frac{L}{\pi})^\gamma \tag{83}$$

Secondly we consider the intermediate time regime for the equilibrium time correlation function,

$$< h(x,t_1), h(x,t_2) >= \frac{L}{\pi}\int_0^\infty < h_q(t_1)h_q(t_2) > dq \tag{84}$$

Provided we consider an equilibrated system, so that $t_1 >> t_2 - t_1 = t$, then from Eq. (76), we have

$$G(t) =< h^2(x,t_1) > - < h(x,t_1), h(x,t_1+t) >= \frac{1}{\pi\tilde{s}}\int_0^\infty \frac{dq}{q^2}(1-e^{-Cq^2I(q)t}) \tag{85}$$

Now note that the right hand side of (85) is the same as the expression (81)(with $t \to 2t$), so that within Langevin theory, the equilibrium fluctuations G(2t) have the same time dependence as the non-equilibrium growth of fluctuations $w^2(t)$.

Appendix B - Sticking Coefficients

We assume that an atom sees the energy profile of Fig. 5. as it approaches as step edge.

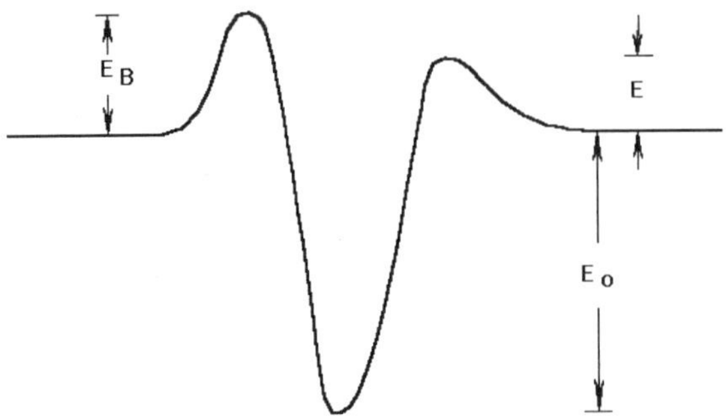

Figure 5. Key energy barriers at a step edge. Here we have ignored the energy undulations as the atom diffuses across a terrace, or along the step edge.

In terms of these step-edge energies, the "lattice" sticking coefficient on approach to the step-edge from the lower terrace is,

$$\alpha_L^l = e^{-\beta E}, \tag{86}$$

$(\beta = 1/(k_B T))$ while the sticking coefficient on approach to the step edge from the upper terrace is,

$$\alpha_U^l = e^{-\beta E_B}. \tag{87}$$

Where the l superscript indicates that there are lattice sticking coefficents. The probability that an atom detaches to the lower terrace from the step edge, is

$$p_L = \frac{e^{-\beta E}}{e^{-\beta E} + e^{-\beta E_B}} = \frac{\alpha_L^l}{\alpha_U^l + \alpha_L^l}, \tag{88}$$

while the probability that an atom detaches from the step edge onto the upper terrace, is,

$$p_U = \frac{e^{-\beta E_B}}{e^{-\beta E} + e^{-\beta E_B}} = \frac{\alpha_U^l}{\alpha_U^l + \alpha_L^l}. \tag{89}$$

Note that we have the identity,

$$p_U \alpha_L^l = p_L \alpha_U^l \tag{90}$$

In terms of the energies defined in Fig. 5, the rate Γ_h appearing in the Langevin theory, is given by,

$$\Gamma_h = a_\perp^2 a_{||}/\tau_h, \tag{91}$$

where the time τ_h is,

$$\frac{1}{\tau_h} = \nu(e^{-\beta(E_0+E)} + e^{-\beta(E_0+E_B)}) \tag{92}$$

where ν is an attempt frequency. Thus,

$$\Gamma_h = a_\perp^2 a_{||} \nu e^{-\beta E_0}(\alpha_L^l + \alpha_U^l) \tag{93}$$

Thus, for the terrace diffusion case, the fundamental activation energies are E_0, E and E_B, as defined in Fig. 5, while the "natural" parameters appearing in the dynamics are Γ_h, α_U^l and α_L^l. The lattice sticking coefficients α_U^l and α_L^l are dimensionless and lie in the interval $[0, 1]$.

The continuum sticking coefficients are defined somewhat differently. A key boundary condition in the continuum models is,

$$\frac{\partial P}{\partial x}|_{x=0_\pm} = \pm \alpha P(0), \tag{94}$$

in which the continuum sticking coefficient α has the dimensions of inverse length and it lies in the interval $[0, \infty)$. As we shall show later in the section to ensure detailed balance, we require,

$$\alpha^l = \frac{\alpha a_\perp}{1 + \alpha a_\perp} \quad or \quad \alpha = \frac{\alpha^l}{(1 - \alpha^l) a_\perp} \tag{95}$$

which provides the relationship between the lattice and continnum sticking coeffients.

Appendix C - Calculation of diffusion kernels

Consider the general case of arbitrary sticking coefficients and two steps. Atoms may diffuse from a step (labelled 1) to itself, and to a step (labelled 2) at distance d from step 1. P_1 is the probability that an atom which is on the lower terrace (with respect to step 1) and at distance $a_\perp$ from step 1 is absorbed by step 1 at distance l from it starting position. P_2 is the probability that the same atom is absorbed by step 2. We start with an atom which has detached from the step labelled 1, and hence is at one atomic spacing from the step edge. In this case, we solve,

$$\nabla^2 P(x, y) = -F\delta(x)\delta(y). \tag{96}$$

Thus the atom is modelled using a source of magnitude, F, at the origin, and we ask for the probability that absorption occurs on the step edges at $x = -a_\perp$ (step 1) and at $x = b$ (step 2), so that $d - a_\perp + b$. Intuitivcly it is clear that $F = 1$, as will be confirmed later. Using the Fourier transform,

$$P(x, y) = \frac{1}{2\pi} \int_{-\infty}^{\infty} P_k(x) e^{iky} dk, \tag{97}$$

and,

$$\delta(y) = \frac{1}{2\pi} \int_{-\infty}^{\infty} e^{iky} dk, \tag{98}$$

we find that,

$$\frac{d^2 P_k(x)}{dx^2} - k^2 P_k(x) = -F\delta(x). \tag{99}$$

We take the solutions to this equation to be,

$$\begin{aligned} P_k^+(x) &= a_k e^{-kx} + d_k e^{kx} \quad for \quad b \geq x > 0 \\ P_k^-(x) &= A_k e^{kx} + D_k e^{-kx} \quad for \quad -a_\perp \leq x < 0. \end{aligned} \tag{100}$$

We must ensure that the following boundary conditions are satisfied,

$$\begin{gathered} P^+(0, y) = P^-(0, y), \\ \frac{\partial P^-}{\partial x}(-a_\perp, y) = \alpha_L P^-(-a_\perp, y) \\ -\frac{\partial P^+}{\partial x}(b, y) = \alpha_U P^+(b, y) \\ \int_{0^-}^{0^+} \frac{d^2 P_k(x)}{dx^2} dx = -F = -\frac{\partial P_k^+}{\partial x}(0^+) + \frac{\partial P_k^-}{\partial x}(0^-) \end{gathered} \tag{101}$$

The first of these equations ensures continuity, while the second and third equate the rate at which atoms stick to the step edge to the current into the edge. The fourth equation ensures flux conservation near the source. Finally, we want to normalize the probabilities and this leads to $F = 1$, as will be demonstrated explicity at the end of the calculation. From the boundary conditions (Eqs. (10)), we get equations for the coefficients A_k, Dk, a_k and d_k,

$$A_k + D_k = a_k + d_k,$$

$$A_k(k - \alpha_L)e^{-ka_\perp} = D_k e^{ka_\perp}(k + \alpha_L),$$

$$a_k(k - \alpha_U)e^{-kb} = d_k e^{kb}(k + \alpha_U)$$

$$k(A_k - D_k + a_k - d_k) = F \tag{102}$$

Solving these equations yields,

$$A_k = \frac{F}{2k}\frac{X_L(1+X_U)}{X_U X_L - 1}, \qquad D_k = \frac{F}{2k}\frac{1+X_U}{X_U X_L - 1},$$

$$a_k = \frac{F}{2k}\frac{X_U(1+X_L)}{X_U X_L - 1}, \qquad d_k = \frac{F}{2k}\frac{1+X_L}{X_U X_L - 1}, \tag{103}$$

where

$$X_L = \frac{k+\alpha_L}{k-\alpha_L}e^{2ka_\perp}, \qquad X_U = \frac{k+\alpha_U}{k-\alpha_U}e^{2kb}. \tag{104}$$

The probabilities we seek, namely $P_1(l)$ and $P_2(l)$ are equal to the rate at which atoms stick to the step edge, so that

$$P_1(l) = \alpha_L P^-(-a_\perp, l) = \frac{dP^-}{dx}(-a_\perp, l) = \frac{1}{2\pi}\int_{-\infty}^{\infty}(kA_k e^{-ka_\perp} - kD_k e^{ka_\perp})e^{iky}dk,$$

$$P_2(l) = \alpha_U P^+(b, l) = -\frac{dP^+}{dx}(b, l) = \frac{1}{2\pi}\int_{-\infty}^{\infty}(ka_k e^{-kb} - kd_k e^{kb})e^{iky}dk. \tag{105}$$

Simplifying Eqs. (103-105), we obtain the key results,

$$P_1(l, \alpha_L, \alpha_U) = \frac{F\alpha_L}{2\pi}\int_{-\infty}^{\infty}\frac{(kcosh(kb) + \alpha_U sinh(kb))e^{ikl}dk}{k^2 sinh(kd) + k(\alpha_U + \alpha_L)cosh(kd) + \alpha_U\alpha_L sinh(kd)}$$

$$P_2(l, \alpha_L, \alpha_U)) = \frac{F\alpha_U}{2\pi}\int_{-\infty}^{\infty}\frac{(kcosh(ka_\perp) + \alpha_L sinh(ka_\perp))e^{ikl}dk}{k^2 sinh(kd) + k(\alpha_U + \alpha_L)cosh(kd) + \alpha_U\alpha_L sinh(kd)} \tag{106}$$

Finally, we require that

$$\int_{-\infty}^{\infty}(P_1(l) + P_2(l))dl = 1, \tag{107}$$

to normalize the probability of absorption on either one step edge or the other. Eq. (107) with Eq. (106) implies $F = 1$, as claimed at the beginning of this calculation.

We also require that the probability that an atom detaches from step 1 and reattaches at step 2, is the same as the probability that an atom detaches from step 2 and reattaches at step 1. That is,

$$P_L\int_0^{\infty}P_2(l, \alpha_L, \alpha_U)dl = P_U\int_0^{\infty}P_2(l, \alpha_U, \alpha_L)dl \tag{108}$$

The identity Eq. (90) with Eq. (106) ensures that Eq. (108) holds. Note however that Eq. (108) is *NOT* true if the integral is removed, and this is an important aspect of the mathematical form of the theory. Finally, we give results in some limiting cases which are relevant to the discussion in the text.

(i) An isolated step with arbitrary sticking coefficient
Take $d \to \infty$, then $P_2 \to 0$, and (from Eq. (106)),

$$P_1(l) = \frac{\alpha_L}{\pi} \int_0^\infty \frac{e^{-ka_\perp} cos(kl) dk}{(k + \alpha_L)} \tag{109}$$

There are two limiting behaviors. If $l >> 1/\alpha_L$, then $P_1(l) \sim 1/l^2$, while if $l << 1/\alpha_L$, then $P_1(l) \sim constant - ln(l)$.

(ii) Both steps perfectly absorbing $\alpha_L \to \infty$, $\alpha_U \to \infty$
Then the integrals Eq. (106) and Eq. (107) can be evaluated with the results,

$$P_1(d,l) \propto \frac{1}{d^2} csch^2(\frac{\pi l}{2d}) \tag{110}$$

$$P_2(d,l) \propto \frac{1}{d^2} sech^2(\frac{\pi l}{2d}).$$

For l small compared to the step separation, we have,

$$P_1(d,l) \propto \frac{1}{2l^2}, \quad and \quad P_2(d,l) \propto \frac{\pi^2}{8d^2} \quad , l << d, \tag{111}$$

while when l is large compared to the step separation, we have,

$$P_1(l) \approx P_2(l) \propto \frac{1}{2d^2} e^{-\frac{\pi l}{d}} \quad , l >> d. \tag{112}$$

(iii) Step 1 perfectly absorbing $\alpha_L \to \infty$, step 2 repelling ($\alpha_U = 0$)
Then

$$P_1(d,l) \propto \frac{1}{d^2} csch(\frac{\pi l}{2d}) coth(\frac{\pi l}{2d}) \tag{113}$$

and,

$$P_2(d,l) = 0. \tag{114}$$

REFERENCES

1. B. Jóos, T.L. Einstein and N.C. Bartelt, Phys. Rev. **B43**, 8153 (1991)
2. X.S. Wang, J.L. Goldberg, N.C. Bartelt, T.L. Einstein and E.D. Williams, Phys. Rev. Lett. **65**, 2430 (1990)
3. B. Swartzentruber, Y.W. Mo, R. Kariotis, M.G. Lagally and M.B. Webb, Phys. Rev. Lett. **65**, 1913 (1990)
4. J. Frohn, M. Giesen, M. Poensgen, J.F. Wolf and H. Ibach, Phys. Rev. Lett. **67** 3543 (1991)
5. C. Alfonso, J.M. Bermond, J.C. Heyraud and J.J. Métois, Sur. Sci. **262**, 371 (1992)
6. E.D. Williams, R.J. Phaneuf, J. Wei, N.C. Bartelt and T.L. Einstein, Sur. Sci. **294**, 219 (1993)
7. D.B. Abraham and P.J. Upton, Phys. Rev. **B39**, 9650 (1989)
8. N.C. Bartelt, J.L. Goldberg, T.L. Einstein and E.D. Williams, Surf. Sci. **273**, 252 (1992)
9. M. Poensgen, J.F. Wolf, J. Frohn, M. Giesen and H. Ibach, Sur. Sci. **274**, 430 (1992)
10. L. Kuipers, M.S. Hoogeman, J.W.M. Frenken, Phys. Rev. Lett. **71** 3517 (1993)
11. M. Giesen-Seibert, R. Jentjens, M Poensgen and H. Ibach, Phys. Rev. Lett. **71**, 3521 (1993)

12. N.C. Bartelt, J.L. Goldberg, T.L. Einstein, E.D. Williams, J.C. Heyraud and J.J. Métois, Phys. Rev. **B48**, 15453 (1993)
13. A. Pimpinelli, J. Villain, D.E. Wolf, J.J. Métois, J.C. Heyraud, I Elkinani and G. Uimin, Sur. Sci. **295**, 143 (1993)
14. G.S. Bales and A. Zangwill, Phys. Rev. **B41**, 5500 (1990)
15. M. Giesen-Seibert, F. Schmitz, R. Jentjens and H. Ibach, Sur. Sci. **329**, 47 (1995)
16. N.C. Bartelt, T.L. Einstein and E.D. Williams, Sur. Sci. **312**, 411 (1994)
17. H.P. Bonzel and W.W. Mullins, Sur. Sci. **350**, 285 (1996)
18. W.W. Mullins, J. Appl. Phys. **30**, 77 (1959)
19. W.W. Mullins, in "Metal Surfaces: Structure, Energetics and Kinetics", R. Vanselow and R. Howe (eds.), New York: Springer 1963
20. W. Selke and P.M. Duxbury, Phys. Rev. **52**, 17468 (1995)
21. E.E. Gruber and W.W. Mullins, J. Phys. Chem. Solids **28**, 875 (1967)
22. D. Kandel and J.D. Weeks, Phys. Rev. **B49**, 5554 (1994)
23. A. Rettori and J. Villain, J. Phys. (Paris) **49**, 257 (1988)
24. F. Lançon and J. Villain, Phys. Rev. Lett. **64**, 293 (1990)
25. P. Nozières, in *Solids Far from Equilibrium*, Ed. C. Godrèche (Cambridge University Press, Cambridge, 1991)
26. M. Ozdemir and A. Zangwill, Phys. Rev. **B42**, 5013 (1990)
27. H. Spohn, J. Phys. I (France) **3**, 69 (1993)
28. W. Selke and T. Bieker, Surf. Sci. **281**, 163 (1993)
29. H.P. Bonzel and E. Preuss, Sur. Sci. **336**, 209 (1995)
30. P. Nozières, J. Physique **48**, 1605 (1987)
31. B. Blagojevic and P.M. Duxbury in preparation
32. G. Forgacs, R. Lipowsky, Th. M. Nieuwenhuizen in Vol. 15 of *Phase Transitions and Critical Phenomena* C. Domb and J.L. Lebowitz eds., Academic Press (1992)
33. M.E. Fisher, J. Stat. Phys. **34**, 667 (1984)
34. M.P.A. Fisher, D.S. Fisher and J.D. Weeks, Phys. Rev. Lett. **48**, 368 (1982)
35. M. Abramowitz and I.A. Stegun, *Handbook of Mathematical Functions*, Dover (1972)
36. N. Bartelt, private communication.

ATOMIC STEPS IN THE DECAY OF 1- AND 2- DIMENSIONAL GRATINGS

Jack Blakely, Christopher Umbach and So Tanaka

Materials Science & Eng., Cornell University, Ithaca, NY, 14853

ABSTRACT

This paper contains a brief summary of our experimental results on the characteristics of the evolution during annealing of periodic modulations on surfaces. Most of the work involves 1-dimensional modulations on vicinal surfaces but recent results with 2-dimensional structures are described. Results on Ni and Si are the most extensive. Amplitude-time decay relationships, scaling of rates with fundamental wavenumber and temperature dependencies are discussed. The importance of steps due to the 'miscut' from singular planes is emphasized. For surfaces near Si(001) we summarize observations on both the overall modulation decay and the dynamics of the corresponding step arrays. The role of processes other than surface diffusion, in particular volume diffusion and evaporation, in the decay process is addressed. Some preliminary results on morphology changes at silicate glass surfaces are described; in the dimension range investigated viscous flow appears to dominate.

INTRODUCTION

On crystal surfaces which are close to low index planes, the topography can be described in terms of arrays of atomic steps so that the evolution of the shape of a surface translates to the evolution of the step distribution. Experimental data on shape development can, in principle, yield information on the elementary kinetic processes of atom transport between steps. With high index crystal surfaces the atomic step picture is of less obvious value, and for singular or vicinal surfaces above their temperatures of terrace roughening, the actual densities of atomic steps will differ significantly from the time averaged distributions which define the shape. The theoretical problem of predicting the characteristics of the evolution of periodic surface modulations has received considerable attention since the original work of Mullins[1]. The simplest form of Mullins' analysis did not explicitly involve the dynamics of surface steps and so is more obviously applicable to fluid or glass surfaces and to crystals above their roughening transitions. However

Dynamics of Crystal Surfaces and Interfaces
Edited by Duxbury and Pence, Plenum Press, New York, 1997

considerable experimental work has been performed below roughening, and it is widely recognized that step dynamics must be used in modeling the observed shape changes. One of the main points that we wish to emphasise in this paper is that the details of the real step distributions making up 1- and 2-dimensional modulations can influence the rate-limiting steps of the decay and such characteristic features as the amplitude-time variation and the scaling of rates with feature dimensions.

In a simple 'continuum' model the evolution of any element of a vicinal surface can be expressed as the rate of motion of that element normal to the terraces as

$$dz/dt = \rho\, v \tag{1}$$

where ρ is the average density of steps(# steps /unit length) and v is their average drift velocity normal to the mean step direction. This type of description will really only be reasonable when the spacing of the steps is much less than the lateral dimensions of the shapes whose evolution is being studied. It is however clear from equation (1) that the exact nature of the step distribution plays a primary role in any model description. Step velocities arise from chemical potential differences due to curvature of individual steps and from interactions among nonuniformly spaced steps; the excess chemical potential for atoms at a step with local radius of curvature, R, is $\beta A/R$, with β the step stiffness[2,3] and A the area per atom, while at a step in a nonuniform array interacting with inverse square potential[4] the additional contribution is proportional to ρK where K is the surface curvature perpendicular to the step direction[5].

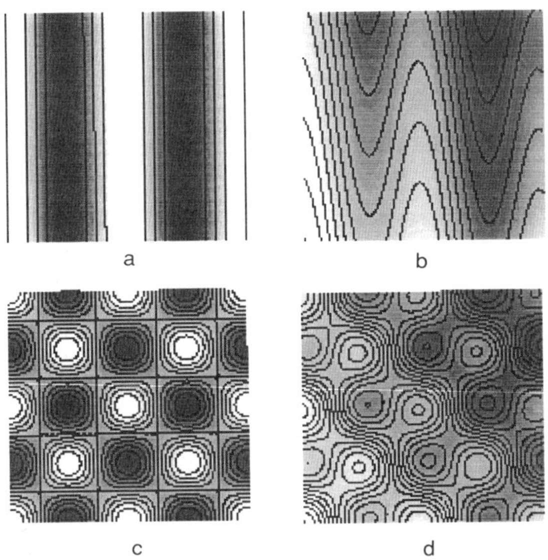

Figure 1. Sets of constant height contours on 1- and 2-D sinusoidal surfaces. The contours correspond to the idealized shapes of the surface steps on modulated crystal surfaces. a) and c) correspond to modulations on singular planes; b) and d) correspond to vicinal surfaces.

1- and 2- dimensional shape modulations on vicinal surfaces allow the creation of special step arrays whose development with time can give fundamental information on atomic exchange processes at steps and the mobility of adatoms on terraces. Figure 1 shows some possible atomic step arrays produced by 1- and 2-D modulations on singular and vicinal surfaces; the correspondence with experimental surfaces will be discussed later. To emphasise the importance of the actual step distribution on relaxation kinetics a simple example is useful. Consider the decay of a 1-D sinusoidal grating, $z(x,t) = A_0 \sin(q x)$, under conditions where the attachment or detachment of adatoms at the steps limits the process. With no steps due to miscut from the exact singular plane, the rate of disappearance of the topmost terrace may be limited by the probability of a fluctuation that causes pinch-off due to collision of the bounding steps. The width of the top terrace is $\propto 1/q$ and according to Pimpinelli et al[6] the frequency of fluctuations large enough to cause pinch-off will then vary as q^4, so that $dz/dt \propto q^4$. On the other hand with crossing steps (assumed non-interacting) attachment/detachment limited kinetics leads to $dz/dt \propto q^2$; this result follows by specializing Mullins'[1] analysis of evaporation-promoted surface smoothing, to the case of exchange of adatoms between steps and adjacent terraces[7,8].

Testing continuum models of shape evolution(or their discrete counterparts) requires careful experiments a) over a range of dimensions to test scaling, b) over a range of orientations to document the effect of step density, c) with 1-D and 2-D structures to explore the effects of step curvature and step density gradients, and d) over ranges of temperature to extract meaningful activation energies.

EXPERIMENTS

Experiments on Metals

The first observations on the development of periodic modulations were made on Pt surfaces annealed in vacuum in the temperature range 920 to 1300C[9]. The 1-dimensional grating structures were made by mechanically ruling portions of gratings on polycrystalline Pt specimens. Annealing caused extensive recrystallization with the formation of large grains with nearly sinusoidal surface sections. Individual crystal surfaces experienced exponential amplitude decay with the average decay constant scaling as q^4 approximately, where q is the wavenumber of the modulation. These measurements were mostly on high index surfaces which have high effective step densities at all temperatures. The 1-dimensional modulation technique was extended to single crystal studies through the use of lithography and etching methods[10] and applied to surfaces near the low index planes (100), (111) and (110) of Ni[11]. The (111) vicinal surfaces exhibited facetting[12] but the surfaces near (100) and (110) remained sinusoidal, at least on an optical scale, and decayed in an exponential fashion as demonstrated in figure 2. The study of the scaling behavior of the decay constant[12] for these surfaces is among the most extensive that has been reported in the literature. Figure 3 shows data on the magnitude of the decay constant as a function of the modulation wavenumber; there is a small contribution from volume diffusion at small q's but otherwise the data follows a q^4 dependence. This is the functional form predicted by Mullins' theory for the surface diffusion contribution. The temperature dependence of the decay constant(corrected for volume diffusion) followed an Arrhenius dependence with an activation energy(dependent on orientation) of 1.5-1.8eV.

The interpretation of this data on metals in terms of microscopic mechanisms of surface atom transport is not totally understood. The original papers[11] proposed that during surface transport the controlling process was adatom terrace diffusion between steps with the adatom concentration being that in local equilibrium with the atomic steps. This may indeed be the case, but in light of other experiments on adatom diffusion[13] and exchange processes at steps[14] the possibility of step attachment/detachment limited kinetics cannot be ruled out.

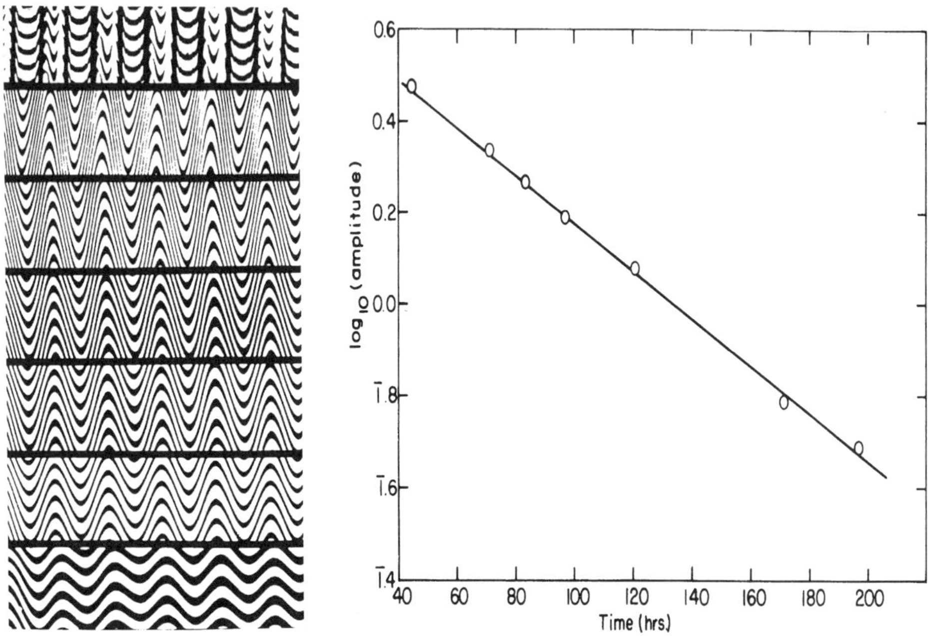

Figure 2. Series of optical interference images showing the decay of a 1-D grating on a Ni(100) surface at 1219C[11]. The interference fringes represent contour intervals of ~0.27μm. The grating repeat spacing is ~25μm. After the initial anneal the amplitude decays exponentially as shown in the corresponding amplitude-time plot.

The Ni surfaces used in these experiments were nominally within 1^0 of (100) and (110) so that they undoubtedly had a fairly large density of the type of background steps described in figure 1. The temperature range used may have spanned that of terrace roughening but the presence of so many steps due to the miscut probably masked any effect of spontaneous step creation on the terraces. Under such experimental conditions it is extremely unlikely that extensive facets at the extrema have strong effects on the mode of decay.

Very recently[7,15] we have carried out a set of experiments on modulated surfaces vicinal to Au(111) using AFM and STM methods to follow the morphology development. Some interesting preliminary results have been obtained for the decay law for both 1- and 2-D modulations and for the scaling behavior. Figure 4 show STM images of both 1- and 2-D grating structures showing the arrays of monoatomic steps which make up the surface shape. The step arrays may be compared to the schematics of figure 1b,d. The decay of the amplitude of these structure when annealed in air has been followed by AFM measurements. In both cases the amplitude decays exponentially with time and with a scaling behavior that is approximately q^2. Due to the relatively low temperature of these experiments it is unlikely that direct evaporation to the vapor

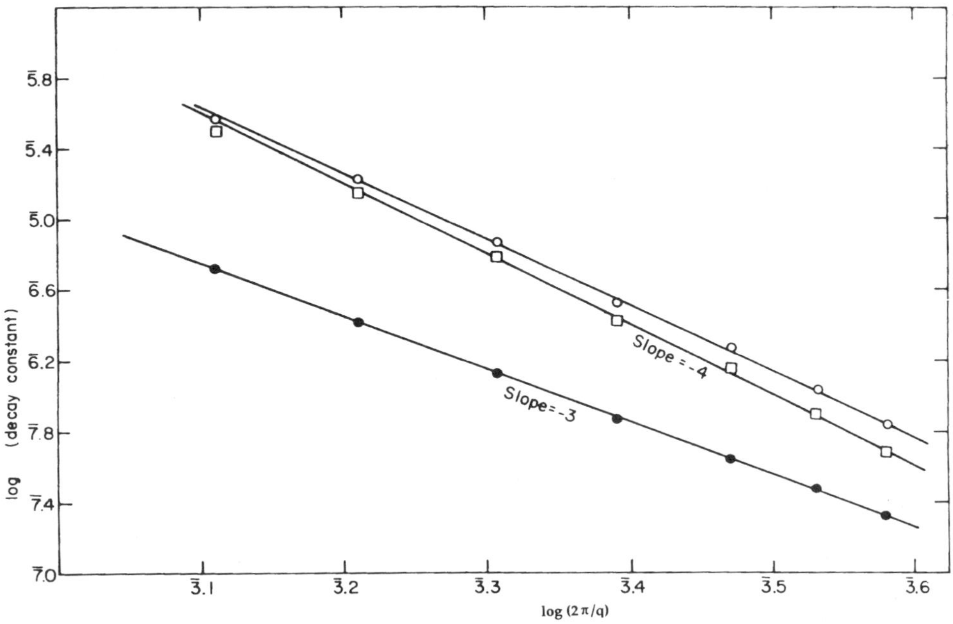

Figure 3. Measured decay constant(top line) as a function of the period (= $2\pi/q$) for 1-D modulations on Ni(100) at 1219C. After correcting[11] for a small volume diffusion contribution(bottom line) the decay constant scales as q^4.

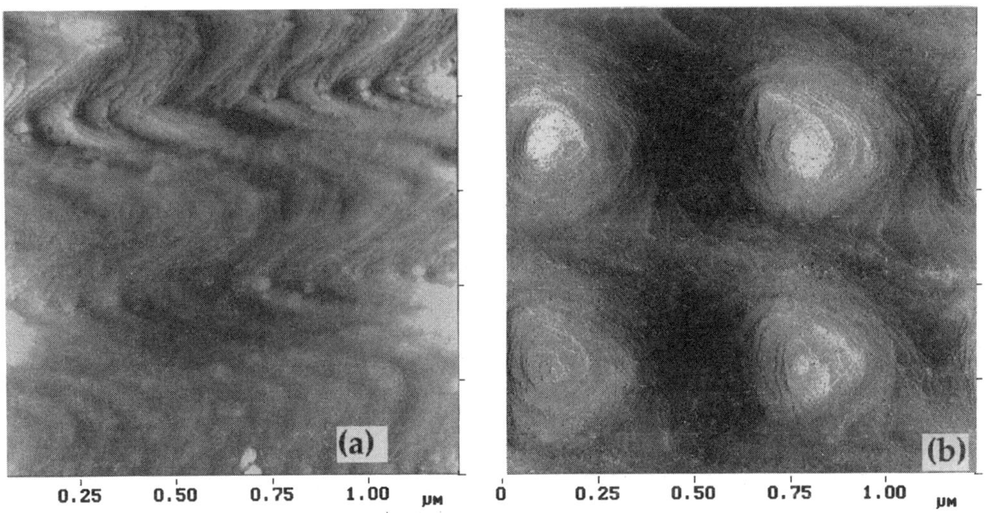

Figure 4. STM images[7,15] showing the arrays of atomic steps making up 1- and 2-D gratings on a surface near Au(111). In a) the steps are grouped into bunches; the general shapes should be compared with figure 1b. In b) the shapes of the closed contours around the maxima are, on average, truncated triangles. Note that the steps due to the 'miscut' remain as open contours. Compare with the schematic of figure 1d.

accounts for this scaling. The observed q dependence can be rationalized in terms of a simple attachment/detachment limited model[7,8] but further data is needed before settling on such a conclusion. It should be noted from a comparison of figures 4a,b that the 2-D modulated structure contains arrays of closed step contours, islands and holes, while the 1-D case has open contours. The average shape of the islands in figure 4b is hexagonal with three long edges and three shorter edges in qualitative agreement with previous observations on island morphology on Au(111). The direction of the steps, due to the miscut, relative to the major axes of the 2-D gratings may be of importance if the step tensions are sufficiently anisotropic; the importance of the direction of the background steps relative to 1-D grating lines has previously been noted and demonstrated by Surnev et al[16].

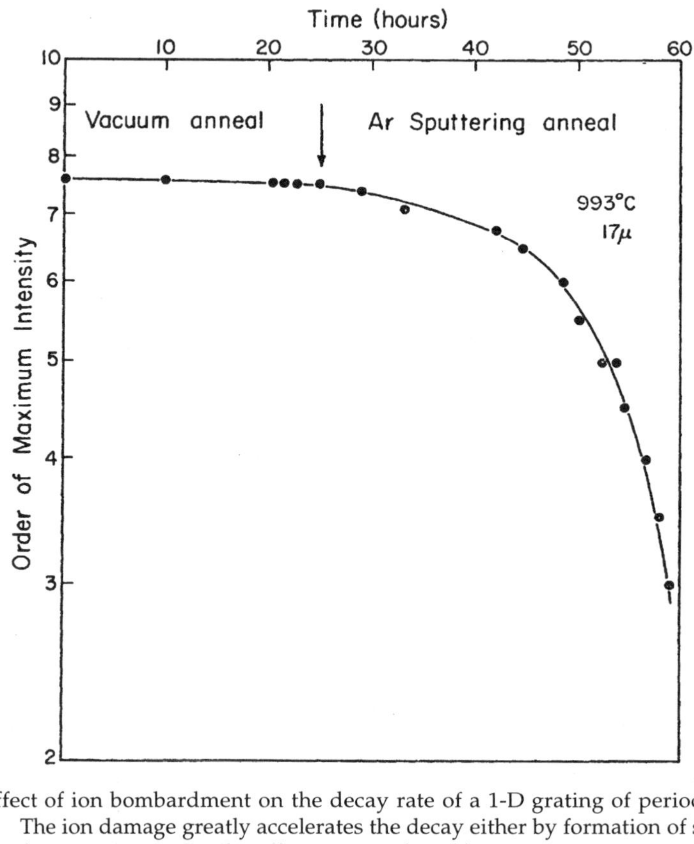

Figure 5. Effect of ion bombardment on the decay rate of a 1-D grating of period 17μm on Au (100) at 993C. The ion damage greatly accelerates the decay either by formation of surface defects or by removing the impurities. Similar effects are produced by oxygen exposure. The ordinate is proportional to the grating amplitude.

How surface adsorbed or segregated impurities may have influenced the results mentioned above is an important question. At the temperatures of the Ni experiments it is likely that desorption or dissolution into the crystal are rapid processes. In fact it was to address this question that we carried out an extensive series of experiments on the surface concentration of C as a function of temperature for various Ni surfaces[18,19]. The data indicates that neither C nor O is likely to remain on the surface but small concentrations of S cannot be ruled out. The effect of impurities in suppressing morphology development on Au surfaces has previously been documented in the work of Olson et al[17,20]. Figure 5 indicates the effects of ion bombardment on the rate of decay of modulated Au surfaces; similar effects were observed with oxygen exposure. Although air or O exposure was found to promote the surface kinetics, a careful study of adsorbate effects on atomic step development would be very desirable.

Recent Experiments on Si

1-D Gratings on Si Over the past several years we have conducted experiments on the development of fabricated grating structures on surfaces near Si(001). In the experiments of Keeffe et al[21-23] the development of 1-D gratings were followed in UHV in the temperature range 800 to 1100C by monitoring the distribution of intensity among the various orders of optical diffraction from the surface; intermittent AFM imaging was used as a check on the indirect diffraction method. As shown in figure 6, the amplitude decay was well described as exponential with the decay constant scaling approximately as q^4; however the range of modulation wavelengths used in the experiments was not extensive. It was found from the optical, AFM and STM studies that a fairly extensive (001) flat formed at the extrema particularly in the intermediate temperature range.

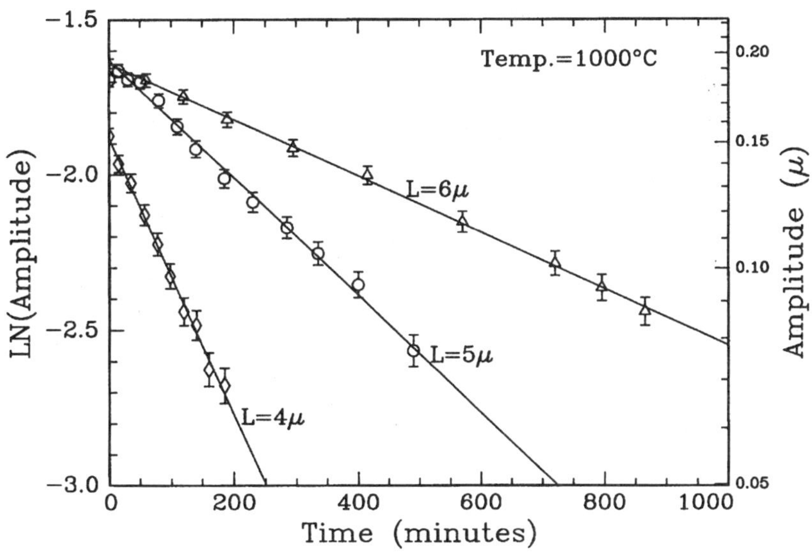

Figure 6. Amplitude-time plots for the decay of 1-D gratings on Si(001). For each of the three periods investigated the decay is well described by an exponential.[22] The dependence of slope on period corresponds approximately to q^4.

The decay constant associated with the exponential decay was found to vary with temperature, at fixed wavenumber, with an effective activation energy of ~2.3 eV. This is in agreement with the value found for step migration producing preferred reconstructed domain growth under applied surface stress[24]. It is also reasonably consistent with the sum of the formation energy for ad-dimers on this surface[22,25] and the activation energy for diffusion of the ad-dimers over the terraces[26]. However the possibility that the rate controlling step is not terrace diffusion but attachment or detachment processes at the steps seems even more likely in view of the results of Tromp et al[27] on inter-island transport and the experimental results discussed below on the kinetics of step motion at the extrema of 2-D modulations; the activation energy would then correspond to the sum of the ad-dimer formation energy and the detachment /attachment activation energy.

Figure 7. STM image showing the crossing steps(due to the 'miscut') at a maximum of a 1-D grating on a surface near Si(001). Due primarily to the difference in free energies of the two step types they occur in pairs[28]. It is believed that the amplitude decay occurs by the motion of such crossing steps along the extrema.

Figure 7 is an STM image of an area near a maximum of a 1-D grating with period of 1.5μm; due to a small 'miscut', steps cross the extrema and the two types of atomic steps of the Si(001) surface are seen to form pairs[28]. It is believed that crossing steps of this type exist on almost all the 1-D grating structures so that the requirement for large amplitude fluctuations to pinch off the extreme terraces is removed[5]. During the past year a collaborative study using low energy electron microscopy(LEEM) with the group of Tromp at IBM, has allowed new insights to be gained on the dynamics of surface step development on periodic structures[5,29,30]. The dynamics of the surface steps at the extrema is illustrated in the (LEEM) images of figure 8a,b [31]; the black-white diffraction contrast is due to the two domains of the (001) surface. These show series of atomic steps sweeping along the extrema as the annealing at 1060C proceeds. At lower temperatures the steps move in opposite directions as expected for the decay of a modulation about the average surface plane. At the higher temperature of 1060C, the steps in fact move in the same direction(figure 8a,b) due to the contribution from evaporation. We have not yet fully analysed the low temperature data on the straightening of these crossing steps and further experiments are planned. Since the fundamental wavenumber of the step modulation is simply related to that of the fabricated structure it should be possible to use the scaling behavior of the step modulation decay[8,15] to isolate the rate determining process in the mass transport; there is some indication from the preliminary data[31] that a significant flux occurs *along* the steps. This possibility certainly has to be considered in trying to understand the approximate q^4 scaling and the measured activation energy for the overall decay; that diffusion along steps must be rapid has been suggested[32] from the observation that small islands and holes are able to maintain their shapes close to equilibrium while their areas change due to inter-island transfer.

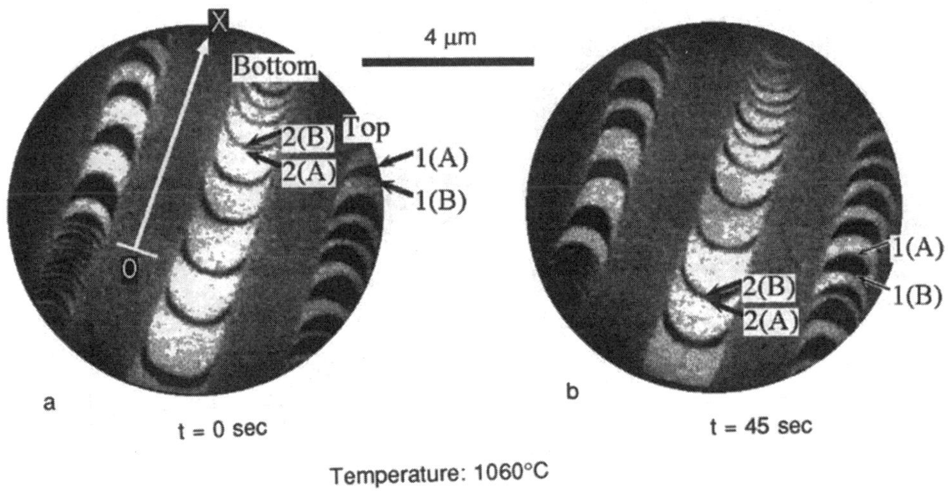

Figure 8. a,b) LEEM images showing monoatomic steps(boundaries of the black-white terrace regions) near maxima and minima of a 1-D grating near Si(001). The difference in spacing at successive extrema is due to long wavelength undulations of the fabricated surface. b) corresponds to the same region as a) after annealing for 45 sec at 1060C; the displacement of particular steps indicated by the arrows is in the *same* direction at the maxima and minima at this temperature.

2-D Gratings on Si(001) There are some fundamental differences in the evolution of 1- and 2-D modulations on singular or vicinal surfaces. As illustrated in figure 1 and discussed by Rettori and Villain[4] the 2-D modulations will generally involve closed step contours so that decay is promoted by differences in curvature of neighbouring steps and also interactions among steps in regions of nonuniform density. Differences in the overall shape of the 1- and 2-D cases is illustrated in the AFM data of figure 9. This shows that in both cases the

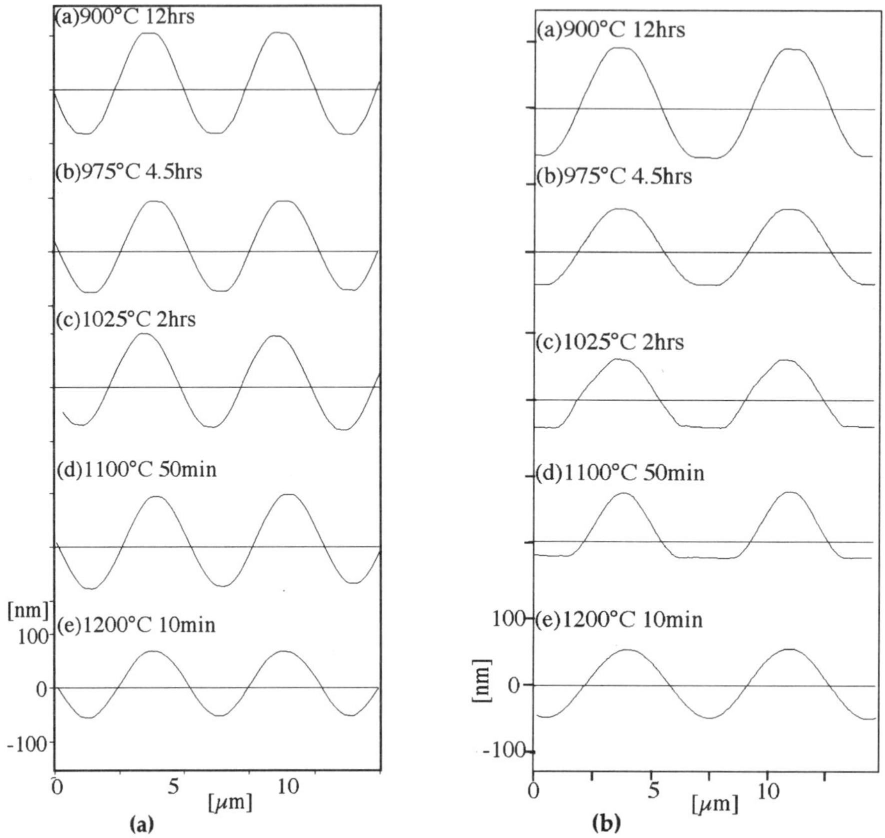

Figure 9. AFM profiles of 1- and 2-D gratings following annealing at sets of temperatures between 900C and 1200C. The 1-D profiles shown in a) have a small flat at the maxima and minima except at 1200C and are nearly symmetrical. The 2-D profiles in b) also show flats at the lower temperatures and at 1100C they are very asymmetrical[30].

annealed profiles are quasi sinusoidal but with a significant flat region at the extrema[30]. At 900C the profiles are each approximately symmetrical, at 1100C the 1-D remains nearly symmetrical while the 2-D becomes highly asymmetrical and at 1200C both profiles are close to sinusoidal. We believe that the flats at the extrema at 900C can be attributed to the surface free energy minimum at (001) and that at this temperature the main mode of decay is surface diffusion; 1200C is above the temperature of terrace roughening[33] so that the (001) singularity is removed; at 1100C very few new steps are nucleated on terraces and the observed morphology is strongly influenced by evaporation. Figure 10 compares LEEM images of 2-D gratings with the same wavenumber following annealing at 980C and at 1060C. The 980C sample shows arrays of islands and holes of similar dimensions at the maxima and minima, while the 1060C sample shows much more extensive (001) facets at the minima. The 'bow-tie' shaped features correspond to the saddle points of the biperiodic structures; note that there is a clear bias for particular (001) domains to be exposed at the saddle points.

Much of the data from these recent LEEM studies is currently being analysed. We give here a brief discussion of i) the kinetics of the island/hole development at the extrema at 950C, ii) the effects of evaporation on the development of the 2-D profiles at 1060C, iii) the saddle point features, and (iv) the motion of the crossing steps on the 1-D modulations.

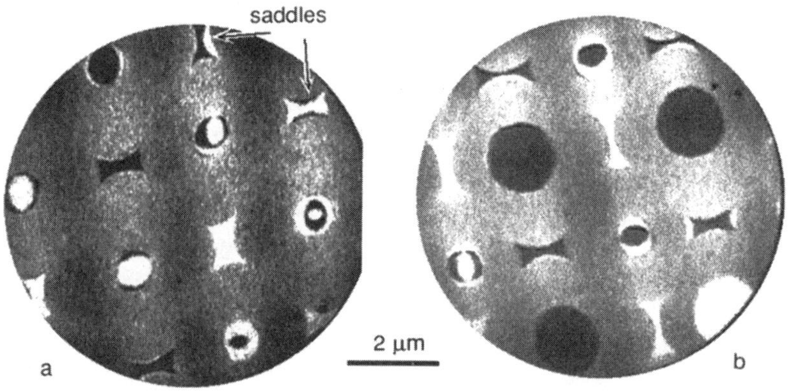

Figure 10. LEEM images of 2-D gratings following annealing at 980C and at 1060C. The 980C sample shown in a) is symmetrical with respect to maxima and minima and the saddle point features('bow-tie' shapes) are approximately 4-fold symmetric hypocycloids. In the 1060C sample the minima have extensive (001) facets and the saddle point features are elongated in the directions between maxima; note also that a particular reconstructed domain is preferred at the saddle points[31,34].

Kinetics of Island and Hole Development at the Extrema of 2-D Gratings. Figure 11 shows a series of snapshots of the development of the stack of islands at a maximum of a 2-D grating during annealing at 950C. By measuring the area as a function of time for sequences of such island decay, the data for a set of successive islands as shown in figure 12 can be obtained[31,35]. The behavior of the holes at the minima is essentially identical. Several features should be noted: a) when the islands are sufficiently far apart they take on the elliptical shape expected for the reconstructed Si(001) surface[31,35], b) in the last stage of island shrinkage the area-time relationship is linear, and c) the islands do not decay independently but in pairs. The linear area-time behavior is characteristic of inter-island transfer that is detachment limited[27]. The pairing of the islands can be modeled extremely well by considering the exchange of fluxes between stacks of interacting islands but only if the steps are taken to be *permeable* to adatoms (or ad-dimers)[35]. The solid curves of figure 12 show the quality of the fit that can be obtained between the data and the simulation. It is worth emphasizing that this involves a very high step permeability, ie an ad-atom/dimer encountering a step is ~30 times more likely to cross to the next terrace than to be incorporated; in light of the observation[36] that terrace-step exchanges seem to involve pairs of dimers steric effects may be responsible for this low incorporation probability.

Effects of Evaporation on the Development of 2-D Grating Profiles. As seen in figures 9 and 10, minima of the 2-D modulated surfaces develop very extensive facets when annealed near 1100C. We believe this to be the result of the evaporative flux from the terraces without the creation of any new steps[30,34]. At the minima evaporation causes the closed steps to expand, thus increasing the lowest terraces, while at the maxima evaporation makes the bounding terraces shrink. We have in fact developed from this observation a technique for producing regular arrays of very large step-free Si(001) regions (>20μm) that may have applications in fundamental surface studies and device fabrication [29].

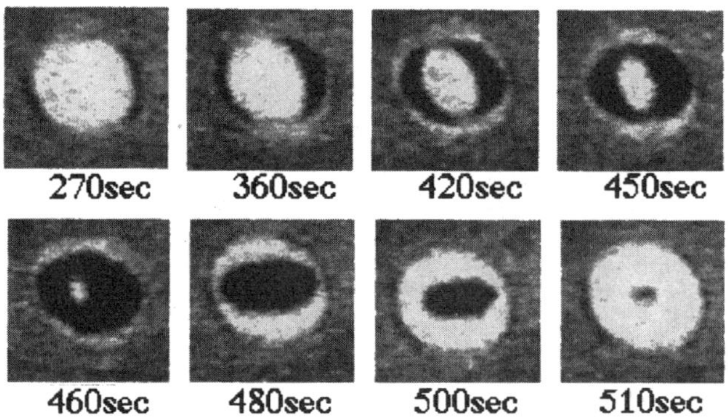

Figure 11. Series of LEEM images[31,35] showing a short sequence in the disappearance of islands at a maximum of a 2-D grating on Si(001) at 950C; the relative annealing times are indicated. At this temperature the behavior of the holes at the minima is essentially the same. Note that the islands are elliptical when separated from the rest of the stack.

The Saddle Point Features of the 2-D Gratings For an ideal 2-D sinewave the saddle point features should appear to have 4-fold symmetry when viewed in LEEM images. From the sketch of figure 13 it can be seen that the hypocycloid shaped terrace at the saddle point has the same type of monoatomic step on all four sides; due to the difference in the free energies of the two steps, S_A and S_B, on Si(001) there should be a strong preference for S_A steps and hence each maximum would prefer to be flanked by two 'white' domains and two "black' ones as is the case in figure 10. (A similar conclusion follows if the the saddle point terrace is surrounded by two S_A steps and two double steps of D_B type[31]).

At a temperature of ~1060C the saddle point features assume an elongated 'bow-tie' shape with the short dimension along a line connecting adjacent minima. We believe this to be the result of the evaporation of islands from above the initial saddle point so that along the line between maxima there are few steps between saddle points compared with the number along the line between minima; inter-step repulsion would then produce the observed elongation.

The relative lifetimes of the two terrace types at any one saddle point location has been measured[31] to differ by a factor of ~ 6 at 1060C. The change in terrace type occurs by the bridging of the short dimension by step fluctuations. Since the probability of a fluctuation of a particular amplitude depends linearly on the step stiffness[8] the observed lifetime ratio is consistent with measured step stiffnesses[37] and the geometrical picture given above[38].

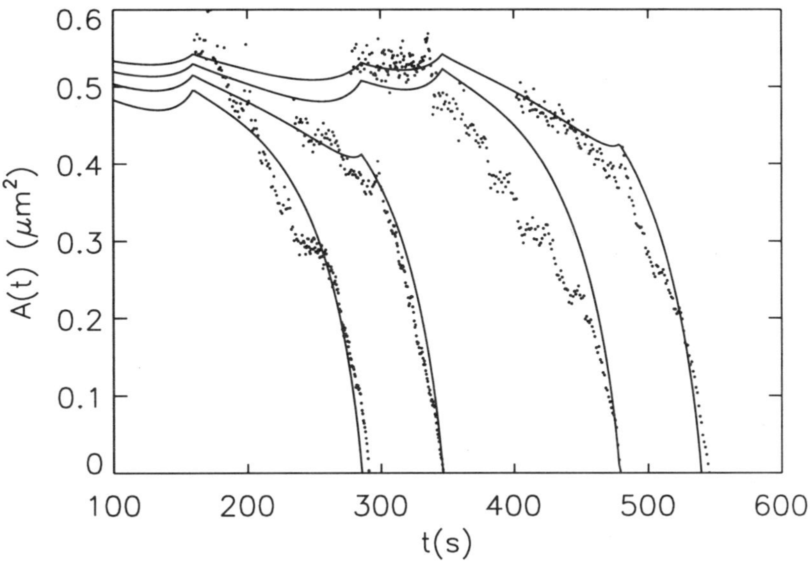

Figure 12. Area vs time for the disappearance of a series of islands at a maximum of a 2-D grating at 950C; data for minima is essentially the same. The individual data points are shown and the solid curves are from a model[35] of the evolution of a stack of concentric islands.

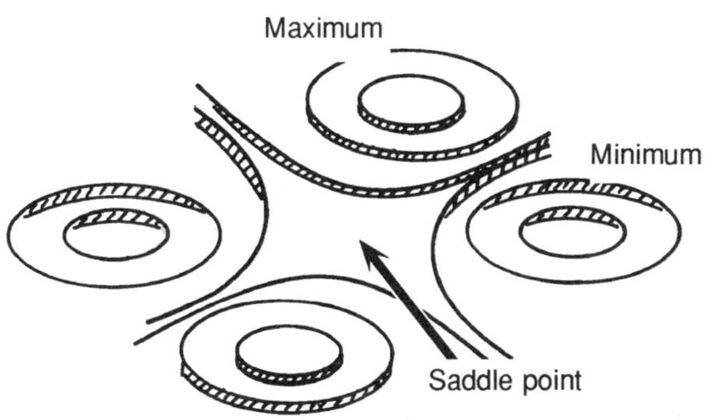

Figure 13. Schematic showing the probable configuration of atomic steps surrounding the (001) terrace at a saddle point of a 2-D grating on Si(001). Note that if the surrounding steps are mono-atomic they are all of the same type and will prefer to be all of the low energy type(S_A); if the saddle point moves up or down by one step unit the surrounding steps will all be of the high energy type. This is believed to be the basic reason for the preference in type of saddle point terrace.

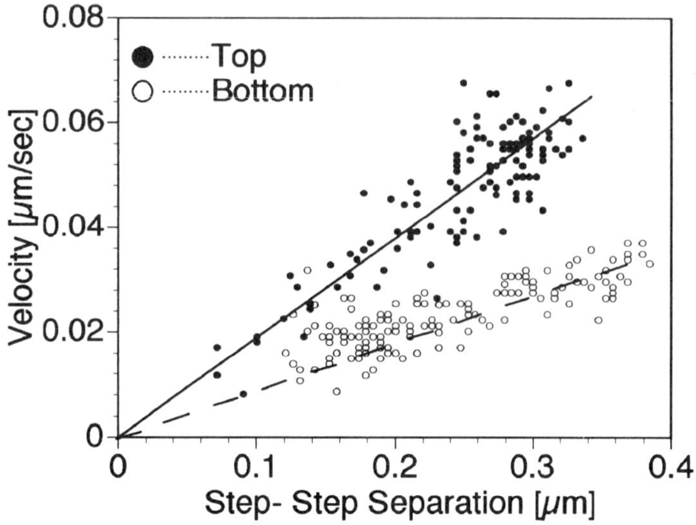

Figure 14. Measured velocity of crossing steps as a function of local step spacing at maxima and minima of a 1-D grating during annealing at 1060C. The data were taken for many different locations on the grating. The velocity-spacing relation is in each case approximately linear. The difference in velocity between maxima and minima at a particular spacing is due primarily to the effect of surface diffusion[5,31,38].

Motion of Crossing Steps on 1-D Gratings. LEEM observations of the type shown in figure 8 allow a direct measurement of the evaporative flux on the development of 1-D profiles. As noted previously, at 1060C the steps at the maxima and minima move in the *same* direction; the velocity depends on the local step density as shown in the data of figure 14. This displays the velocity as a function of step separation for various locations at maxima and minima[5,31]. The variation in step separation arises from 'waviness' of the fabricated surface. The relationship between velocity and step separation is reasonably linear as expected when evaporation of ad-atoms/dimers controls the process[39]. At any location on the surface it is the difference in the average step velocities at the maxima and minima that contributes to the decay of the modulation. The measured difference from figure 14 can be attributed to two sources, viz the effect of step curvature on the local density of ad-atoms and the contribution of surface diffusion transfer. The first of these can be shown to be quite negligible in comparison to the observed difference for any reasonable value of step stiffness[38]; the main contribution to the amplitude decay is thus surface diffusion as was deduced from the experiments of Keeffe et al[22] in this temperature and grating spacing range.

Other Materials

The periodic modulation decay method can in principle be applied to any type of material, metals, semiconductors, ceramics, polymers and to both crystalline and non-crystalline systems. It may also be applied to interfaces including solid-solid boundaries[40]. We have, in the past carried out experiments on Al_2O_3 where transport within extended defect charged layers seems to play an important role[41]. Some very recent work in our group on silicate glasses is very different from all the crystal studies that have been reported and will be summarised here; it is also a system to which the original form of Mullins' continuum analysis[1] is expected to apply quite well.

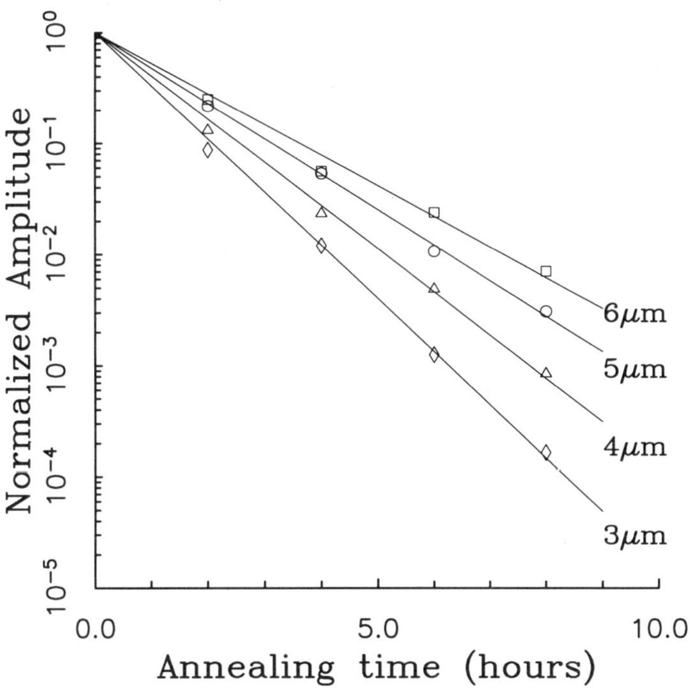

Figure 15. Amplitude -time plots obtained from AFM measurements for the decay of sinusoidal 1-D gratings with the periods indicated on the surface of a silicate glass(Corning 1737) annealed in air at 827C.[42]. The exponential decay constant scales approximately as q as expected for a viscous flow controlled process[1].

1- and 2-D grating structures can be fabricated on silicate glasses in much the same fashion as with crystalline substrates. We have made such structures and examined them with AFM techniques following annealing [42]. Figure 15 shows an amplitude-time plot for 1-D gratings of spacings 3, 4, 5, and 6μm on a silicate glass (Corning 1737) annealed in air. The profiles remain sinusoidal as expected during the decay, the decay is exponential with time, and the decay constant scales with wavenumber approximately as q. (The experimental value obtained so far was actually 0.8±0.02; we believe this deviation from unity to be due to a systematic error introduced by plasma cleaning the surface between anneals). This scaling behavior suggests that viscous flow processes are responsible for the observed mass transport; the measured decay constants are of the correct order of magnitude expected from independent determinations of viscosity and surface tension. These experiments are part of a fundamental study of the composition, morphology and mass transfer at surfaces of glass particularly those used in large area electronics. The orientation of the silicate units and how they are transported at the surface are fundamental to the understanding of silicate glass surface properties.

FUTURE DEVELOPMENTS

As indicated in this and other papers in this volume, the connection between the overall development of crystal surface shape and the details of atom transport among non-uniform step arrays is becoming clearer. Many fundamental questions remain to be answered such as the rate controlling process and the temperature domains over which particular processes dominate. The arrays of steps created by fabricating 1- and 2-D modulated structures are reasonably well defined and are amenable to modeling. For a more basic understanding we need to investigate the same phenomena on sets of similar materials eg C, Si, Ge in order to make connections between the bonding characteristics and the energies of defect formation and motion. The extension of the techniques described here to systems under irradiation, deposition or gas exposure should lead to more fundamental understanding of aspects of those processes. Mass transport by atom and molecule motion near the surface of glassy materials is still relatively unexplored; the methods discussed in this volume promise to further that area of study.

ACKNOWLEDGMENTS

This work is currently supported by NSF grant # DMR 9313371 and by the Materials Science Center at Cornell University under NSF Grant # DMR 9121654. We are particularly grateful to the staff of the Cornell Nanofabrication Facility for their assistance in creating the periodic starting structures. The LEEM experiments on Si were carried out in collaboration with Ruud Tromp and Marion Mankos at IBM, Yorktown Heights, NY. Norm Bartelt of Sandia Labs, Livermore, CA has modeled the kinetics of the island and hole decay on the 2-D gratings.

REFERENCES

[1] W.W. Mullins, J. Appl. Phys., **30,** 77-83 (1959).
[2] C. Herring, in The Physics of Powder Metallurgy , edited by W.E. Kingston (McGraw-Hill, New York, 1951) p. 143.
[3] N.C. Bartelt, T.L. Einstein and E.D. Williams, Surface Science Letters, **240,** L591 (1990).
[4] A. Rettori and J. Villain, J. Phys., **49,** 257-67 (1988).
[5] S. Tanaka, C.C. Umbach, J.M. Blakely, R.M. Tromp and M. Mankos, MRS Symp. Proc. Thin Films and Surfaces: Structure and Morphology, Fall 1996, (1997, to be published).
[6] A. Pimpinelli, J. Villain, D.E. Wolf, J.J. Metois, J.C. Heyraud, I.Elkinani and G.Uimin, Surf. Sci., **295,** 143 (1993).
[7] S.L. Pellathy, M.S. Thesis, Cornell University, (1997).
[8] N.C. Bartelt, T.L. Einstein and E.D. Williams, Surf. Sci., **312,** 411 (1994).
[9] J.M. Blakely and H. Mykura, Acta Met., **10,** 565 (1962).
[10] P.S. Maiya and J.M. Blakely, Appl. Phys. Lett., **7,** 60 (1965).

[11] P.S. Maiya and J.M. Blakely, J. Appl. Phys., **38,** 698 (1967).
[12] J.M. Blakely, Appl. Phys. Lett., **11,** 335 (1967).
[13] J.A. Meyer, J. Vrijmoeth, H.A. van der Vegt, E. Vlieg and R.J. Behm, Phys. Rev. B, **51,** 14798 (1995).
[14] G. Ehrlich, Surf. Sci., **331/333,** 865 (1995).
[15] S.L. Pellathy, A. Judy and J.M. Blakely, (1997, to be published).
[16] S. Surnev, B. Voigtlander, H.P. Bonzel and W.W. Mullins, Surf. Sci., **360,** 242 (1996).
[17] D.L. Olson H.P. Patil and J.M. Blakely, Scripta Met., **6,** 229 (1972).
[18] L.C. . Isett and J.M. Blakely, Surf. Sci., **58,** 397 (1976).
[19] M. Eizenberg and J.M. Blakely, J. Chem. Phys., **71,** 3467 (1979).
[20] D.L. Olson, Ph.D. Thesis, Cornell University, (1970).
[21] M.E. Keeffe, Ph. D. Thesis, (Materials Science Center Report #7910), Cornell University, (1994).
[22] M.E. Keeffe, C.C. Umbach and J.M. Blakely, J. Chem. Phys. Solids, **55,** 965 (1994).
[23] C.C. Umbach, M.E. Keeffe and J.M. Blakely, . Vac. Sci. Technol. B, **9,** 721 (1991).
[24] F.K. Men, W.E. Packard and M.B. Webb, Phys. Rev. Lett., **61,** 2469-71 (1988).
[25] C. Roland and G.H. Gilmer, Phys. Rev. B, **46,** 13428 (1992).
[26] Y.W. Mo, J. Kleiner, M.B. Webb and M.G. Lagally, Phys. Rev. Lett., **66,** 1998 (1991).
[27] W. Theis, N.C. Bartelt and R.M. Tromp, Phys. Rev. Lett., **75,** 3328 (1995).
[28] C.C. Umbach, M.E. Keeffe and J.M. Blakely, J. Vac. Sci. Technol. A, **10,** 1000 (1993).
[29] S. Tanaka, C.C. Umbach, J.M. Blakely, R.M. Tromp and M.Mankos, Appl. Phys. Lett, **69,** 1235 (1996).
[30] S. Tanaka, C.C. Umbach, J.M. Blakely, M. Mankos and R.M. Tromp, J. Vac. Sci. Technol. A, (1997, in press).
[31] S. Tanaka, Ph.D. Thesis, Cornell University, (1997).
[32] N.C. Bartelt, private communication.
[33] R.M. Tromp and M.C. Reuter, Phys. Rev. Lett., **68,** 820 (1992).
[34] S. Tanaka, C.C. Umbach, J.M. Blakely and R.M. Tromp, Cornell University Materials Science Center Report, #7995, (1995).
[35] S. Tanaka, N.C. Bartelt, C.C. Umbach, R.M. Tromp and J.M. Blakely, (1997, to be published).
[36] B.S. Swartzentruber and M. Schacht, Surf. Sci., **322,** 83 (1995).
[37] N.C. Bartelt, R.M. Tromp and E.D. Williams, Phys. Rev. Lett., **73,** 1656 (1994).
[38] S. Tanaka, C.C. Umbach, J.M. Blakely and R.M. Tromp, (1997, to be published).
[39] J.J. Metois and D.E. Wolf, Surf. Sci., **298,** 71 (1993).
[40] J.M. Blakely and P.S. Maiya, in Surfaces and Interfaces I, edited by J.J. Burke, N.L. Reed, and V. Weiss, (Syracuse Univ. Press, Syracuse, 1967) p. 325.
[41] S.M. Mukhopadhyay and J.M. Blakely, in Surfaces and Interfaces of Ceramic Materials , edited by L.-C. Dufour, C. Monty and G. Petot-Ervas (Kluwer Acad., Dordrecht, 1989) p. 285.
[42] C.C. Umbach, J. Schaeffer and J.M. Blakely, (1997, to be published).

MORPHOLOGIES OF PERIODIC SURFACE PROFILES AND SMALL PARTICLES: A SOURCE OF STEP AND STEP INTERACTION ENERGIES

H. P. Bonzel and S. Surnev

Institut für Grenzflächenforschung und Vakuumphysik
Forschungszentrum Jülich
D-52425 Jülich, Germany

1. INTRODUCTION

Equilibrium surface morphologies can only be observed for isolated or adsorbed small crystalline particles [1-4]. To obtain the equilibrium crystal shape (ECS), single crystal particles have to be prepared, e.g. by vapor deposition, and annealed at high temperature for a sufficiently long time under clean surface conditions. For particles adsorbed on a (single crystal) substrate, a truncated form of the ECS is observed, where the contact angle between particle and substrate is given by the surface and interfacial energies of the materials [5-7]. The ECS is a reliable source of information on the orientation dependence of the surface free energy via the well known Wulff construction. Together with an absolute value of the surface free energy, step energies can be evaluated from the orientation dependent surface energy near a cusp orientation. Also, the shape of the particle near a facet can be fitted by theoretical expressions, to yield critical exponents and energetic parameters. We will discuss here some new results obtained for Pb particles on Cu(111) imaged by scanning tunneling microscopy (STM).

Another interesting surface morphology is a periodic surface profile etched into a single crystal surface. This is a non-equilibrium structure which may transform into a flat surface at elevated temperature [2]. On the other hand, such a profile has also a well defined shape during smoothing, influenced by the orientation dependent surface free energy, $\gamma(\theta)$ (θ being a polar angle relative to a cusp orientation), along a particular crystallographic zone, which is selected by the orientation of the profile [8-18]. In general, the rate of assuming this well defined shape is fast compared to the rate of profile decay [9,16], and it can therefore also provide information on the relative surface energy anisotropy. At the same time, profile decay can be measured at elevated temperature to yield kinetic data, such as surface self-diffusion coefficients [19-21]. The kinetics of profile decay has preferably been studied on surfaces where the anisotropy of the surface energy can be neglected, e.g. on surfaces which are above the roughening temperature [2]. Another important case are periodic profiles on vicinal surfaces, i.e. near a cusp orientation, which behave substantially different from those on "rough" surfaces [18,22]. The anisotropy of $\gamma(\theta)$ is found to play an essential role in determining the shape as well as the decay kinetics of the profile. The theory and experiment of profile decay on vicinal surfaces will therefore be reviewed in this paper.

2. ANISOTROPIC SURFACE FREE ENERGY

The phenomenon of the orientation dependent surface free energy of metals is theoretically and experimentally well established [1-4]. An example from the experimental work of Heyraud and Métois for Pb is shown in fig. 1 [5]. Here the relative anisotropy of $\gamma(\theta)$ is derived from the ECS of Pb particles on graphite measured by scanning electron microscopy. In fig. 1 $\gamma(\theta)$ is plotted for part of the $[1\bar{1}0]$ zone at T = 473 K [5]. The minimum is characteristic of the low-index (111) orientation. The origin of the anisotropy of $\gamma(\theta)$ is due to the presence of steps and kinks on vicinal surfaces [2,23-25]. Since these elementary defects have their own specific energies, the surface energy per unit area of a vicinal surface will increase proportionally to the density of these defects. For a surface z(x,y) with the step density defined as $\rho = (z_x^2 + z_y^2)^{1/2}$, where z_x and z_y are derivatives, the surface free energy per unit area can be written as an expansion in the step density [2,18,23]:

$$f(\rho) = 1 + a\rho + b\rho^2 + c\rho^3 + \ldots \qquad (1)$$

The constant a is proportional to the step formation energy, $f^{(1)}$, and b and c account for step-step interactions. Entropic, dipole and elastic interactions between steps give rise to the ρ^3 term [24,25] whereas the ρ^2 term may be due to electronic effects [26]. The experimental data of fig. 1 can be fitted by $\gamma(\theta) = \gamma_0 \cos\theta \; f(\rho)$ over the whole range of orientations when all terms of eq.(1) are allowed [27]. More about this point in section 6. In principle, it is a matter of great interest to test the validity of eq. (1) and to determine the step as well as step interaction energies.

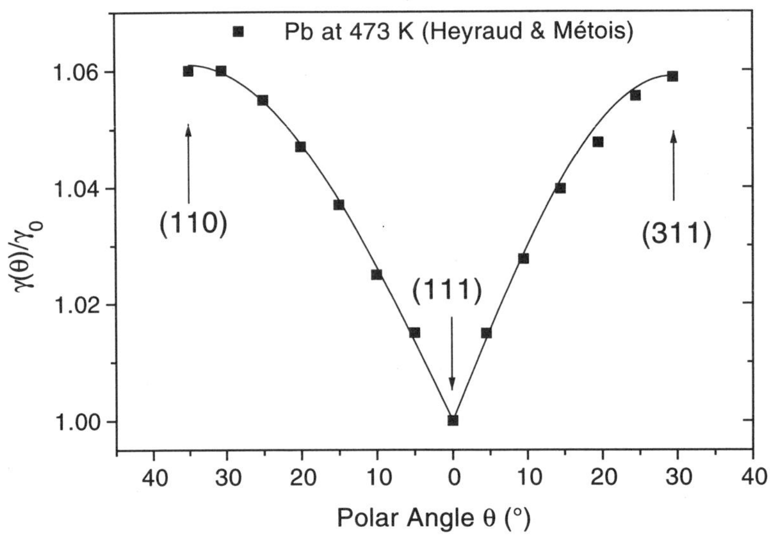

Figure 1. Relative surface energy versus polar angle at 473 K for Pb [5].

3. MODULATED VICINAL SURFACES

Periodic surface profiles on vicinal surfaces have received considerable attention in the past, both from a continuum as well as an atomistic point of view [8-18]. Here we describe briefly some recent work for surfaces of miscut α (about 3-10°) based on continuum mechanics specifically designed to take the anisotropy of $\gamma(\theta)$ into account [18]. The approach is based on eq.(1) and the excess chemical potential given by [2]

$$\mu(z_x, z_y) = -\Omega\,(f_{z_x z_x} z_{xx} + 2f_{z_x z_y} z_{xy} + f_{z_y z_y} z_{yy}) \tag{2}$$

where the subscripts denote derivatives and Ω is the atomic volume. Assuming surface diffusion as the kinetic mechanism for profile smoothing, the combined use of Fick's first law and the continuity equation lead to a differential equation which can be solved analytically under the assumption of small slopes everywhere (slope $<< \alpha$). Then the profile shape should be sinusoidal and the decay of amplitude exponential with time. The small slope approximation also has the advantage that steps of only one sign are present on the surface. Here the orientation of the profile relative to that of the intrinsic steps of the flat vicinal surface is important. If the modulation wave vector makes an angle ψ with the step direction, the decay constant B of profile smoothing can be written as [18]

$$B(\psi) = \frac{N_0 \Omega^2 D(\psi) E(\psi)}{kT} \tag{3}$$

where D and E are the ψ dependent diffusion coefficient and energy factor, respectively. N_0 is the number of surface atoms per unit area. Based on the assumption that step interaction is accounted for by the ρ^3 term only [18], eq. (1), D and E are given by

$$D(\psi) = D_p \cos^2\psi + D_n \sin^2\psi \tag{4}$$

$$E(\psi) \approx \frac{f^{(1)}}{\sin\alpha}\cos^2\psi + \frac{6\varepsilon\sin\alpha}{\cos^4\alpha}\sin^2\psi, \quad \alpha > 0. \tag{5}$$

D_p and D_n are surface diffusion coefficients parallel and normal to steps, respectively, and ε is the step interaction energy. Obviously, two simple cases arise for $\psi = 0$ and $\psi = \pi/2$, i.e. for the profile modulation parallel and perpendicular to the intrinsic steps. Since the step energy is in general larger than the step interaction energy [24,28] and the diffusion parallel to steps faster than normal to steps [29,30], the decay rate of such profiles is expected to be much faster when the modulation is parallel to the steps. The dependence of B, D and E on ψ for a miscut $\alpha = 15°$, $D_p/D_n = 2$ and $f^{(1)}/\varepsilon = 2$ is illustrated in fig. 2. Smaller miscuts lead to much greater anisotropies, as seen from the (approximate) ratio for $\alpha < 15°$:

$$\frac{B(0)}{B(\pi/2)} \approx \frac{D_p f^{(1)}}{D_n \varepsilon}\,\frac{1}{6\tan^2\alpha}, \quad \alpha > 0. \tag{6}$$

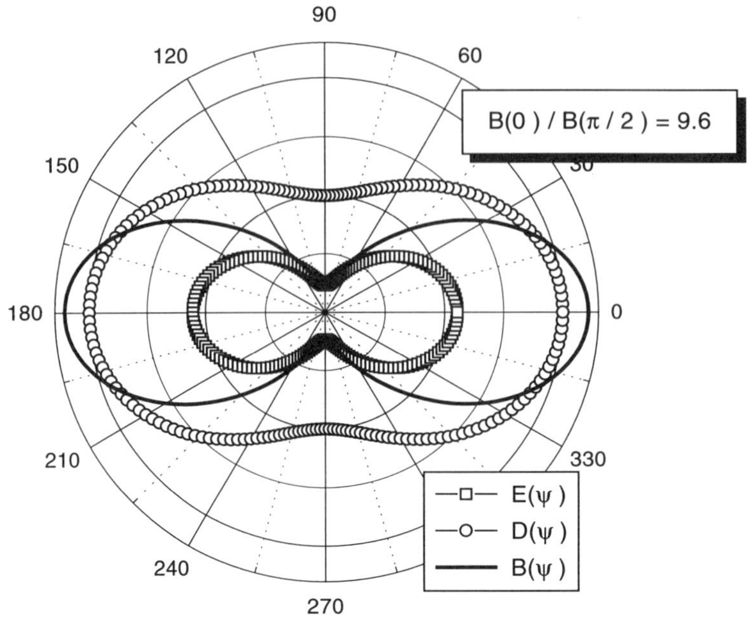

Figure 2. Polar plot of energy factor E, diffusion coefficient D and decay constant B for a vicinal surface of miscut $\alpha = 15°$.

Eqs. (3), (5) and (6) depend on the particular form of eq.(1). If the main step interaction term in eq.(1) is proportional to ρ^2 rather than ρ^3, the expressions for $E(\psi)$ would be different, particularly for the $\psi = \pi / 2$ case. Whereas the variation of $B(\pi/2)$ with miscut α is strong ($\propto \tan\alpha / \cos^3\alpha$) for ρ^3 dependence, it is weak ($\propto 1/\cos^3\alpha$) for ρ^2 dependence. Thus a measurement of $B(\pi/2)$ for several miscut surfaces, although tedious, would be a means to distinguish between those dependences. The anisotropy factor, eq.(6), would also exhibit a weaker variation with miscut for ρ^2 dependence.

When the small slope approximation is not fulfilled, the profile shape is expected to deviate from a sine wave and the decay kinetics are not necessarily exponential. Numerical calculations for $\psi = 0$ orientations and for not so small slopes show profiles with flattened maxima and minima as well as non-exponential decay behavior [18]. Examples of amplitude decay for several miscuts α are plotted in fig. 3. Calculations for ψ near $\pi / 2$ are also possible but have not been carried out as yet.

Clearly, the anisotropy arises mainly from eq.(5) because the smoothing for the $\psi = 0$ modulation is driven by the step energy, while for $\psi = \pi / 2$ it is driven by the step interaction energy. This can also be understood on geometric grounds. The $\psi = 0$ modulation introduces an oscillatory step configuration while for $\psi = \pi / 2$ steps remain straight but their separation is modulated [18,22]. Hence it is clear that the total surface energy can be lowered by reducing the total step length in the first case, and by a reduction of step interaction in the second.

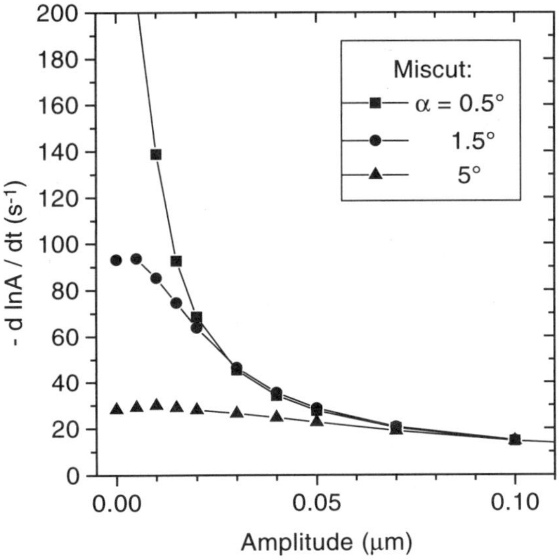

Figure 3. Calculated decay rate of profile orientation $\psi = 0$ versus amplitude for several miscuts α.

4. EXPERIMENT ON VICINAL Au(111)

An experimental test of the theory outlined above was carried out with two vicinal Au(111) crystals which had miscuts of 1.5° and 5° in the $[\bar{1}\bar{1}2]$ azimuth [22]. Each crystal had been cleaned in an ultra-high vacuum (UHV) chamber by extended annealing and Ar sputtering cycles, where cleanliness was controlled by Auger and x-ray photoelectron spectroscopies. Subsequently the crystals were patterned lithographically with two periodic line gratings of 4.3 μm (or 5.4 μm) wavelength, oriented parallel and perpendicular to the intrinsic steps. The crystal was mounted to a heater and profiles were imaged by STM under ultra-high vacuum conditions.

Initially crystals with prepared gratings were annealed at 1070 K to reshape the profiles to a quasi-steady-state shape [10,11]. This procedure eliminated surface roughness and sharp edges due to preparation and caused also a certain decrease in amplitude. Once brought into shape, crystals were annealed for periods of time at 1023 K and analyzed for shape changes by STM, after cooling to room temperature. The temperature during the anneal was measured by an infrared pyrometer. The STM with extra long piezo-electric scanning elements allowed large area scans (up to 25 μm per dimension). Amplitudes were measured from line scans across the profile image. Certain sections of the profile were also imaged at higher magnification to examine the shape and density of steps.

We observed a very different behavior of the two profiles during their decay, both with regard to shape and rate of decay. Figure 4 shows STM images from the 5° vicinal crystal after annealing at 1023 K. The $\psi = 0$ modulation (4a) is well rounded and nearly sinusoidal after 270 min of annealing while the $\psi = \pi/2$ modulation (4b) exhibits sharp edges and flat parts even after 720 min annealing. Thus the $\psi = 0$ profile behaves as expected for small slopes. Figure 5 shows an STM image of the top (or bottom) part of a $\psi = 0$ profile on a crystal of much smaller miscut (< 1°), with strongly curved steps emerging from the highly sloped area. This is a direct verification of the oscillatory shape of steps for the $\psi = 0$ modulation of a vicinal surface [2,18,22]. The condition for the small slope approximation requires that the amplitude be smaller than 0.06 μm (if the shape were

a. $\psi = 0$

b. $\psi = \pi/2$

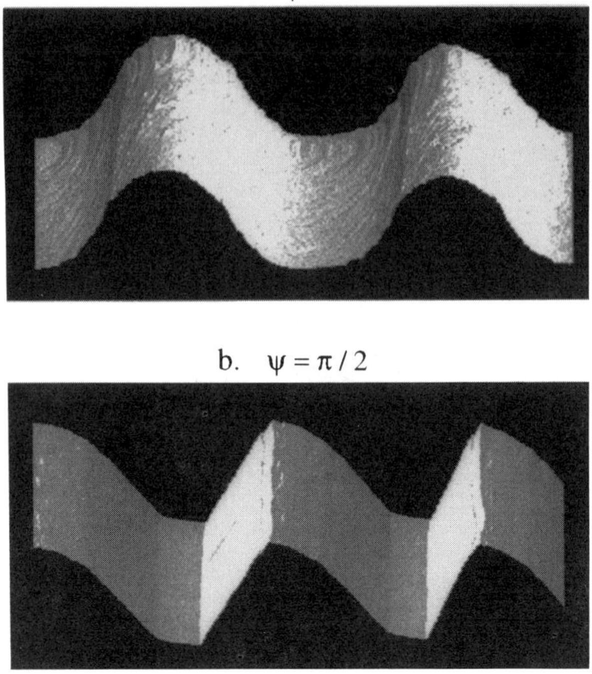

Figure 4. STM images of profiles after annealing at 1023 K for both modulation directions.

sinusoidal). This is not fulfilled for the $\psi = \pi/2$ profile because its shape is non-sinusoidal. Sharp edges also imply missing orientations on the equilibrium shape, a feature which is not accounted for by the theory.

On the other hand, the rates of decay for the two profiles are significantly different as seen in fig. 6. The ratio turns out to be 25 which is almost equal to the factor $1/6 \tan^2 \alpha$ in eq. (6). This is qualitatively consistent with theory. It follows that the ratio of surface diffusion coefficients and energies in eq.(6) is near unity at this temperature. Unravelling this ratio is not possible without additional assumptions [18]. For example, selecting a

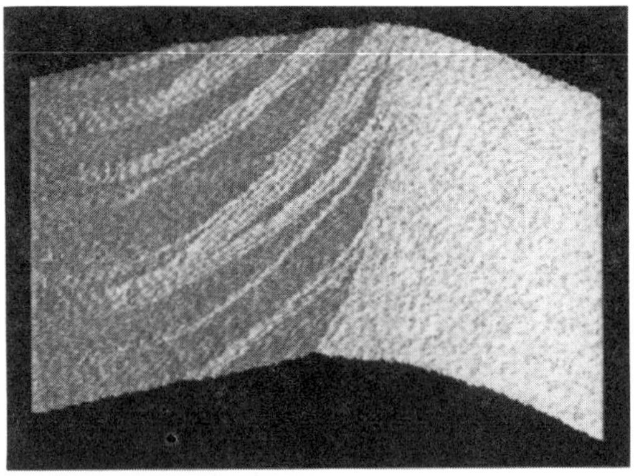

Figure 5. STM image of the top part of a profile with $\psi = 0$ modulation showing individual steps.

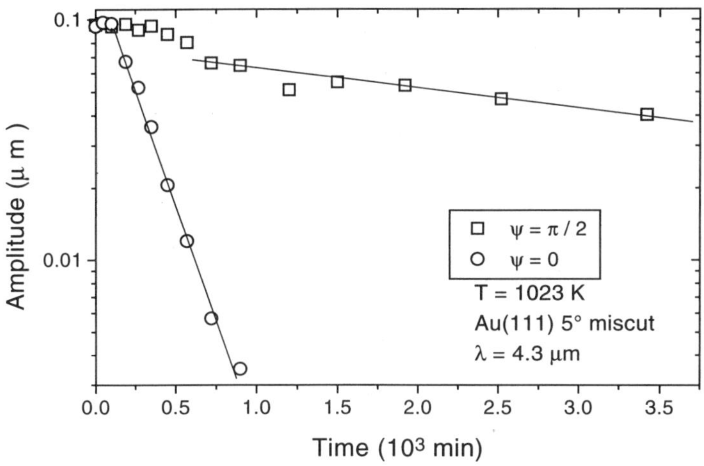

Figure 6. Decay of profile amplitude versus time at 1023 K for 5° vicinal Au(111).

"typical" experimental surface self-diffusion coefficient for Au from the literature [31] for the $\psi = 0$ rate constant in eq.(3) would in principle provide a value for the step energy but reliable measurements of D_p at this temperature are not available. A second possibility is to estimate the ratio $\varepsilon/f^{(1)}$ via the following expression [24]

$$\frac{\varepsilon}{f^{(1)}}(T) = \frac{2\pi^2}{(2+e^{\beta})(\beta - 2e^{-\beta})}, \qquad \beta = \frac{f^{(1)}(0)}{kT} \tag{7}$$

Using a value of the step formation energy at 0 K as 224 meV/atom [32] yields $\varepsilon/f^{(1)} =$ 0.8 such that the ratio of diffusion coefficients is near unity at this temperature. Again, nothing quantitative is known about the temperature dependence of D_p/D_n although it is generally believed to be greater than unity [29,30]. The general dilemma with measurements of this kind is that they always yield products of diffusion coefficient × energy. Additional measurements are necessary to decouple these numbers.

In context with the observations reported here it is of interest to point out a previous study of scratch smoothing and grain boundary grooving on Au surfaces by Gjostein [31]. In this investigation of the anisotropy of surface self-diffusion of Au he found that scratches oriented parallel and perpendicular to the [110] and [0$\bar{1}$1] zones annealed at much different rates. This effect was correctly interpreted as mainly stemming from the orientation dependence of the surface free energy ("curvature") near cusps. Hence the observed "anomously" low diffusion rates parallel to these zone directions, i.e. perpendicular to the steps of the vicinal surface in question. It is exactly the same effect that was studied here and attributed to the different role of step and step interaction energies.

Several similar experiments of scratch healing on polycrystalline Au had been published by Geguzin et al. [33,34]. They noted that the healing rate depended on the orientation of the imposed scratch relative to the "natural roughness steps" of the surface. When surface diffusion was mainly perpendicular to these, a substantial retardation was observed. In a later paper [34] the azimuthal dependence was measured and a plot similar to that in Fig. 2 was shown. Since the "natural roughness steps" may well have the same orientation as the intrinsic steps on a well annealed vicinal surface, a certain consistency between Geguzin's and our present results seems very likely.

The results of this chapter are also of some technical significance. Consider a poorly polished polycrystalline surface which shall be improved by a heat treatment. The scratches will generally not heal at the same rate, but depending on the orientation of the grains, those scratches running perpendicular to the [110] and [$0\bar{1}1$] zones will stay while those more parallel will anneal out fast. The closer the orientation of the grain is to (111) or (100), the more drastic the effect is going to be. An evenly smooth surface can therefore not be obtained by simple annealing (aside from grain boundary grooves which will always appear after some time).

6. STM OF ECS OF SMALL PARTICLES

The investigation of the exact ECS of small particles adsorbed on a flat (monocrystalline) substrate appears to be an excellent way to determine the relative orientation dependence of $\gamma(\theta)$ and from there the relative step and step interaction energies, provided the ECS exhibits all orientations, such that the conversion to $\gamma(\theta)$ is unique [1-4]. In that sense ECS studies are an important supplement to studies of transient morphological shapes which normally yield the product of mobility × energy.

The preferred tool for obtaining the ECS of small particles has been the scanning electron microscope (SEM) [5,35,36]. Here the shape along a certain (high symmetry) crystallographic direction is obtained by viewing the particle exactly perpendicular to this direction. This is experimentally difficult and the result is limited by the resolution of the instrument, typically 6-10 nm. By comparison, the resolution of the scanning electron tunneling microscope (STM) is atomic, and therefore it should be ideally suited for exact shape imaging. However, the probing tip at the end of a wire, oriented normally to the substrate surface, is of finite sharpness. Therefore only a finite slope of any morphological feature can be measured, probably about 35-50°. Hence truncated particles with contact angles of not more than this slope can be imaged completely. Facets on the ECS will be imageable with atomic resolution. Under these conditions, their flatness and step structure are accessable, if they appear on the truncated ECS and if the angle they make with the macroscopic surface is less than about 30°.

We have tested the applicability of STM to imaging the ECS for Pb particles deposited on a Cu(111) substrate [37]. The preparation of particles for this system has been described before [36]. They were annealed at 570 K for 15 hours and then cooled to room temperature. With the aid of an optical microscope, a Pb particle of several μm in diameter was approached by the STM tip. An image of the complete particle was taken at an ambient temperature of about 285 K to avoid surface-to-tip contacts [38]. Figure 7a shows a 3D representation of such a Pb particle with a single (111) facet parallel to the Cu(111) surface. The complete shape allows (computer) line scans in any azimuthal direction across the particle to be drawn for an exact analysis of the shape. This is one of the advantages of STM. An example is presented in fig. 7b. Furthermore, a line scan can be arbitrarily expanded in the z-direction normal to the facet. Such an expanded scan illustrates the flatness of the facet and the transition from the facet to the curved region of the shape which in this case is continuous. This is another important advantage of the STM and its superior vertical z-resolution. Since no orientations in this part of the shape are missing, a transformation to $\gamma(\theta)$ can be performed if the center of the particle shape is known [3]. Using a simplified geometrical construction to obtain the center, the line scan in fig. 7b is converted into $\gamma(\theta)$ by the formalism of the Legendre transform [3,4]. The result is

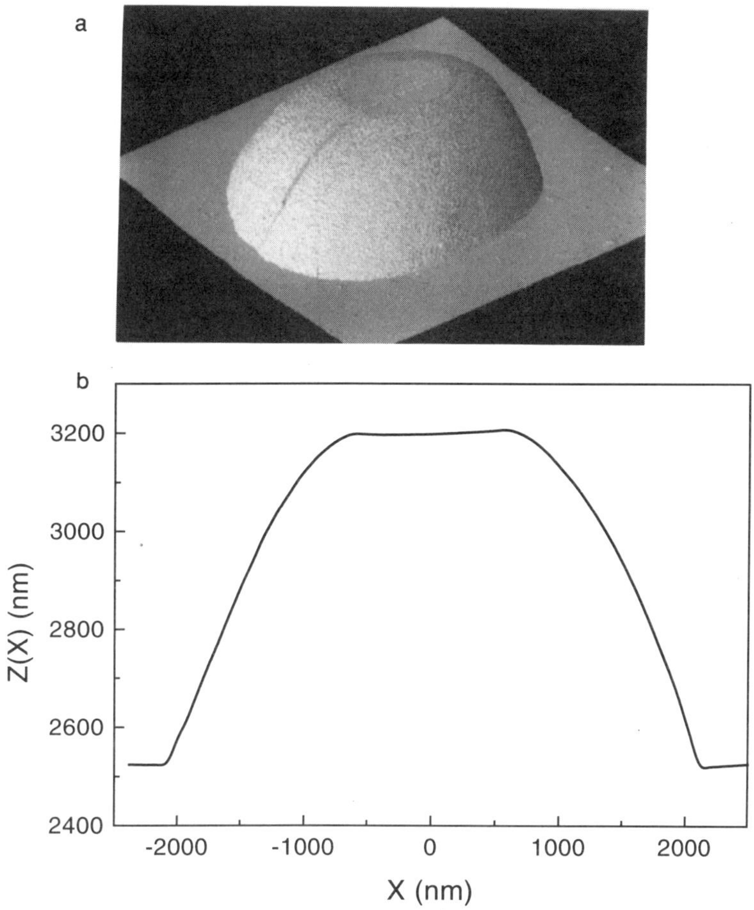

Figure 7. (a) ECS of truncated Pb particle on Cu(111) imaged by STM. (b) Line scan across particle.

presented in fig. 8a. Because of the accuracy of the shape, the corresponding $\gamma(\theta)$ function is accurate for the whole range of orientations, but especially near the cusp orientation. Hence it is possible to determine the relative step energy from the limit of the derivative:

$$\left[\frac{d\gamma}{d\theta}\right]_{\theta=0} = \left[\frac{d}{d\theta}(f(\rho)\cos\theta)\right]_{\theta=0} = \frac{f^{(1)}}{\gamma_0 h}. \tag{8}$$

A comparison of the present data for $\gamma(\theta)$ obtained on the basis of STM shape imaging, fig. 7a, with those previously published by Heyrauld and Métois for Pb [5] shows good agreement. This data as well as the experimental data of fig. 1 can be fitted by $\gamma(\theta) = \gamma_0 \cos\theta \; f(\rho)$ over the whole range of orientations allowing all terms of eq.(1) to be present [27]. An equally good fit can also be obtained by using a phenomenological expression for $\gamma(\theta)$ [32]. The step free energy can then be evaluated from the derivative $d\gamma/d\theta|_{\theta=0}$, eq.(8), and with the surface energy of Pb as 38 meV/Å^2 [39]. Since the two branches of $\gamma(\theta)$ in fig. 1 correspond to the two different close-packed steps on the fcc

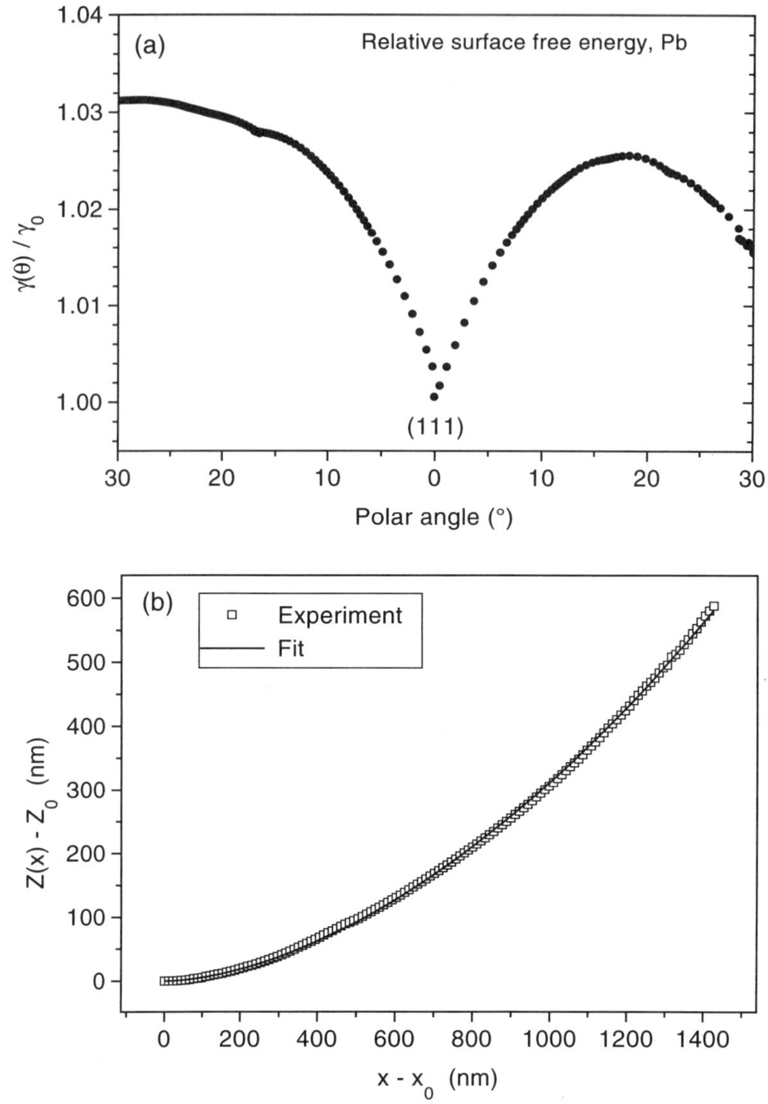

Figure 8. (a) Relative surface energy versus polar angle obtained by Legendre transform of shape in fig.6a. (b) Shape fitting of experimental data by eq. (9).

(111) surface, one can evaluate both step energies which are 17.0 and 19.6 meV/Å for the [110](111) and [110](100) steps, respectively, at T = 473 K. The more open step has the higher energy. The step energy ratio is thus 0.87. This ratio can be compared with 0.935 calculated for Al(111) [40], with 0.89 calculated for Pt(111) [41] and 0.87 found experimentally for Pt(111) from the shape of small two-dimensional islands at 625 K by Michely and Comsa [42]. It is remarkable that such very similar ratios are found by these different approaches.

Furthermore, the functional form of the experimental $\gamma(\theta)$ can be compared with known analytical expressions based on step density expansions [18,23-26]. The same can in principle be done via an analysis of the particle shape itself. Figure 8b shows a fit of the measured shape according to the expression [43-46]

$$z(x) - z_0 = \text{const}\,(x - x_0)^n \quad (9)$$

where z_0, x_0 are the coordinates of the edge of the facet and n is the universal exponent, predicted to be 3/2 if step-step interactions are described by the cubic term in step density only [44,45]. From the fit we deduce n = 1.72 near the facet as well as rather far away from it, corresponding to an angular range of about 35°. This finding, which deviates from the prediction, and other related issues will be the subject of future work in this area [37].

6. CONCLUSIONS AND OUTLOOK

The study of periodic surface profiles and of equilibrium shapes of small crystals provide complementary information on basic kinetic and energetic surface quantities at a given temperature. Fitting shapes with an orientation dependent surface free energy can provide numbers for the step and step interaction energies. Using those, the rate of profile decay on vicinal surfaces can in addition yield surface self-diffusion coefficients parallel and perpendicular to steps. Measuring the dependence of decay rate on miscut angle offers in principle a way to check the analytical expression of the surface free energy as a function of step density [18]. Determining the ECS of particles by STM seems to be a new promising approach yielding the orientation dependent relative surface free energy, step and step interaction energies, and the functional dependence of the surface free energy on step density (universal exponent in shape). STM imaging is an excellent way to determine the structure of facets on the ECS, in particular the edge of a facet which is important for fitting the shape by theoretical expressions [37].

ACKNOWLEDGEMENT

Our sincere thanks are due to P. Coenen and Dr. B. Voigtländer for discussions and substantial help with the experiments, and Profs. W. Mullins and P. Wynblatt for a very fruitful and pleasant collaboration and comments on the manuscript. S. Surnev thanks the A. von Humboldt Foundation for a Fellowship.

REFERENCES

[1] C. Herring, in: *Structure and Properties of Solid Surfaces* (edited by R. Gomer and C. S. Smith). Univ. of Chicago Press, 1952. P. 5.
[2] W. W. Mullins, in: *Metal Surfaces:Structure, Energetics and Kinetics,* (edited by N. A. Gjostein and R. W. Roberts). Am. Soc. Metals, Metals Park, OH, 1963. P. 17.
[3] M. Wortis, in: *Chemistry and Physics of Solid Surfaces, Vol. VII* (edited by R. Vanselow and R. Howe). Springer Verlag, New York, 1988. P. 367.
[4] H. van Beijeren and I. Nolden, in *Topics in Current Physics*, Vol. 43, p. 259. Springer Verlag, Berlin 1987.
[5] J. C. Heyraud and J. J. Métois, Surf. Sci. 128 (1983) 334.
[6] A. Pavlovska, D. Dobrev and E. Bauer, Surf. Sci. 326 (1995) 101.
[7] G. Rao, D. B. Zhang and P. Wynblatt, Acta Metall. Mater. 41 (1993) 3331; U. Gangophadyay and P. Wynblatt, Metall. Mater. Trans. 25A (1994) 607.
[8] K. Yamashita, H. P. Bonzel and H. Ibach, Appl. Phys. 25 (1981) 231.
[9] H. P. Bonzel, E. Preuss and E. Steffen, Appl. Phys. A35 (1984) 1.
[10] H. P. Bonzel, E. Preuss and B. Steffen, Surf. Sci. 145 (1984) 20.
[11] A. Rettori and J. Villain, J. Phys. France 49 (1988) 257; L. Lancon and J. Villain, Phys. Rev Lett. 64 (1990) 293. See also: J. Villain and A. Pimpinelli: *Physique de la Croissance Cristalline*. Editions Eyrolles, Paris 1995.

[12] M. Ozdemir and A. Zangwill, Phys. Rev. B 42 (1990) 5013.
[13] W. Selke and Oitmaa, Surf. Sci. 198 (1988) L346; W. Selke and Duxbury, acta phys. slovaca 44 (1994) 215; Phys. rev. B 52 (1995) 17468.
[14] H. Spohn, J. Phys. I France 3 (1993) 63.
[15] H. P. Bonzel and E. Preuss, Surf. Sci. 336 (1995) 209.
[16] U. Breuer and H. P. Bonzel, Surf. Sci. 273 (1992) 219.
[17] H. P. Bonzel, U. Breuer, B. Voigtländer and E. Zeldov, Surf. Sci. 272 (1992) 10.
[18] H. P. Bonzel and W. W. Mullins, Surf. Sci. 350 (1996) 285.
[19] N. A. Gjostein, in: *Metal Surfaces:Structure, Energetics and Kinetics,* (edited by N. A. Gjostein and R. W. Roberts). Am. Soc. Metals, Metals Park, OH, 1963. P. 17.
[20] J. M. Blakely, Progr. Materials Sci. 10 (1963) 395.
[21] H. P. Bonzel, in *Structure and Properties of Metal Surfaces,* Vol. 1, p- 248. Honda Memorial Series on Materials Science (ed. by S. Shimodaira). Maruzen, Tokyo 1973.
[22] S. Surnev, B. Voigtländer, H. P. Bonzel and W. W. Mullins, Surf. Sci. 360 (1996) 242.
[23] N. Cabrera, Surf. Sci. 2 (1964) 320. See also: C. Herring, Phys. Rev. 82 (1951) 87.
[24] E. E. Gruber and W. W. Mullins, J. Phys. Chem. Solids 28 (1967) 875.
[25] C. Jayaprakash, C. Rottman and W. F. Saam, Phys. Rev. B30 (1984) 6549.
[26] N. Garcia and P. A. Serena, Surf. Sci. 330 (1995) L665.
[27] H. P. Bonzel and P. Wynblatt, Z. Physik. Chemie (in print).
[28] M. S. Hoogemann, D. C. Schlösser, J. B. Sanders, L. Kuipers and J. W. M. Frenken, Phys. Rev. B53 (1996) R13299.
[29] R. Butz and H. Wagner, Surf. Sci. 87 (1979) 85. N.-J. Wu, H. Yasunaga and A. Natori, Surf. Sci. 260 (1992) 75.
[30] G. L. Kellog, Surf. Sci. 359 (1996) 237. Surf. Sci. Reports 21 (1994) 1.
[31] N. A. Gjostein, Trans. Metall. Soc. AIME 239 (1967) 785.
[32] H. P. Bonzel, Surf. Sci. 328 (1995) L571.
[33] Ya. E. Geguzin and N. N. Ovcharenko, Sov. Phys.-Sol. State 4 (1963) 2281.
[34] Ya. E. Geguzin, Yu. S. Kaganovskiy and N. V. Stoychev, Fiz. Metal. Metalloved. 26 (1968) 298.
[35] J. C. Heyraud and J. J. Métois, J. Crystal Growth 50 (1980) 571; Surf. Sci. 180 (1987) 647.
[36] G. Rao, D. B. Zhang and P. Wynblatt, Acta metall. mater. 41 (1993) 3331.
[37] S. Surnev, P. Coenen, P. Wynblatt and H. P. Bonzel, to be published.
[38] L. Kuipers and J. W. M. Frenken, Phys. Rev. Lett. 70 (1993) 3907.
[39] A. R. Miedema and J. W. F. Dorleijn, Surf. Sci. 95 (1980) 447.
[40] R. Stumpf and M. Scheffler, Phys. Rev. Lett. 72 (1994) 254.
[41] J. Jacobsen, K. W. Jacobsen and J. K. Norskov, Surf. Sci. 359 (1996) 37.
[42] T. Michely and G. Comsa, Surf. Sci. 256 (1991) 217.
[43] A. F. Andreev, Zh. Eksp. Teor. Fiz.80 (1981) 2042.
[44] C. Rottman and M. Wortis; C. Rottman et al., Phys. Rev. Lett. 52 (1984) 1009.
[45] C. Jayaprakash, W. F. Saam and S. Teitel, Phys. Rev. Lett. 50 (1983) 2017; C. Jayaprakash and W. F. Saam, Phys. Rev. B 30 (1984) 3916.
[46] J. J. Sáenz and N. Garcia, Surf. Sci. 155 (1985) 24.

ANISOTROPY OF WETTING OF PB CRYSTALS BY THEIR OWN MELT AND BY LIQUID GA-PB ALLOYS

Dominique Chatain,[1] and Paul Wynblatt[2]

[1]Centre de Recherche sur les Mécanismes de la Croissance Cristalline*
CNRS, Campus de Luminy, Case 913, 13288 Marseille Cedex 9, France
[2]Department of Materials Science and Engineering
Carnegie Mellon University, Pittsburgh PA 15213, USA

INTRODUCTION

This paper addresses two different sets of observations on the anisotropy of wetting of Pb crystals by its own melt and by Ga-Pb alloys. The observed anisotropies in these cases are due to the anisotropy of the surface free energy of solid Pb and to the intervention of surface phase transitions.

With the exception of the work described below, few reports of anisotropic wetting of solids by liquids are to be found in the literature, even though the anisotropy of the surface free energy of solids, and of the solid-liquid interfacial energy, may engender such behavior. The only other experiments reporting anisotropy of wetting were performed on pure Ge[1] and ice[2]. However, no investigations of such effects in equilibrated multi-component systems appear to have been reported.

Our purpose is to review recent work on the wetting of solid Pb by metallic liquids as a function of the surface orientation of the solid, and of temperature, since the latter changes the surface energy anisotropy of Pb. The simplest case is liquid Pb on solid Pb, which coexist at a single temperature. Thus, in order to explore the effect of temperature, another degree of freedom has to be introduced without changing the behaviour of the solid surface. This may be accomplished by introducing a second component, namely Ga, which melts at a lower temperature than Pb, without segregating to the Pb surface.

The wetting experiments to be described were performed on Pb single-crystals. Interpretation of the results requires a knowledge of the surface energy anisotropy of solid Pb, as well as the atomic scale physical state of Pb surfaces of different orientations. These features of Pb surfaces have been studied by several authors in great detail and are described in the following section.

SOLID SURFACE OF A PB CRYSTAL

In the temperature range explored (between 540K and 600K) the equilibrium shape of Pb exhibits {111} and {100} facets connected by curved regions. Below 560K the connection is smooth, and all the orientations of the solid exist on the crystal. Above this temperature, the curved regions of the crystal undergo premelting and the connection

---

* Laboratoire associé aux Universités d' Aix-Marseille 2 et 3

between the facets and the rounded parts of the crystal becomes sharp, corresponding to the disappearance of some of the orientations on the surface[3,4].

Between 560K and the melting point, premelting begins at {110} orientations[5] and propagates to other surface orientations. At 599K, 5 to 6 monolayers of Pb are premelted on the curved regions of the crystal[6]. The {111} facets remain dry up to the melting point. On the {100} facets a liquid-like disorder develops up to a maximum thickness of about one monolayer ("incomplete premelting") as temperature approaches the bulk melting point[7-10].

The temperature dependence of the anisotropy of solid surface energy of Pb has also been determined, from studies of the equilibrium form of crystals, in the temperature range 540 to 600K[11].

WETTING OF SOLID PB BY ITS OWN MELT

Wetting of the different orientations of a Pb single-crystal by its own melt has been observed in experiments performed on a Pb crystal of millimetric size. Such crystals are large enough for solid-liquid equilibrium to be maintained by imposing a small thermal gradient across their length. This is achieved by passing a current through a vitreous carbon plate which supports the crystal. The experimental device used for preparation of the Pb crystal as well as the optical method used for measuring its wetting by its own melt have been described in detail elsewhere[12].

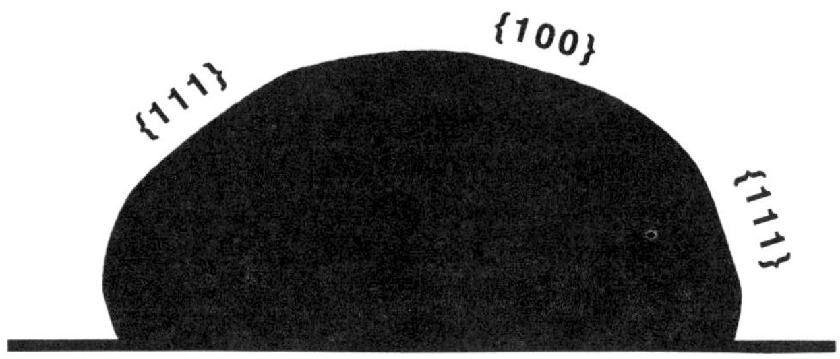

Figure 1. Binary image of a Pb crystal of 7 mm of diameter supported by a vitrous carbon plate.

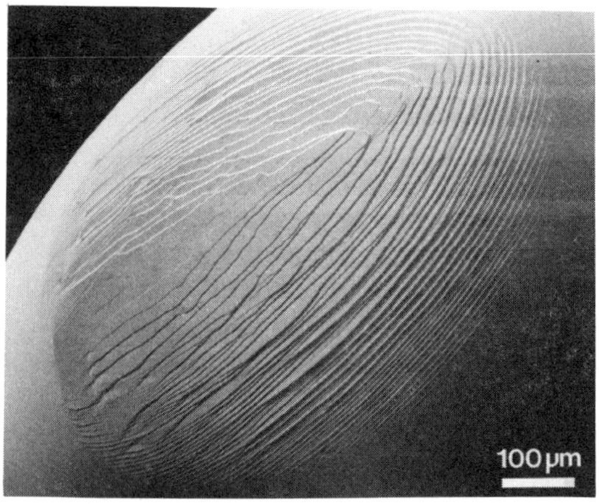

Figure 2. SEM image of the topography of a macroscopic {111} facet.

A typical profile of the solid single-crystal is shown in figure 1. It is a slightly flattened sphere showing plane facets. The angles between the facets allow their orientations, {111} and {100}, to be identified, and prove that the solid is a single-crystal. These planes are the same as those observed on the equilibrium shape of micron-sized crystals near the melting point[11]. Nevertheless, they are not smooth on the atomic scale, but contain some steps of macroscopic size as shown in figure 2.

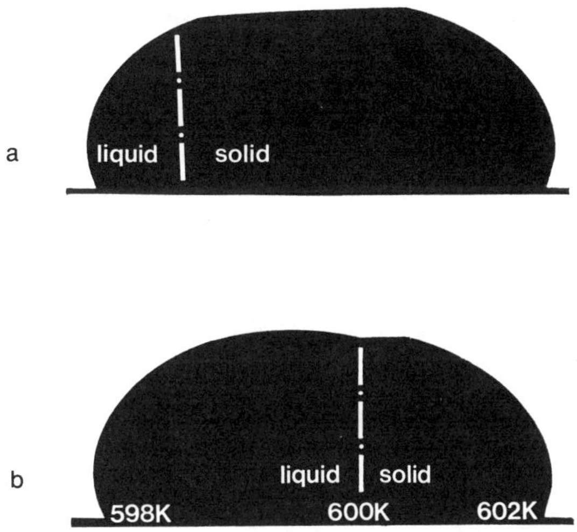

Figure 3. Wetting behaviour of the melt on the crystal. (a) rounded part (the dashed white line corresponds to the position of the external solid-liquid-vapor line seen during the experiment). (b) {111} facet.

Partial melting and freezing cycles allow observation of the manner in which the facets and the curved regions of the crystal form or melt. The curved regions melt and solidify reversibly without any change in curvature on the scale of the observations (see figure 3a). One may therefore conclude that perfect wetting of the melt takes place on the curved regions. This behaviour is to be expected on these premelted surfaces, as the thickness of the surface melted zone is large enough (5 to 6 monolayers at 599 K[6]) to have a surface which is similar to that of the macroscopic liquid.

Different behaviour is observed when crystallization or melting of {111} and {100} facets occurs. When the liquid front reaches the edge of a facet, the liquid swells over the facet in order to increase its contact angle with the facet up to a thermodynamic advancing contact angle. This convex meniscus then invades the facet (see figure 3b). When the liquid reaches the end of the facet, the liquid again perfectly wets the connected curved region. During solidification, the reverse process occurs, with the receding contact angle being identical to the advancing one. This process is consistent with a mechanism proposed by Mutaftschiev and Zell[13] which assumes that the facet already exists under the melt. The partial wetting observed on {111} and {100} facets is characterized by different contact angles, equal to $16.2\pm0.4^\circ$ and $11.4\pm0.8^\circ$, respectively.

These contact angles can be related to the physical state of the surface. The {100} facet is better wetted than the {111} one because the {100} surface is partly premelted. But, the liquid-like disordered monolayer is too thin to have the properties of the macroscopic liquid, and this "adsorbed liquid layer" coexists with a non-wetting macroscopic liquid. This so-called "incomplete surface melting" has also been observed on a pure single crystal of ice[2].

The partial wetting of a melt on a low-index plane of its solid has already been predicted[14-16] and observed on Cd[13], Ga[17], Ge[1], NaCl[18] and KCl[19]. However, as mentioned earlier, indications of anisotropic wetting have been limited to Ge[1] and ice[2]. In the case of Pb, the wetting angle is shown to increase as the surface free energy, or the atomic density of the surface plane, decreases. Germanium was quoted as having a similar behaviour[1].

WETTING OF SOLID PB BY GA-PB ALLOYS

As the temperature decreases below the melting point, premelting eventually disappears at 560K, and the anisotropy of the surface energy of a Pb crystal increases. The equilibrium shape of the crystal then displays {111} and {100} facets which are smoothly connected to the curved regions. The anisotropy of wetting of such a solid has been studied with Ga-Pb liquid alloys which do not modify the surface properties of solid Pb, because the solubility of Ga in Pb is low and the surface energy of Ga is higher than that of solid Pb. The temperature range studied extended from 540 to 586K, in which a Pb-rich solid solution co-exists with a Ga-rich liquid solution. At 586K, a monotectic reaction takes place which produces the melting of the solid Pb-rich phase[20]. For the sake of brevity, henceforth we refer to the Pb-rich solid and the Ga-rich liquid simply as the Pb solid and Ga liquid.

Two-phase particles ranging from 10 to 20 microns in size, supported on a graphite substrate, were observed in-situ in the UHV chamber of a scanning Auger microprobe. Both surface composition analysis and imaging of the particles could be undertaken. The preparation of the samples has been described in detail elsewhere[21].

The particles are formed from compositionally uniform liquid Pb-Ga alloy drops which exhibit a Pb-rich liquid surface. Because of supercooling effects, it is necessary to cool about 25K below the monotectic temperature in order to solidify the Pb solid solution from the liquid, and produce a two-phase particle consisting of a Pb crystal and a small amount of liquid Ga. At this temperature (~560K), a large number of Ga droplets are observed to form at the surface of the Pb crystal. Immediately after solidification of the Pb, Ga droplets tend to form in an irregular pattern which avoids the {111} and {100} orientations of the crystal. This is consistent with the wetting behaviour observed for pure Pb: as the contact angle is the highest on the densest facets, liquid droplets prefers to avoid these locations so as to minimize the total energy of the two-phase system.

Figure 4 shows the equilibrium shapes of liquid Ga-solid Pb particles obtained after annealing at different temperatures. At a temperature of 560K, Ga droplets distribute themselves on the surface of the Pb in a three-fold pattern around the (111) facet, as shown in Fig. 4a. The Pb orientations that are wetted by Ga are close to {210}. From simple considerations of surface free energy anisotropy, based on the number of broken bonds per unit area, the {210} orientations of face centered cubic solids are expected to have the highest surface free energy[22]. In particular, in the case of pure Pb, Heyraud and Metois[11] have demonstrated experimentally that there is a maximum in surface free energy in the vicinity of {210} along the {100} zone, at temperatures below about 573K. Thus, we conclude that wetting of the solid Pb surface by Ga is highly selective, and only occurs where the surface free energy of Pb is the highest. There may also be a contribution to this anisotropic wetting that arises from the anisotropy of the solid Pb-liquid Ga interface. In order for the effects at the solid-liquid interface to favor wetting of {210} orientations, the interfacial energy would need to be relatively low at {210} substrate orientations, a conclusion that appears unlikely. Rather, we suspect that the contribution to wetting associated with the removal of the high energy solid {210} surface must overwhelm any effects due to solid-liquid interface energy anisotropy. In addition, it should be noted that the energy of the solid-liquid interface could be lowered by segregation effects, thus further diminishing its overall role in the observed wetting behavior.

At a temperature of about 570K, the Ga droplets begin to scatter on the rounded parts of the crystal away from the {210} orientations. This is illustrated for somewhat higher temperature in Fig. 4b. The motion of the droplets indicates that {210} ceases to be the orientation with the maximum surface free energy; this conclusion is consistent with the observations of Heyraud and Métois[11] who show that the maximum in surface free energy of Pb disappears somewhere between 550 and 573K. This change is related to the surface premelting of Pb which begins at about 560K on the {110} orientations[5], and gradually spreads to orientations of lower packing density with increasing temperature[23]. Premelting erases the surface energy anisotropy of the curved regions and produces the scattering of droplets away from {210}.

Figure 4c shows the disappearance of Ga droplets when the temperature exceeds the monotectic temperature at which the solid Pb phase melts. The Ga liquid is no longer visible either by Auger spectroscopy or by the secondary electron imaging. This means that the Pb layer covering the surface of the liquid Ga is thick enough (at least 2nm) for concluding that perfect wetting of liquid Ga by liquid Pb prevails. This feature has been studied in more detail in a recent experimental paper[24].

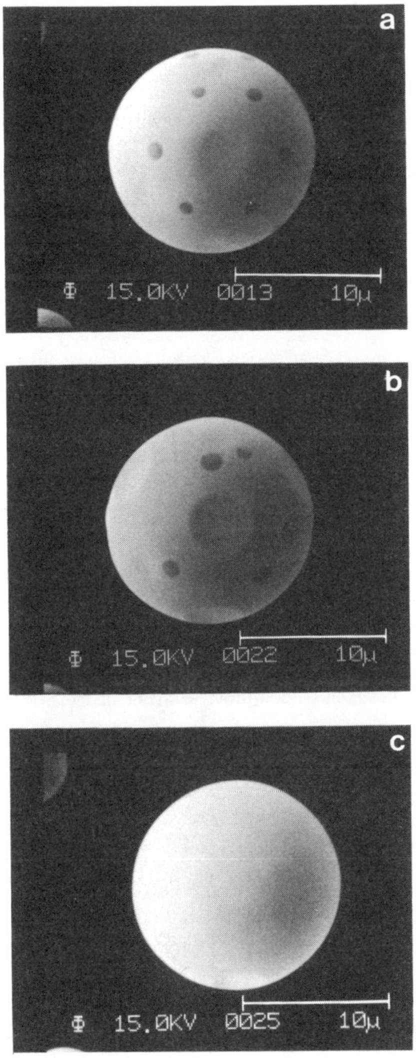

Figure 4. Evolution of the distribution of Ga liquid on a Pb crystal as temperature is increased from (a) 560K, to (b) 582K, and to (c) 586K.

SUMMARY

In a first set of experiments, observations of anisotropic wetting of different orientations of a macroscopic (milimeter-sized) solid single crystal of Pb by pure liquid Pb have been made at the melting point. It was found that the surface orientations which undergo premelting transitions below the bulk melting point are perfectly wetted by the liquid, whereas the {100} and {111} facets, which do not premelt, are only partially wetted. On those surfaces, wetting improves with decreasing atomic density of the surface.

In a second set of observations on Pb-Ga alloys, wetting of a mezoscopic (micron-sized) solid single crystal of Pb by liquid Ga was studied in a scanning Auger microprobe. At relatively low temperatures, below the premelting of the solid Pb surfaces, liquid Ga droplets are found to reside only at the {210} orientations of Pb surface. As the temperature is raised above that where most Pb surface orientations undergo premelting, the anisotropic distribution of Ga droplets disappears.

In both cases, the anisotropy of wetting is driven by the anisotropy of surface energy of the solid. The degree of wetting is shown to be strongly correlated to the degree of premelting at the solid surface.

Acknowledgment

This paper has drawn heavily on materials published previously in references 12 and 21, which involved the co-authorships of Drs. J.J. Métois and W.C. Cheng. The authors acknowledge with thanks support of their research by a joint grant from CNRS and the International Program Division of NSF (grant INT9313994). PW also wishes to acknowledge the support of the NSF under grant DMR 9530469.

REFERENCES

1. Yu.V. Naidich, N.F. Grigorenko and V.M. Perevertailo, Interphase and Capillary Phenomena in Crystal Growth and Melting Processes, *J. Crystal Growth*, 53:261 (1981).
2. M. Elbaum, S.G. Lipson and J.G. Dash, Optical Study of Surface Melting on Ice, *J. Crystal Growth*, 129:491 (1993).
3. J.C. Heyraud, J.J. Métois and J.M. Bermond, Surface Melting and Equilibrium Shape; the Case of Pb on Graphite, *J. Crystal Growth*, 98:355 (1989).
4. A. Pavlovska, K. Faulian and E. Bauer, Surface Roughening and Surface Melting in the High Temperature Equilibrium Shape of Small Crystal, *Surface Science*, 221:233 (1989).
5. J.W.M. Frenken and J.F. van der Veen, Observation of Surface Melting, *Phys. Rev. Lett.*, 54:134 (1985).
6. B. Pluis, A.W. Denier Van der Gon, J.W.M. Frenken and J.F. Van der Veen, Crystal-Face Dependence of Surface Melting, *Phys. Rev. Lett.*, 59:2678 (1987).
7. B. Pluis, A.W. Denier Van der Gon, J.F. Van der Veen and A.J. Riemersma, Surface-Induced Melting and Freezing I. Medium-Energy Ion Scattering Investigation of the Melting of Pb {hkl} Crystal Faces, *Surface Science*, 239:265 (1990).
8. M. Van Pinteren and J.W.M. Frenken, Incomplete Melting of Pb (100) and Vicinal, *Surface Science*, 275:383 (1992).
9. H.N. Yang, K. Fang, G.C. Wang and T.M. Lu, Vacancy-Induced Disordering in the Pb (100) Surface, *Phys. Rev. B*, 44:1306 (1991).
10. E.A. Murphy, H.E. Elsayed-Ali, K.T. Park, J. Cao and Y. Gao, Angle-Resolved X-Ray-Photoemission Study of the Surface Disordering of Pb (100), *Phys. Rev. B*, 43:12615 (1991).
11. J.C. Heyraud and J.J. Métois, Equilibrium Shape and Temperature : Lead on Graphite, *Surface Science*, 128:334 (1983) .
12. D. Chatain and J.J. Métois, A New Procedure for the Determination of the Free Energies of Solid-Fluid Interfaces from the Anisotropy of Wetting of a Melt on its Solid, *Surface Science*, 291:1 (1993)
13. B. Mutaftschiev, J. Zell, Energie d'Interface et Croissance du Cadmium en Bain Fondu, *Surface Science*, 12:317 (1968).
14. I.N. Stranski, Uber den Schmelzvorgang bei Nichtpolaren Kristallen, *Die Naturwissenschaften*, 30:425 (1942).
15. B. Mutaftchiev, L'Interface Cristal-Bain Fondu, *Bull. Soc. Fr. Mineral. Cristallogr.*, 92:558 (1969).
16. G.A. Chadwick, in "*Interfaces*", R.G. Grifkin Ed., Butterworths, Melbourne, 101 (1969).
17. O. Züger, U. Dürig, STM Investigation of Gallium Single-Crystal Surfaces Close to Bulk Melting, *Ultramicroscopy*, 42-44:520 (1992).
18. G. Grange, B. Mutaftschiev, Méthode de Mesure de l'Angle de Contact à l'Interface Cristal-Bain Fondu, *Surface Science*, 47:723 (1975).
19. G. Grange, R. Landers, B. Mutaftschiev, Contact Angle and Surface Morphology of KCl Crystal-Melt Interfaces Studied by the "Bubble Method", *Surface Science*, 54:445 (1976).
20. I. Ansara and F. Ajersch, in: *Binary Alloy Phase Diagrams*, 2nd Ed., T. B. Massalski Editor, ASM International, Materials Park, 1837 (1990).
21. W.C. Cheng, D. Chatain and P. Wynblatt, Observations of Anisotropic Wetting in Solid-Liquid Pb-Ga Alloys, *Surface Sience Lett.*, 327:501 (1995).
22. J.K. Mackenzie, A.J.W. Moore and J.F. Nicholas, Bonds Broken at Atomically Flat Crystal Surfaces - I Face-Centred and Body-Centred Cubic Crystals, *J. Phys. Chem. Solids*, 23:185 (1962).
23. H. Hakkinen and M. Manninen, Computer Simulation of Disordering and Premelting of Low-Index Faces of Copper, *Phys. Rev. B*, 46:1725 (1992).
24. D. Chatain and P. Wynblatt, Experimental Evidence for a Wetting Transition in Liquid Ga-Pb Alloys, *Surface Science*, 345:85 (1996).

RELAXATION OF NANOMETER-SCALE SURFACE MORPHOLOGY

S. Jay Chey and David G. Cahill

Department of Materials Science and Engineering, and
Materials Research Laboratory
University of Illinois at Urbana-Champaign
Urbana, IL, 61801

ABSTRACT

We review recent research on the relaxation of microscopic surface roughness in ultra-high vacuum, including our scanning tunneling microscopy experiments on the smoothing of Ge(001). For nanometer-scale roughness on Ge(001), the dependence of the smoothing kinetics on the in-plane length scale of the roughness shows that the mass transport is "non-conserved, step-mobility-limited." Step mobilities extracted from our experiments on Ge(001) are compared to data from several experiments on Si(001). The range of step-mobilities spans a factor of 10^{10}; most of the data are consistent with an activation energy of 1.8 eV and an attempt-rate comparable to the frequency of atomic vibrations.

INTRODUCTION AND REVIEW

The evolution of surface morphology during the deposition and etching of thin films is controlled by the thermodynamics and kinetics of mass transport on surfaces. One important example is the roughening of surfaces during low temperature processing: reduced temperatures are often desirable in the fabrication of thin film devices, but at low temperatures, kinetic limitations to the motion of adatoms and surface vacancies can produce rough surface morphologies [1, 2]. These morphologies have direct consequences for the abruptness of interfaces in heteroepitaxy and can also play a critical role in controlling the microstructure of thin film materials [3]. Recent work on the self-assembly of "quantum dots" provides a second example: during deposition of lattice-mismatched layers, compressive strain drives the formation of nanometer-scale islanding with narrow distributions of island sizes and spacings. In this case,

Dynamics of Crystal Surfaces and Interfaces
Edited by Duxbury and Pence, Plenum Press, New York, 1997

lattice strain modifies the surface energies and kinetics in manner that produces self-organization.

The complete description of morphology evolution during thin film growth is an extremely complex problem. But we can make significant progress by studying the relaxation of a surface morphology during thermal annealing. In other words, we first ask: what is the correct description of how a non-equilibrium surface morphology approaches equilibrium?

The initial answer to this question was provided by Mullins [4], building on the work of Herring [5]. For surfaces orientations at temperatures above their thermodynamic roughening temperature T_R (the free energy for step formation becomes zero at $T = T_R$), the relaxation is driven by the stiffness [6] of the surface $\tilde{E} = E + dE^2/d\theta^2$, where E is the surface energy and θ is the orientation of the surface. For mass transport by surface diffusion, the dynamics of the surface at $T > T_R$ are described by

$$\frac{\partial z}{\partial t} = -\frac{D\nu\Omega^2\tilde{E}}{kT}\nabla^4 z , \tag{1}$$

where z is the height of the surface as function of time t and position $\mathbf{r}$, Ω is the volume of the surface species that controls mass transport on the surface (e.g., Ω is the volume per adatom [7]), D is the surface diffusivity of adatoms, and ν is the density of adatoms per unit area.

Review of Experiments

Surprisingly, experiments on the relaxation of *micron-size* surface structures are generally in good agreement with Eq. 1 despite the fact that, in most cases, the annealing temperatures are below T_R [8]; and, therefore, the derivation of Eq. 1 cannot be justified because of the finite free energy of surface steps. In Fig. 1, we summarize selected data for the relaxation of one dimensional (1-D), gratings on single crystal surfaces [9, 10, 11, 12]. The grating period L is typically on the order of 10 μm, and the decay rate τ^{-1} is derived from the exponential decay of the grating amplitude, $z(t) = z_0 \exp(-t/\tau)$.

Although, Eq. 1 is routinely applied in models of morphology evolution during thin film deposition and etching, to the best of our knowledge, no experiment has demonstrated the validity of Eq. 1 at the small length scales and relatively low temperatures relevant to these models. The lack of grating relaxation data for $L \ll 1$ μm is likely due to the difficulty of the experiments: high temperatures are typically needed to clean surfaces and eliminate defects in the near-surface region but high temperature annealing would also rapidly relax the grating.

Consequently, recent experiments on the dynamics of microscopic surface morphologies have employed kinetically roughened surfaces produced by low temperature crystal growth or ion-etching [13, 14, 15, 16]. High-resolution diffraction is used to measure the evolution of the average terrace size l; the time dependence of l is interpreted in terms of a power law, $l \propto t^\alpha$. Experimental values for α can then be compared to theory to gain insight on the relative importance of various atomic-scale mechanisms.

Diffraction experiments have discovered values of α that span nearly the entire range $0 < \alpha < 1$, see Fig. 2. Theories for the coarsening of self-similar morphologies predict $\alpha = 1/i, i = 3, 4, 5$; the value of i depends on assumptions about the geometry of the roughness (e.g., 1-D vs. 2-D) and mechanisms for mass-transport; many of the experiments have indeed found $1/5 < \alpha < 1/3$. We note however, that in our opinion, the interpretation of these diffraction experiments is greatly complicated by the fact that the variation of the step density in the plane of the surface, the lateral characteristic

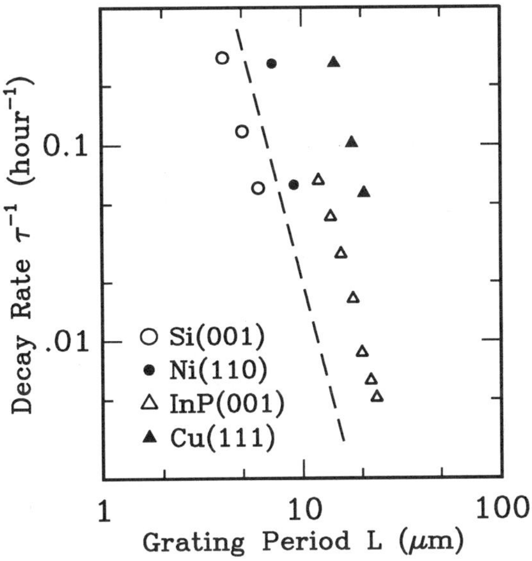

Figure 1. Exponential decay rates τ^{-1} of one-dimensional gratings with periodicity L. The dashed line shows the dependence $\tau \propto L^4$ predicted by Eq. 1. The annealing temperatures of the Si(001) [12], Ni(110) [10], InP(001) [11], and Cu(111) [9] samples are 1000, 900, 780, and 970°C, respectively. For Si(001), $T_R \simeq 1200$°C [30]. For Ni(110), $T_R \simeq 1200$°C ; for Cu(111), T_R exceeds the melting point [8].

length scale of the roughness, is essentially unknown. In other words, the experiments do not tell us over what range of length scales the surface morphology is truly self-similar nor do the experiments tell us what range of length scales is involved in the relaxation of the surface roughness.

Review of Theory

At $T > T_R$, the relaxation of a non-equilibrium surface morphology by surface diffusion can be described by Eq. 1: the thermodynamic driving force for smoothing smoothing is the surface stiffness $\tilde{E}$ and the kinetics of the smoothing is determined by the concentration and mobility of the surface point defects that provide the mass transport, e.g. adatoms. At $T < T_R$, on the other hand, we must consider a more microscopic description of the dynamics that is based on the thermodynamics of the interactions between steps, and the kinetics of step motion [17].

First, we consider the *thermodynamics* of a stepped surface. The step density, $n = 1/l$, is related to the surface orientation θ by $\tan|\theta| = na$, where a is the step height. To third order in n, the surface free energy $E(n)$ is given by [17]

$$E(n) = E_0 + \beta n + \frac{\gamma}{3} n^3 \quad , \tag{2}$$

where β is the free energy of a step, and γ measures the strength of the interactions between steps [17] created by step entropy or elastic deformations. For entropic interactions between steps [18],

$$\gamma = \frac{\pi^2 (kT)^2}{2\tilde{\beta}} \quad , \tag{3}$$

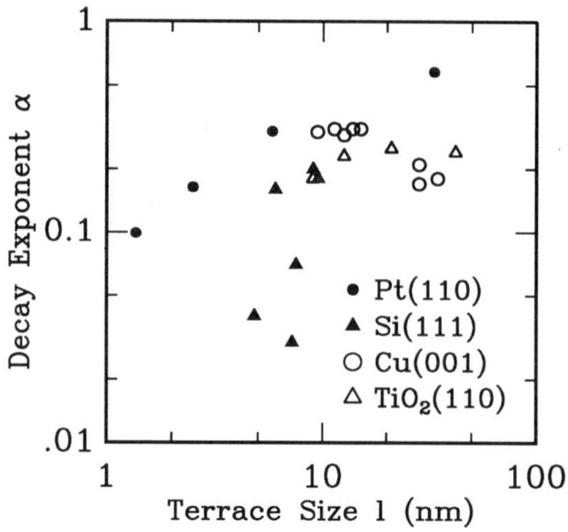

Figure 2. Summary of data for the exponent α in the power-law growth of average terrace size $l \propto t^{\alpha}$ measured by x-ray diffraction from Pt(110) [13] and electron diffraction from Si(111) [15], Cu(001) [14], and $TiO_2(110)$ [16]. Data points for α are plotted at approximately the geometric mean $l_m = \sqrt{l_{min} l_{max}}$ of the range of l over which the power law is assumed to describe the data. In most cases, $l_{max}/l_{min} \sim 3$.

where $\tilde{\beta}$ is the step edge stiffness $\tilde{\beta} = \beta + d^2\beta/d\theta_s^2$ and θ_s is the orientation angle of the step. And for elastic strain interactions [8],

$$\gamma = \frac{6(1-\nu^2)\mathbf{d_1} \cdot \mathbf{d_2}}{\pi E_Y}, \tag{4}$$

where ν and E_Y, are the Poisson's ratio and Young's modulus that describe the elastic deformation of the solid. For identical steps, $\mathbf{d_1} \cdot \mathbf{d_2} = d^2$, where d is the magnitude of the stress dipole moment of the step.

Next, we consider the *kinetics* of mass transport on the surface. Descriptions of surface mass transport can be classified into two limiting cases i) diffusion-limited; and ii) step-mobility-limited models. Rettori and Villain [19]; and Ozdemir and Zangwill [6] analyzed the diffusion-limited relaxation of a surface described by Eq. 2; the kinetics of surface mass transport were assumed to be the same as in the Mullins model—determined by the concentration and mobility of surface point-defects. For a 1-D surface morphology, this assumption [6] predicts that the lifetime τ of non-equilibrium surface structure should scale as $\tau \propto L^5/z_0$, where L is the lateral length scale of the morphology and z_0 is the initial amplitude of the surface roughness.

Step-mobility-limited models can be further separated into two limits: "conserved" and "non-conserved" [20]. This terminology refers to the *local* conservation of mass; transport is said to be conserved if a surface defect generated at a step edge eventually annihilates at the same step or at one of the two adjacent steps. Thus, the motion of adjacent steps is coupled. The 1-D "conserved" model of Nozières [21] predicts $\tau \propto L^4$, independent of z_0. On the other hand, in a "non-conserved" model the motion of adjacent steps is uncorrelated: surface defects generated at a step edge can annihilate at *any* step edge on the surface. Uwaha [22] has considered this case and found $\tau \propto L^2(L/z_0)^2$. In the discussion below, we will use these two limiting cases of step-mobility-limited models [21, 22] to extract the step-mobilities on Si(001) and Ge(001) surfaces from experiments on relaxation kinetics.

DYNAMICS OF THE Ge(001) SURFACE

To obtain a more complete description of the smoothing of microscopic surface roughness, we have used ultra-high vacuum scanning tunneling microscopy (STM) to precisely measure the structure of rough Ge(001) surfaces with high spatial resolution over a wide range of lateral length scales. Furthermore, using ion-etching, we have produced starting surfaces with different characteristic in-plane length scales L but nearly the same average step densities [23, 24]. These starting surfaces give us a way of quantifying the dependence of smoothing kinetics on L for $L \sim 100$ nm.

Production and Characterization of Nanometer-scale Roughness

We first prepare a smooth surface by ion-etching the Ge crystal for 1 hour at 430°C. Subsequently, this flat surface is roughened by ion-etching at reduced temperature, 270°C [23]. Etching times of 10, 42, or 180 minutes are used to produce rough surface morphologies with characteristic in-plane length scale L of 37, 65, and 118 nm, respectively, see Fig. 3.

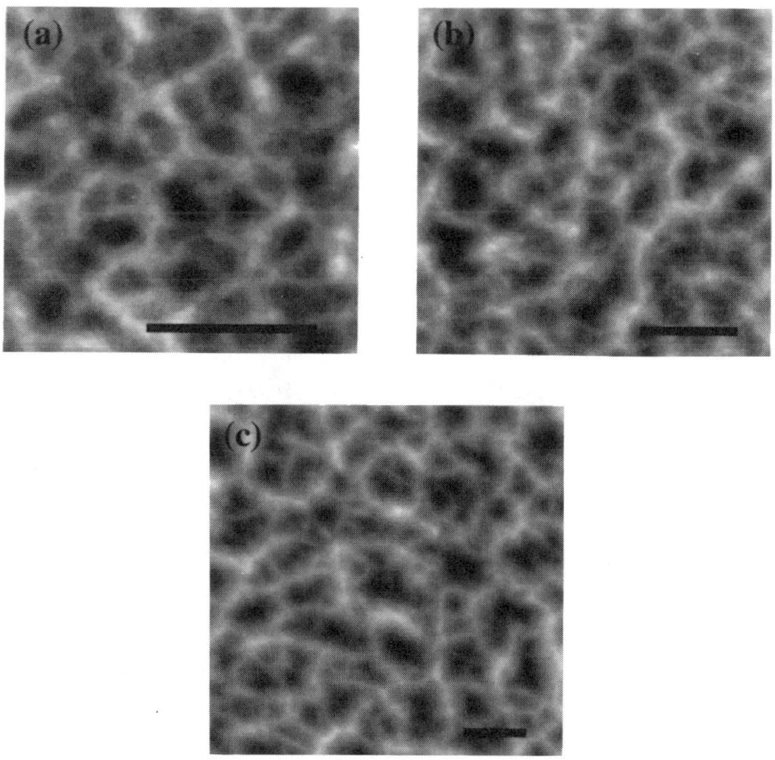

Figure 3. STM images of surface morphologies produced by etching of Ge(001) by 240 eV Xe ions at 270°C: ion-etching time, L, and scan area are: (a) 10 min, 37 nn, 200×200 nm^2; (b) 42 min, 65 nm, 360×360 nm^2; (c) 180 min, 118 nm, 650×650 nm^2. The scale bar is 100 nm.

To quantify the roughness, we evaluate the height-difference correlation function,

$G(\rho) = \langle (h_j - h_i)^2 \rangle$, and the height-height correlation function $H(\rho) = \langle h_j h_i \rangle$ [8]; h_j and h_i are the heights of the surface at two locations labelled by i and j separated by a distance ρ. The brackets signify an average over pairs of points i, j. The small ρ limit of $G(\rho)$ shows that the average step densities of the three surfaces shown in Fig. 3 are nearly identical [24]. We define a characteristic in-plane length scale L by the position ρ of the first local maximum in $H(\rho)$ [24]; L corresponds to the separation or repeat distance between dominant features in the surface morphology. Consistent with our observation that the average step density is nearly constant, the amplitude of the surface roughness as measured by $G^{1/2}(L/2)$ increases nearly linearly with L. $G^{1/2}(L/2) = 0.32$, 0.50, and 0.80 nm for $L = 37$, 65, and 118 nm respectively, giving a power law of $G^{1/2}(L/2) \propto L^{0.8}$. Not only is the *average* step density constant, the *distribution* of step densities is also nearly independent of L, see Fig. 4.

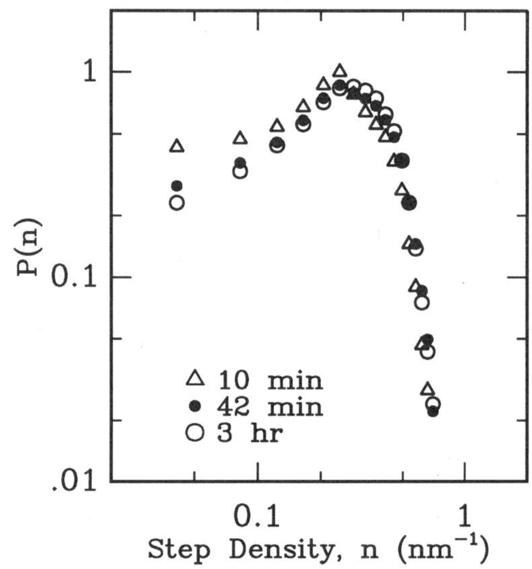

Figure 4. Histogram of step densities for the three Ge(001) surfaces shown in Fig. 3. The relative probabilities $P(n)$ of step density n are determined by measuring the local surface orientation θ in an area 4×4 nm^2 centered on each image pixel; and using $na = \tan|\theta|$.

Relaxation of Surface Roughness

Empirically, we find that the time evolution of $G(\rho, t)$ can be approximated as $G(\rho, t) = f^2(t/\tau)G(\rho, 0) + G_f(\rho)$ where G_f describes the roughness of a surface following a relatively high temperature, long time anneal, 325°C for $t = 6$ hours. Since the largest ratio of $G(\rho, 0)/G_f(\rho)$ occurs at a lateral length scale near $\rho = L/2$, we use the time dependence of $G(L/2, t)$ to extract $f(t/\tau)$. We find that a power law of the form $f(t/\tau) = (1 + t/\tau)^{-1/2}$ describes the decay of the roughness quite well, see Fig. 5a. Each data point in Fig. 5a constitutes a complete cycle of experiments: preparation of a flat surface at elevated temperature, roughening the surface by etching at 270°C, annealing, and imaging by STM at room temperature [24].

To determine the dependence of τ on the in-plane length scale of the surface roughness L, we study the relaxation of the three rough starting surfaces with $L = 37$, 65 and 118 nm at a fixed annealing temperature $T = 270$°C, see Fig. 5a. Here, we assume that the time constant of the smoothing process scales as $\tau \propto L^n$ and adjust n to obtain the best fit to the data. We find $n = 2.2 \pm 0.4$, consistent with the non-conserved step-mobility-limited kinetics for the relaxation of a 1-D morphology [22, 24]. Fig. 5b shows a typical step structure on an annealed surface.

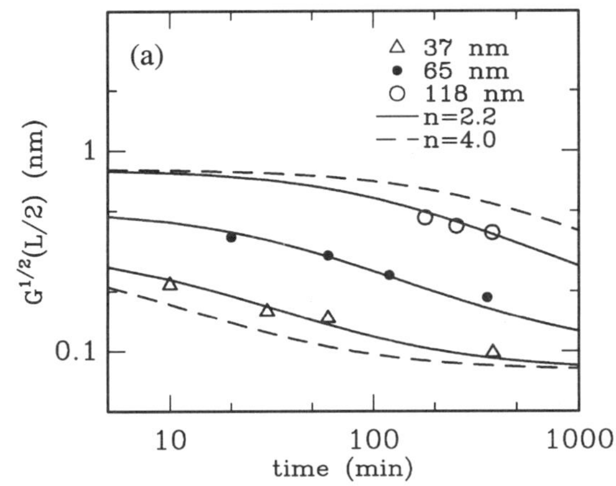

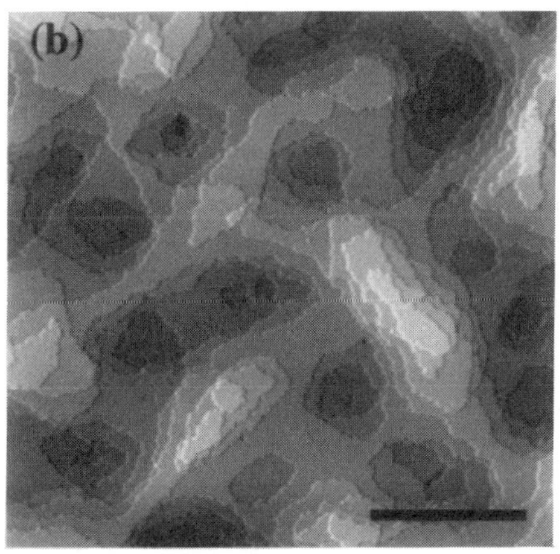

Figure 5. (a) Decay of Ge(001) surface roughness as a function of lateral length scale L for an annealing temperature $T = 270°C$; $L = 37$ nm (open triangles), $L = 65$ nm (filled circles), $L = 118$ nm (open circles). Assuming $\tau \propto L^n$, the solid line shows the calculated decay using $n = 2.2$; the dashed line is calculated using $n = 4.0$. (b) STM image of the $L = 118$ nm morphology, see Fig. 3c, following a 6 hour anneal at 270°C. The scale bar is 100 nm.

STEP MOBILITY

In this section, we analyze experiments on the relaxation of non-equilibrium Si(001) [12, 25] and Ge(001) [24] morphologies to extract values for the step-mobility as a function of temperature. Mobilities derived from the relaxation experiments are compared to more direct measurements of step-mobilities using low energy electron microscopy (LEEM) [26] and STM [27, 28].

Grating Relaxation on Si(001)

Keefe et al. [12] observed that the relaxation of micron-sized 1-D gratings on Si(001) is consistent with Eq. 1. But as discussed above, the derivation of Eq. 1 is not strictly valid at $T < T_R$. We show here that these experiments are also in agreement with dynamics of the conserved, step-mobility-limited model derived by Nozières [21]:

$$z_t = -\frac{a\eta}{2}\frac{\partial}{\partial x}\left(\frac{1}{n}\zeta_{xx}\right) , \tag{5}$$

where η is the step-mobility, and ζ is the step chemical potential $\zeta = dE/dn$; To second order in n, $\zeta = \beta + \gamma n^2$ (see Eq. 2). The subscripts t and x denote partial derivatives with respect to time and position. Using $z_x = an$, we find

$$z_t = -\eta\gamma\left(\frac{2z_{xx}z_{xxx}}{z_x} - \frac{z_{xx}^3}{z_x^2} + z_{xxxx}\right) . \tag{6}$$

We analyze Eq. 6 by examining the time evolution of the shape-preserving solution. Separating variables, we write $z(x,t) = u(x)\omega(t)$, where $u(x)$ is the profile of the shape preserving solution, and $\omega(t)$ describes the time evolution. The ordinary differential equation for $\omega(t)$, $\omega_t = -\tau^{-1}\omega$, shows that the shape preserving profile decays exponentially, $z(x,t) = z_o(x)\exp(-t/\tau)$. The differential equation for $u(x)$ is probably unsolvable analytically, but dimensional analysis shows that $\tau^{-1} \simeq 2\eta\gamma(2\pi/L)^4$ where L is a characteristic in-plane length scale of the surface morphology. We see that the Nozières model [21], Eq. 6, produces essentially the same behavior as the Mullins model, Eq. 1: the grating amplitude decays exponentially in time with a decay rate $\tau^{-1} \propto L^{-4}$.

To extract a value of the step-mobility η from the grating relaxation experiments [12], we must evaluate the strength of the step-step interaction γ. Computational work suggests that γ due to elastic interactions between Si(001) steps is $\simeq 0.2$ eV nm [29], while, we estimate that the entropic interaction is ~ 10 times larger. (We use a step stiffness $\tilde{\beta}$ calculated from the geometric mean of $\tilde{\beta}$ for S_A and S_B steps given in Ref. [30]: $\tilde{\beta} \simeq 0.03$ eV nm^{-1}.) Therefore, entropic repulsion should dominate, and

$$\tau^{-1} \simeq \eta\,\frac{\pi^2(kT)^2}{\tilde{\beta}}\left(\frac{2\pi}{L}\right)^4 . \tag{7}$$

Figure 6 includes η derived (using Eq. 7) from data for the relaxation of a $L = 5$ μm Si(001) grating at 900 and 1100°C [12].

LEEM Measurements of Step Dynamics

Bartelt and Tromp [26] have recently described direct measurement of step-mobilities using low energy electron microscopy (LEEM). We include their data for Si(001) in Fig. 6.

Relaxation of Step Spacing

Next, the step-mobility of Si(001) is estimated at lower temperatures, $T \simeq 500$°C, from the experiments of Webb et al. [25] on the relaxation kinetics of non-equilibrium step-spacings. In this experiment, the average terrace size was large, and therefore, due to the stress anisotropy of the 2×1 surface reconstruction, a long range interaction of the form

$$E = C_2 n \ln(\pi b n) \tag{8}$$

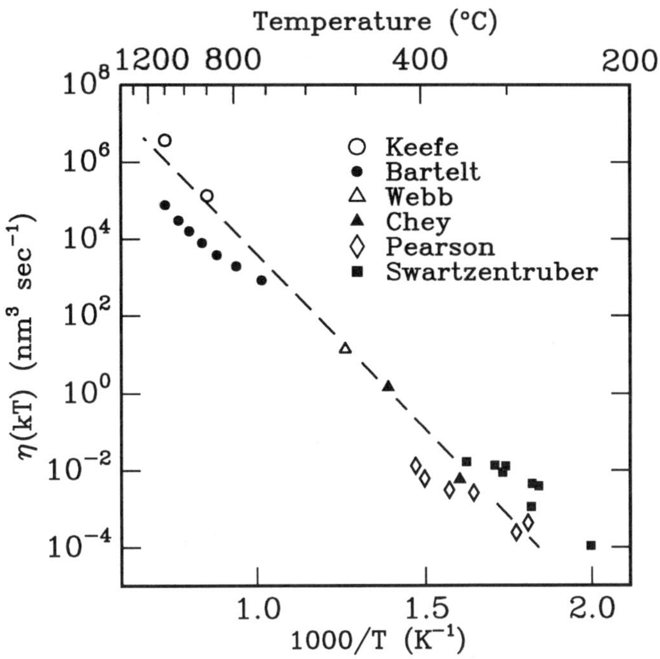

Figure 6. Compilation of step-mobilities derived from several experiments on Si(001) and Ge(001). The temperatures for the two data points for Ge(001) (filled triangles) have been scaled by the ratio of the cohesive energy of Si to Ge, 1.20. The dashed line shows a thermally activated process with an activation energy of 1.8 eV and a prefactor $b^3k\Theta/\hbar$; Θ is the Debye temperature of Si, 650 K, and $b = 0.38$ nm.

is thought to dominate the step-step interactions [25]. For Si(001), $C_2 \simeq 0.008$ eV nm^{-1} [31].

We again apply the conserved step-mobility-limited model of Nozières (Eq. 5); but, because the change in the step-spacing is small, or at least comparable to the average step-spacing, we can linearize Eq. 5 using $n(x) = n_0 + \delta n(x)$. The change in the step chemical potential $\delta\zeta(x)$ due to the surface stress anisotropy is related to the change in the step density $\delta n(x)$ by

$$\delta\zeta(x) = C_2 \frac{\delta n(x)}{n_0} \quad , \tag{9}$$

leading to the equation of motion

$$z_t = -\frac{\eta C_2}{2} \frac{\delta z_{xxxx}}{n_0^2} \quad . \tag{10}$$

In Ref. [25], the periodicity of the modulation in step spacing is $2l$. Therefore, $\delta z_{xxxx} \simeq (\pi/l)^4 \delta z$ and

$$\tau^{-1} = \eta \frac{C_2 \pi^4}{2l^2} \quad . \tag{11}$$

For a 0.05 degree miscut surface ($l = 156$ nm), the time constant of the relaxation of the step-spacing was $\tau = 315$ sec at $T = 520°$C [25]. The step-mobility calculated using these data and Eq. 11 is included in Fig. 6.

Ge(001) Nanometer-scale Roughness

In our previous work [24], we modeled the relaxation of our Ge(001) surfaces using the 1-D model of Uwaha [22] because this approach is the simplest consistent model that provides good agreement with the data. We do not know of a way to quantitatively justify the use of a 1-D model for these complex morphologies, see for example Fig. 5b, but we can examine the step chemical potential $\zeta = \beta + \gamma n^2$ to gain some insight on when a 1-D model is appropriate. We suggest that the ratio of the two terms, $f = \gamma n^2/\beta$, should give an approximate measure of when a 1-D model is valid. Choosing appropriate parameters at $T = 600$ K, $n = 0.3$ nm^{-1} (see Fig. 4), $\tilde{\beta} = 0.030$ eV nm^{-1} and $\beta = 0.050$ eV nm^{-1} [30], gives $f \sim 1$; therefore, we cannot easily rule out a contribution from the step curvature (2-D aspect of the morphology) to the driving force for relaxation.

Nevertheless, the 1-D, non-conserved, step-mobility-limited model [22, 24] does appear to fit the data and we will apply it here to extract the step-mobility. For non-conserved transport, the step velocity v is simply related to the gradient in the step chemical potential by the step-mobility η:

$$v = -\eta \frac{d\zeta}{dx} \quad . \tag{12}$$

Since $z_t = -van$ and $z_x = an$,

$$z_t = \frac{2\eta\gamma}{a^2} z_x^2 z_{xx} \quad . \tag{13}$$

As before, we examine the time evolution of the shape-preserving solution by separating variables. The ordinary differential equation for the time evolution, $d\omega/dt = -\alpha\omega^3$, yields $\omega(t) = (1 + t/\tau)^{-1/2}$, which is the functional form we use to fit the relaxation data, see for example Fig. 5a. Dimensional analysis for the decay time constant—combined with Eq. 3 expressing the entropic repulsion between steps— gives

$$\tau^{-1} \simeq \eta \, \frac{\pi^2 (kT)^2}{\tilde{\beta}} \left(\frac{2\pi}{L}\right)^4 \left(\frac{z_0}{a}\right)^2 \quad , \tag{14}$$

where z_0 is the initial amplitude of the surface roughness.

Step-mobilities for Ge(001) using Eq. 14 are included in Fig. 6 by scaling the annealing temperature by the ratio of the cohesive of energy of Si to Ge, 1.20 [32]. In other words, to enable a comparison between the Ge and Si experiments, we assume that activation energies for Si(001) are 1.20 times larger than the equivalent activation energies for Ge(001).

STM Measurements of Step Dynamics

Using "hot" STM [27, 28], atomic-scale fluctuations in the step configurations have been measured in real-time. Because of the symmetry of the surface reconstruction, the length of a S_B step is only seen to change in units of $2b$, where b is the size of the surface 1×1 unit cell, $b = 0.38$ nm. Therefore, we convert the measured rates to step-mobilities using

$$\tau^{-1} = \frac{\eta(kT)}{(2b)^3} . \tag{15}$$

To add the data from Ref. [28] to Fig. 6, we use τ^{-1} for kink diffusion. From Ref. [27], we use τ^{-1} for the total "event rate."

CONCLUSION

Data for step-mobilities shown in Fig. 6 span an impressively large range: a factor of 10^{10} separates step-mobilities measured by STM from the step-mobilities extracted from the relaxation of micron-sized gratings. Some discrepancies exist, but most of the step-mobilities are consistent with a single activation energy of 1.8 eV and an attempt rate given by the frequency of atomic vibrations. We hope that this initial comparison of step-mobility data will help motivate more detailed theoretical analysis and experiments on the connections between step-mobility and the evolution of surface morphology.

ACKNOWLEDGMENTS

This work was supported by the U.S. Department of Energy grant DEFG02-91-ER45439 through the University of Illinois Materials Research Laboratory. We thank Norm Bartelt for helpful discussions and for sending a preprint of his work prior to publication.

REFERENCES

[1] M. D. Johnson, C. Orme, A. W. Hunt, D. Graff, J. Sudijono, L. M. Sander, and B. G. Orr, Phys. Rev. Lett. **72**, 116 (1994).

[2] J. E. Van Nostrand, S. Jay Chey, M.-A. Hasan, D. G. Cahill, and J. E. Greene, Phys. Rev. Lett. **74**, 1127 (1995).

[3] The break-down in epitaxial growth of Si(001) is thought to be intimately connected to the kinetic roughening of the surface, see for example, D. P. Adams and S. M. Yalisove, J. Appl. Phys. **76**, 5185 (1994).

[4] W. W. Mullins, J. Appl. Phys. **30**, 77 (1959).

[5] C. Herring, in *The Physics of Powder Metallurgy*, edited by E. Kingston (McGraw-Hill, New York, 1951), p. 143.

[6] M. Ozdemir and A. Zangwill, Phys. Rev. B **42**, 5013 (1990).

[7] In general, the identity of the surface species that dominates surface transport (adatom, vacancy, dimer, dimer-vacancy, etc.) is not known.

[8] J. Lapujoulade, Surface Science Reports **20**, 191 (1994).

[9] H. P. Bonzel and N. A. Gjostein, Phys. Stat. Sol. **25**, 209 (1968).

[10] H. P. Bonzel and N. A. Gjostein, J. Appl. Phys. **39**, 3480 (1968).

[11] Z. L. Liau and H. J. Zeiger, J. Appl. Phys. **67**, 2434 (1990). The decay of the InP(001) grating amplitude was not exponential; decay constant is extracted from the initial slope. The decrease in the decay rate as a function of time was attributed to surface contamination.

[12] M. E. Keefe, C. C. Umbach, and J. M. Blakely, J. Phys. Chem. Solids **55**, 965 (1994).

[13] K. Kern, I. K. Robinson, and E. Vlieg, Surf. Sci. **261**, 118 (1992).

[14] J.-K. Zuo and J. F. Wendelken, Phys. Rev. Lett. **70**, 1662 (1993).

[15] H.-N. Yang, G.-C. Wang, and T.-M. Lu, Phys. Rev. Lett. **74**, 2276 (1995).

[16] B. Grossman and P. Piercy, Phys. Rev. Lett. **74**, 4487 (1995).

[17] E. D. Williams, Surf. Sci. **299/300**, 502-524 (1994).

[18] N. C. Bartelt, T. L. Einstein, and Ellen D. Williams, Surf. Sci. **312**, 411 (1994).

[19] A. Rettori and J. Villain, J. Physique **49**, 257 (1988).

[20] E. S. Fu, M. D. Johnson, D.-J. Liu, J. D. Weeks, and E. D. Williams, Phys. Rev. Lett. **77**, 1091 (1996).

[21] P. Nozières, J. Physique **48**, 1605 (1987).

[22] M. Uwaha, J. Phys. Soc. Japan **57**, 1681 (1988).

[23] S. J. Chey, J. E. Van Nostrand, and D. G. Cahill, Phys. Rev. B **52**, 16696 (1995).

[24] S. Jay Chey, J. E. Van Nostrand, and D. G. Cahill, Phys. Rev. Lett. **76**, 3995 (1996).

[25] M. B. Webb, F. K. Men, B. S. Swartzentruber, R. Kariotis, and M. G. Lagally, Surf. Sci. **242**, 23 (1991).

[26] N. C. Bartelt and R. M. Tromp, Phys. Rev. B, in press.

[27] B. S. Swartzentruber and M. Schacht, Surf. Sci. **322**, 83 (1995).

[28] C. Pearson, B. Borovsky, M. Krueger, R. Curtis, and E. Ganz, Phys. Rev. Lett. **74**, 2710 (1995).

[29] T. W. Poon, S. Yip, P. S. Ho, and F. F. Abraham, Phys. Rev. B **45**, 3521 (1992).

[30] N. C. Bartelt, R. M. Tromp, and E. D. Williams Phys. Rev. Lett. **73**, 1656 (1994).

[31] O. L. Alerhand, D. Vanderbilt, R. D. Meade, and J. D. Joanopolous, Phys. Rev. Lett. **61**, 1973 (1988); this reference gives $C_2 \simeq 0.03$ eV nm^{-1} but using data for the stress anisotropy from F. Wu and M. G. Lagally, Phys. Rev. Lett. **75**, 2534 (1995); we calculate $C_2 \simeq 0.008$ eV nm^{-1}.

[32] C. Kittel, *Introduction to Solid State Physics*, 5th edition, (John Wiley & Sons, New York, 1976), p. 74.

SMOOTHING OF A GROOVED SINGULAR SURFACE WHOSE NEIGHBORING ORIENTATIONS ARE UNSTABLE

C. Duport [1], Anna Chame [1,2], W. W. Mullins [3] and J. Villain [1]

1 CEA, Départment de Recherche Fondamentale sur la Matière Condensée, SPSMS, 38054 Grenoble Cedex 9, France
2 Permanent address: Universidade Federal Fluminense, Instituto de Fisica, 24210-340, Niteroi RJ, Brazil
3 Carnegie-Mellon University, Department of Materials Science and Engineering, Pittsburgh, PA 15213, USA

1. INTRODUCTION

The smoothing of grooves artificially made in a crystal surface is a classical problem whose solution for non-singular surfaces is well known[1-5] and experimentally substantiated. The case in which the final (or average) surface orientation is singular and the neighboring orientations are stable has been treated by a number of authors[6-14]. The results depend on how the effect of the singularity (i.e. the cusp in gamma plot) on the kinetics is modeled or interpreted in terms of step dynamics.

In this paper, we treat the decay of a grooved singular surface under the assumption that the orientations neighboring the singular one are unstable. The discussion reviews and supplements a recent paper[15] based on this assumption. The assumption means that the singular surface meets adjacent surfaces on the equilibrium form at sharp angles which we assume also to be true on the decaying profile. The assumption also implies that steps on a given unstable surface attract each other (see appendix A), since these surfaces spontaneously facet into step bundles alternating with terraces of the singular surface.

Consequently, for a profile to decay as depicted in Fig. 1, segments of the outermost steps of the side walls must break away from their attracting neighbors, cross the terraces and meet steps of opposite sign thereby initiating a step annihilation process. As a result,

Dynamics of Crystal Surfaces and Interfaces
Edited by Duxbury and Pence, Plenum Press, New York, 1997

consecutive atomic layers peel away from the top terrace and build up on the bottom terrace. The treatment applies to the decay of fully faceted profiles and, as discussed later, gives results in sharp contrast to previous treatments of this problem[14, 16] that do not explicitly consider the layer mode of facet motion.

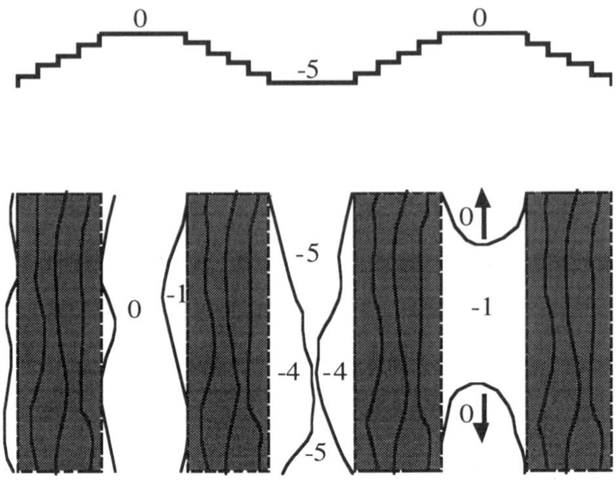

Fig. 1 Decay of grooved surface by fluctuations of step segments away from the side walls and subsequent annihilation; numbers indicate level of surface

The treatment is restricted to the transport process of evaporation/condensation[4] which also applies[14], in principle, to the case of surface diffusion in which the rate limiting step is the attachment/detachment kinetics of surface mobile atoms to surface sites (e.g. kinks in steps).

2. KINEMATICS OF LOOP MOTION

As indicated in Fig. 1, when step segments break away from the side walls and cross over the terraces, they contact steps of opposite sign and annihilate a section forming pairs of step loops. These loops then move along the terraces by evaporation (top) or condensation (bottom) until they meet a loop of opposite sign and annihilate. In this way, layers on the top terraces peel away and those on the bottom ones build up causing the profile amplitude progressively to decrease. Let P be the contact rate of steps of opposite sign per unit length, each contact producing a loop pair, ρ be the linear density of loop pairs and v be the loop velocity. Then equating the production rate of loop pairs (contact rate) to the annihilation rate, we have

$$P = \rho^2 v \quad . \tag{1}$$

Our central goal is the estimate of the time τ required for one atomic layer to peel away from the top terrace (or the time to build up one layer on the bottom terrace); in terms of the preceding quantities, τ is estimated by

$$\tau = 1/\rho v = (1/Pv)^{1/2} \quad . \tag{2}$$

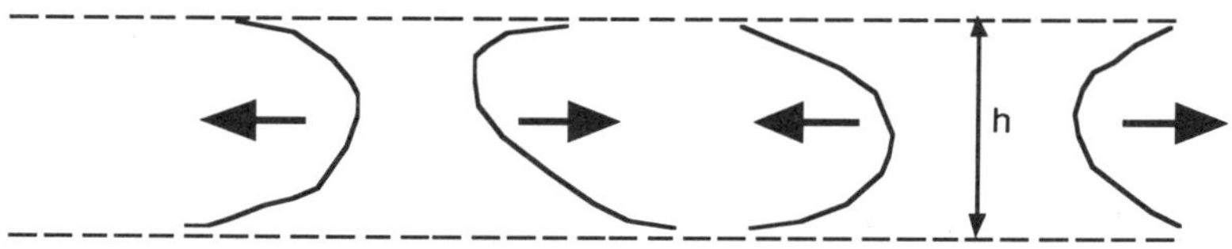

Fig. 2 Loop motion along top or bottom facet

We turn to the estimate of P in the next sections; v is easily estimated in terms of the effect of curvature on the evaporation rate.

3. EQUILIBRIUM SYSTEM USED AS A BASIS FOR ESTIMATE OF P

The contact rate P depends on the probability that the outside step will elongate, by a fluctuation, across the terrace to meet a step of opposite sign. To analyze this probability, we use the known result[17-20] of the following equilibrium problem: a single step made up of x and y links of size a (atomic distance) and energy W_1, in a semi-infinite space $x>0$, depicted in Fig. 3, is attracted to a line at $x=0$ with a binding energy W_o per link.

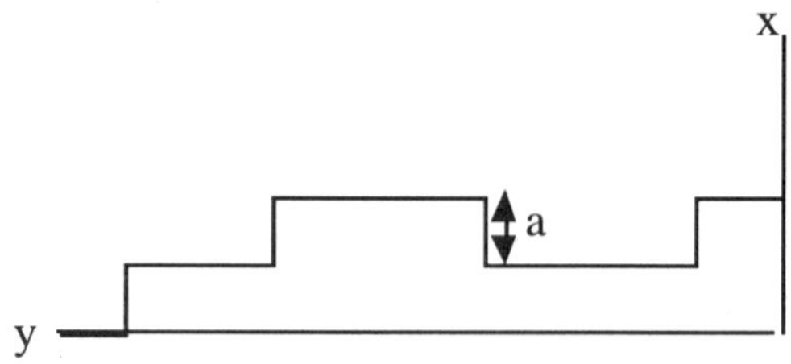

Fig. 3 Part of a semi-infinite step fluctuating away from an attracting line at x=0

Then under the simplifying assumptions that $\beta W_1 >> 1$ ($\beta = 1/kT$, in which T is the absolute temperature and k Boltzmann's constant), so that multiple links parallel to the x axis are negligible, and a continuum assumption stated below, the probability $p^+(h)$ that, at any given y and time t, an elongation exceeding h (subsequently called an h-exceedence) is observed can be shown to be

$$p^+(h) = \exp[-2\kappa h] \qquad (3)$$

where κ is given by the solution of the equation

$$\exp[2\kappa a] - (D/\varepsilon)\exp[\kappa a] - D = 0 \qquad (4)$$

in which $D = (\exp[\beta W_0] - 1)$ and $\varepsilon = \exp[-\beta W_1]$. From (3) it is clear that the probability of an h-exceedence drops off exponentially with h and that a typical elongation distance is $1/\kappa$; the continuum assumption referred to above is $\kappa a << 1$ which means that the typical elongation distance is much larger than the link distance a.

Fig. 4 shows a plot of κa as a function of βW_0 for several values of βW_1. For any given value of W_1, κa decreases with increasing temperature until, at a critical value of $T = T_W$, called the wetting temperature and given from Eq. (4) by the solution of $D = \varepsilon/(1+\varepsilon)$, κ vanishes and the step is no longer bound to the attracting line; for $T > T_W$, the step bundle dissociates and faceting no longer occurs. We assume T is below T_W but close enough so that the continuum assumption is valid. Under these conditions,

$$\exp[\kappa a] \cong 2W_0\zeta a/(kT)^2 = (T_w/T)^2, \qquad (5)$$

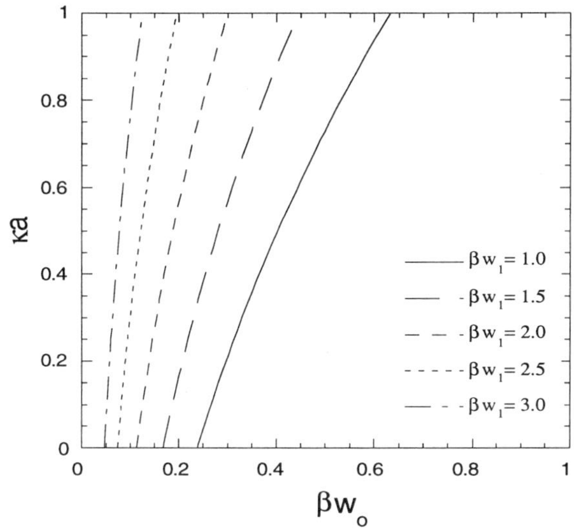

Fig. 4 Plot of κa vs. βW_0 for several values of βW_1

where ζ, the step stiffness, and T_W are given in our model by

$$\beta\zeta a = 1/2\varepsilon \ , \tag{6}$$

$$kT_w = \sqrt{2W_0\zeta a}\,. \tag{7}$$

4. FORMAL ESTIMATE OF LOOP PAIR PRODUCTION P

In this section, we give a formal estimate of the loop pair production or contact rate P per unit length (along y) by equating it to P', the rate at which h-exceedences (elongations exceeding h) are produced per unit length in the unrestricted space x>0. To estimate P', we define τ_0 and Y respectively to be the typical duration and spatial extent in the y direction of an h-exceedence. Fig. 5a shows a surface x(y,t) representing a step position as a function of y and t, with a portion exceeding h, and 5b is a view of the plane x=h over time with the exceedences (x>h) shaded; P' is the number n of shaded areas per unit area of the plane and their typical dimensions are the quantities τ_0 and Y.

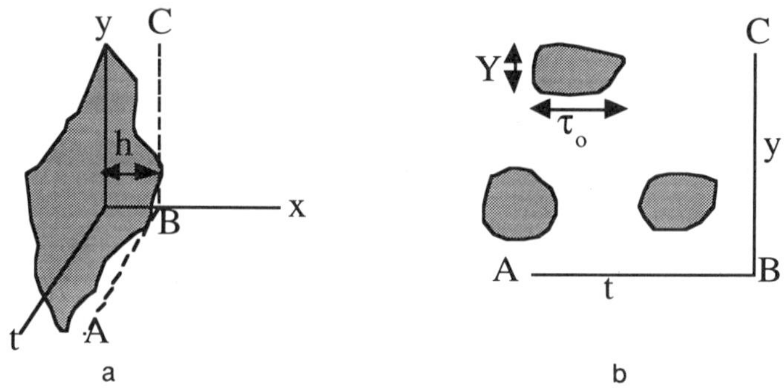

Fig. 5: a. A fluctuating step represented as a surface x(y,t);
b. The extent and duration of h-exceedences (x>h)

Clearly, the probability $p^+(h)$ of finding an h-exceedence at an arbitrary y and t is estimated by

$$p^+(h) = \frac{nY\tau_0}{LT} = P'\,Y\tau_0 \quad . \tag{8}$$

Equating P to P' and substituting Eq. (8) into Eq. (2), we obtain the formal estimate of τ as

$$\tau = \left[\frac{\tau_0 Y}{p^+ v}\right]^{1/2} \quad . \tag{9}$$

It remains to estimate Y, τ_0 and v; the latter two quantities are kinetic ones that depend on the transport mechanism.

5. ESTIMATE OF Y

We estimate Y as follows: the mean square difference $< \delta x^2 >$ of the x coordinates of a pair of points whose y coordinates differ by y is calculated for two cases: y<<Y and y>>Y. Y is then estimated as the value of y for which these two estimates are equal. A more rigorous calculation based on the autocorrelation of x coordinates as a function of y, described in ref. 12, gives essentially the same result. It is assumed that the elongation is very large compared with the typical value so that $\kappa h >> 1$.

If y<<Y, the step may be approximated as free and[21]

$$<\delta x^2> = y/\beta\zeta \qquad (10)$$

If y>>Y, the points may be approximate as independent so that

$$<\delta x^2> = \int_0^\infty dx p(x) \int_0^\infty dx' p(x')(x-x')^2 = 1/2\kappa^2 \quad . \qquad (11)$$

where $p(x)=-dp^+(x)/dx$ in which $p^+(x)$ is given by Eq. (3). Equating Eqs. (10) and (11) gives the desired estimate of Y,

$$Y = \frac{\beta\zeta}{2\kappa^2} \quad . \qquad (12)$$

6. ESTIMATE OF τ_0 AND v

A typical elongation that has already reached $x=h>>1/\kappa$, exceeds h by about $1/\kappa$. We estimate τ_0 as the time required for the elongation to recede from $h+1/\kappa$ to h, using macroscopic equations for the motion of the step. It may be shown from the statistics of the step that the typical base λ at x=0 of an elongation of amplitude x=A is given by

$$\lambda(A) = \frac{4\zeta\beta}{\kappa} A \quad . \qquad (13)$$

Approximating the elongation as the arc of a circle of base λ and height A gives a radius of curvature R of

$$R = \lambda^2/8A = 2(\beta\zeta)^2 A/\kappa^2 \qquad (14)$$

The rate of shrinkage dA/dt of the circular arc is given from standard capillary theory as

$$dA/dt = \sigma_0 \beta\zeta a^3/R \quad , \qquad (15)$$

where σ_0 is the evaporation rate per site of a straight step. Hence, using Eq. (14) for R with A=h, since the radius is about constant during the shrinkage of the step from $h+1/\kappa$ to h, we obtain

$$\tau_0 \approx (1/\kappa)/(dA/dt) \approx 2\beta\zeta h/\sigma_0(\kappa a)^3 . \qquad (16)$$

Finally, we estimate v for a loop moving along the terrace as

$$v = 2\sigma_0 \beta \gamma a^3 / h \qquad ; \qquad (17)$$

the radius of the loop is taken to be h/2 and the step energy γ rather than the step stiffness ζ is used since the disappearance of the step sides is the driving force for the loop motion along the terrace.

7. RESULTS AND NUMERICAL ESTIMATE

Substituting Eqs. (3), (6), (12), (16) and (17) into Eq. (9) gives our main result

$$\tau \approx C(h/a)\exp[\kappa h] \qquad (18)$$

where

$$C = \left[\frac{1}{8\beta(\kappa a)^5 \varepsilon^2 (\gamma a) \sigma_0^2} \right]^{1/2} \qquad (19)$$

This is the same result, apart from a numerical factor, given by Eq. 7.1 of reference 15. Alternatively, we can obtain the density of loops on the top or bottom terrace by combining these results with Eq. (2) and (17) to get

$$\rho a \approx \left[\frac{2\varepsilon^2 (\kappa a)^5}{\beta \gamma a} \right]^{1/2} \{\exp[-\kappa a]\}^{h/a}; \qquad (20)$$

note this result is independent of σ_0.

We use Eq. (20) as the basis of a numerical estimate with the following assumed values: T=1000°K, $\beta W_1=2$ (corresponding to $\varepsilon=0.1353$ and a surface energy of $\approx 0.55 J/m^2$), $\beta W_0=0.1387$ (corresponds to $T_W=1085$°K), $\kappa a=0.2$, $\gamma a = w_1 = 2/\beta$. The evaluation gives

$$\rho a \approx 2.4 \times 10^{-3} [0.8187]^{h/a} \qquad (21)$$

This formula gives the following values: for h/a=10, 20 and 30 the value of ρa respectively is 3.2×10^{-4}, 4.4×10^{-5} and 6.0×10^{-6}. In the latter case, for example, there is about one loop pair in every 1.7×10^6 atom spacings.

8. DISCUSSION

It is clear from the form of Eq. (18) or (20) that the difficulty for a fluctuating step to cross a terrace grows exponentially with the terrace width. In fact, the numerical example indicates that this process effectively stops for terrace widths of more than a few tens of atom spacings so that the idealized system we have discussed is stuck in the grooved state with no decay in reasonable times. The situation is analogous to the negligible growth rate of a singular crystal surface at low supersaturations where the nucleation rate is negligible.

In practice, one expects defects to control the rate of advance or retreat of a singular surface under most experimental conditions. In our case, important defects are the built-in steps produced by a slight miscut of the original crystal. For example, a miscut of 0.1°, almost certainly a minimum, produces a vicinal surface with a step density ρ_m given by $a\rho_m=\theta_m=1.7\text{x}10^{-3}$. Depending on the details of how the grooves are oriented with respect to the built-in steps, the density of built-in loops on the groove terraces may be as high as ρ_m. Comparison of this value with those estimated from thermal fluctuations in section 7, shows that decay rate of grooves would almost always be dominated by the built-in steps or other defects of comparable density. Step fluctuations may, however, be the rate controlling step in the motion of singular surfaces in other related processes; for example, in the approach to the equilibrium shape of small defect-free faceted particles.

We have assumed $\varepsilon<<1$, corresponding to a large step stiffness ($\zeta>>1$), so that multiple kinks of a step in the x direction are negligible. Less stiff steps would cross the terraces more easily and lead to a larger decay rate of the grooved surface.

The average chemical potential reduction (driving force) per atom for decay of the grooved profile we have considered is the order of $\Delta\mu=\gamma\Omega/h$, where Ω is the atomic volume. The decay rate we have calculated is very non-linear in $\Delta\mu$ (e.g. it decreases exponentially with h) and contrasts markedly with theories[11c,16] based on a linear dependence of the decay rate on $\Delta\mu$.

APPENDIX A: FACETING AND STEP ATTRACTION

A simple model that illustrates the connection between faceting and step attraction is provided by the following hypothetical linear function F(n) giving the free energy of a bundle of n steps per unit length in local equilibrium on the vicinal surface

$$F(n) = f_0 + nf_1 \quad . \tag{22}$$

with $f_0 > 0$ and $f_1 > 0$. Eq. (22) evidently describes a nearest neighbor attraction between steps in which f_0 is the work per unit length required to separate two steps since $2F(1) - F(2) = f_0$; similarly f_0 is the work per unit length required to separate an outer step from a bundle. The free energy per step of a macrosurface composed of these steps is $f_1 = \lim_{n\to\infty} F(n)/n$.

Eq. (22) gives a reduction of energy upon faceting of the vicinal surface with N separate steps into two surfaces, the singular one and the one composed of the single bundle of N steps, since

$$F(N) - NF(1) = -(N-1)f_0 \quad . \tag{23}$$

Note, further, that Eq. (22) implies an edge energy of the bundle of steps given by $f_0/2$ since the work required to separate a bundle of N+M steps into two bundles of N and M steps respectively is

$$F(N) + F(M) - F(N+M) = f_0 \quad . \tag{24}$$

Eq. (22) is meant only as an example to illustrate that an unstable surface which spontaneously facets is composed of attracting steps. We do not address the underlying cause of the attraction. Note, in this model, f_0 would be a function of temperature which would vanish at the faceting transition temperature and become negative above this temperature corresponding to a repulsion between steps.

REFERENCES

1. J. M. Blakely and H. Mykura, Acta Metall. **10**, 565 (1962)
2. H. P. Bonzel, in Surface Physics of Materials, J. M. Blakely Ed., v. 2, p.275 (1975)
3. W.W. Mullins, J. of appl. phys. **28**, 333 (1957)
4. W.W. Mullins, J. of appl. phys. **30**, 77 (1959)

5. H. P. Bonzel and W. W. Mullins, Surface Science **350**, 285-300 (1995)

6. F. Lançon and J. Villain, in Kinetics of Ordering and Growth at Surfaces, M. J. Legally Ed. (Plenum Press, New York, 1990)

7. H. Spohn, J. Phys. I France **3**, 69 (1993)

8. J. Hager and H. Spohn, Surf. Sci. **324**, 365 (1995)

9. H. P. Bonzel, E. Preuss and B. Steffen, Appl. Phys. **A35**, 1 (1984)

10. H. P. Bonzel, E. Preuss and B. Steffen, Surf. Sci. **145**, 20 (1984)

11. A. Rettori and J. Villain, J. Phys. France **49**, 257 (1988)

12. L. H. Tang, in "Dynamics of crystal surfaces and interfaces", edited by P.M. Duxbury and T. Pence, (Plenum, NY) 1997

13. W. Selke and P. M. Duxbury, Phys. Rev. **B52**, 17468 (1995)

14. J. W. Cahn and J. E. Taylor, Acta Metall. Mater. **42**, 1045 (1994)

15. C. Duport, A. Chame, W. W. Mullins and J. Villain, J. Phys. I France **6**, 1095 (1996)

16. W. C. Carter, A. R. Rosen, J. W. Cahn and J. E. Taylor, Acta Met. **43**, 4309 (1995)

17. D. B. Abraham Phys. Rev. Lett. **44**, 1165 (1980)

18. D. B. Abraham Phys. Rev. Lett. **51**, 1279 (1983)

19. S. T. Chui and J. D. Weeks, Phys. Rev. B **23**, 2438 (1981)

20. M. Vallade and J. Lajzerowicz, J. Phys. France **42**, 1505 (1981)

21. A. Pimpinelli, J. Villain, D. E. Wolf, J. J. Métois, J. C. Heyraud, I. Elkinani, G. Uimin Surface Science **295**, 143-153 (1993)

STEP FLUCTUATIONS: FROM EQUILIBRIUM ANALYSIS TO STEP UNBUNCHING AND CLUSTER DIFFUSION IN A UNIFIED PICTURE

T. L. Einstein and S. V. Khare*
Department of Physics
University of Maryland
College Park, MD 20742-4111

INTRODUCTION

In addition to providing information about the energy of kinks, thermal fluctuations of steps on a vicinal surface provide a rich source of insight into the microscopic atomic processes which underlie the fluctuations. In recent years it has become possible to make quantitative measurements of these fluctuations using STM (scanning tunneling microscopy), LEEM (low-energy electron microscopy), and REM (reflection electron microscopy). The fluctuations of the steps can be viewed as a form of Brownian motion (viz. random motion of a heavy entity due to thermal fluctuations of light particles) and can be analyzed using a capillary-wave approach and Langevin formalism. From this analysis one can deduce the key macroscopic parameters of step stiffness, step-step interaction strength, and mobility which govern the macroscopic behavior of the steps. These parameters can then be applied to situations far from equilibrium or in which the steps are driven by some external force. Furthermore, the analysis of nearly-straight steps can be adapted to treat nearly-circular steps and thereby describe the Brownian motion of monolayer clusters of atoms or vacancies on surfaces, for which quantitative experimental data has also been obtained recently. In this short review we gather together, integrate, and in some cases amplify results from several previous papers. We also summarize and catalog the relevant experimental results.

EQUILIBRIUM FLUCTUATIONS OF ISOLATED STRAIGHT STEPS

For a theorist it is easy to model the equilibrium fluctuations of steps by performing Monte Carlo simulations within the SOS (solid-on-solid) model. In this model, one assigns to each site **i** on a square net (2D lattice) an integer h_i, which denotes the height (in units of the lattice constant) at that site. The Hamiltonian is then a characteristic energy ε times the sum over all nearest neighbor pairs **i,j** of the height difference $|h_i - h_j|$. Thus, both the energy of a single-height step per lattice constant along a principal direction and the energy of a simple kink in such a step is ε, while the energy of an isolated adatom or vacancy is 4ε. For specificity, we shall assume that the step runs along the y direction, so that discursions perpendicular to it are in the x direction (sometimes called "Maryland" notation), as depicted in Fig. 1. To create such a step, one can simply apply screw periodic boundary conditions in

*Present address: Department of Physics, Ohio State University, Columbus, OH 43210

Dynamics of Crystal Surfaces and Interfaces
Edited by Duxbury and Pence, Plenum Press, New York, 1997

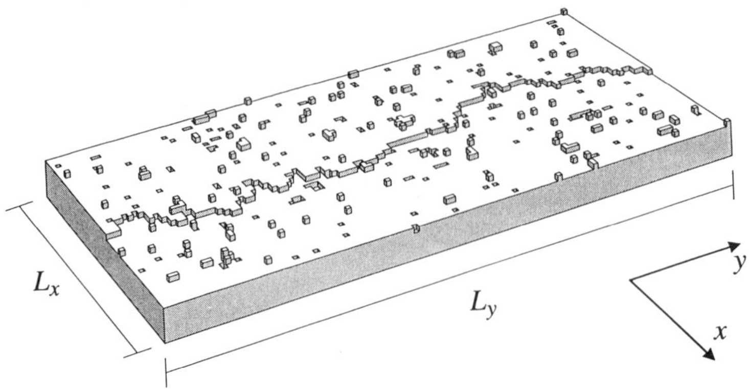

Fig. 1. A step in the SOS model at temperature $k_BT/\varepsilon = 0.9$. The concentration of adatoms and of vacancies is about $0.02/\Omega$, where Ω is the area of the surface unit cell. From Bartelt *et al.* (1994a), with permission.

the $\hat{x}$ direction: $h_{i+L_x} = h_i + 1$, where $\mathbf{L_x}$ is the upstairs-pointing vector spanning the lattice, projected in the terrace plane.

In Monte Carlo propagation of this lattice system, one can graphically watch how the step configuration changes as adatoms or vacancies attach to or detach from the step edge. In LEEM, REM, or STM experiments, one either lacks the resolution to observe atomic events or these events happen so rapidly that they cannot be observed individually. The challenge, then, is to deduce as much as possible about these atomic processes from observations of the step configurations alone. To do so, we apply capillary wave analysis. As the "wavelengths" (characteristic size in the $\hat{y}$ direction) of the equilibrium fluctuations increase, so do their amplitudes (in $\hat{x}$) and their duration, as shown below.

There are three well-characterized limiting cases, denoted hereafter EC, TD, and PD (Bartelt *et al.*, 1992, 1994a). In EC [2D] evaporation/condensation, or attachment/detachment, of atoms and/or vacancies at the step edge limit the production and decay of fluctuations. Once the adatom or vacancy is free of the step, it is assumed to be instantly equilibrated into a 2D "gas" of "carriers" on the terraces. In TD, diffusion across the terrace is the limiting process, leading to slower behavior and a non-uniform distribution of carriers on the terrace that decays exponentially toward the thermal value for a flat surface. The slowest process is PD: periphery (or edge) diffusion. Here motion along the step edge limits the rate of the healing of fluctuations.

To make quantitative progress, we use a Langevin formulation. This amounts to an overdamped harmonic oscillator driven by a noise term (Kardar, 1994). The generic form is

$$\partial x(y,t) / \partial t = -restoring \text{ "}force\text{" } + \eta(y,t) \tag{1}$$

While one might hope that the restoring force [divided by "mass"] is linear in $x(y,t)$ with a prefactor τ^{-1} from dimensional arguments, in general the behavior is more complicated, involving a convolution over y. However, by performing a capillary-wave analysis, i.e. writing $x(y,t)$ as $\Sigma_q \exp(iqy) x_q(t)$, we find

$$\partial x_q / \partial t = -\tau_q^{-1} x_q(t) + \eta_q(t), \qquad \langle \eta_q(t) \eta_{q'}(t') \rangle = L_y^{-1} f(q) \delta_{q,q'} \delta(t-t'), \tag{2}$$

where f(q) is given below. We seek the behavior of the [measurable! (Bartelt *et al.*, 1993; Bartelt and Tromp, 1996; Pai *et al.*, 1996)] autocorrelation function $G_q(t-t')$ of the capillary modes:

$$G_q(t-t') \equiv \left\langle \left| x_q(t) - x_q(t') \right|^2 \right\rangle = 2\left\langle \left| x_q(t) \right|^2 \right\rangle - 2\left\langle x_q(t) x_q(t') \right\rangle = A\left(1 - e^{-|t-t'|/\tau_q}\right) \tag{3}$$

From equipartition arguments, one finds that the prefactor $A = 2k_BT/\tilde{\beta}q^2L_y$ depends simply on known quantities, except perhaps the step stiffness $\tilde{\beta}$, which can thus be determined from this relation or checked with previous independent determinations. [The stiffness is the coefficient of the integral of $(1/2)(\partial x/\partial y)^2$ in the Hamiltonian of the step; $\tilde{\beta}(\theta) = \beta(\theta) + \beta''(\theta)$, where $\beta(\theta)$ is the free energy per length of a step.] The q^2 dependence does not depend on the limiting case. On the other hand, by integrating eqn. (2) and computing $\langle x_q(t)\, x_q(t')\rangle$, we find

$$\tau_q^{-1} = \tilde{\beta}q^2 f(q) / 2k_BT; \qquad f(q) = 2\Gamma_a,\ 4D_{su}c_{su}\Omega^2|q|,\ 2\Gamma_{st}q^2 \tag{4}$$

for the cases EC, TD, and PD, respectively. The subscripts su and st are abbreviations for surface and step, respectively. The lattice spacings $a_\perp$ and $a_\parallel$ are in the unit spacings in the $\hat{x}$ and the $\hat{y}$ directions, respectively, and $\Omega = a_\perp a_\parallel$ is the area per atom (i.e. of the surface unit cell). The mobility or friction coefficients can be related to microscopic times:

$$\Gamma_a = a_\perp^2 a_\parallel/\tau_a \qquad \text{and} \qquad \Gamma_{st} = a_\perp^3 a_\parallel^2/\tau_{st}\,, \tag{5}$$

where τ_a is the mean time between successive attachment (or successive detachment) events, and τ_{st} is the mean time between successive hops in the same direction along the step edge. Similarly, on a square lattice of sites $D_{su} = a^2/4t_h$, where t_h is the time between hops, and c_{su} is the sum of the adatom and vacancy concentrations far from the fluctuating step. [Bartelt *et al.* (1994a) also show how in Monte Carlo simulations one can change TD behavior to EC—except at very small q—by including a [very small] sticking coefficient in the attachment/detachment of carriers to step edges, thereby making that the rate-limiting process. We can now characterize this crossover analytically (Khare and Einstein, 1996, 1997).]

In terms of the models of dynamical critical phenomena (Hohenberg and Halperin 1977), EC and PD are examples of cases A (no conservation law) and B (conserved order parameter), respectively. The behavior $\tau_q^{-1} \propto q^2$ in EC is due to the fact that the relaxation is proportional to the gradient of the local surface free energy, which in this case is just the local curvature of the step edge. The extra factor of q^2 in PD ($\tau_q^{-1} \propto q^4$) arises from the additional $-\partial^2/\partial y^2$ coming from the conservation condition. In spite of some similarities, TD is not (Siggia, 1996) an example of case C (coupling to an auxiliary conserved density). The extra factor of $|q|$ ($\tau_q^{-1} \propto |q|^3$) comes from the exponential decay of the concentration toward the terrace value as one moves away from the step edge. In contrast, the terrace concentration is uniform in EC and effectively zero in PD. From the perspective of the long-studied problem of the decay of sinusoidal gratings in 3D, reviewed/studied from different perspectives by H. Bonzel, W.W. Mullins, L.-H. Tang, W. Selke, J. Erlebacher, and M.V. Ramana Murty in this volume, EC, TD, and PD are called [3D] evaporation/condensation, volume diffusion, and surface diffusion, respectively. Notice from eqn. (3) that in all three cases the early-time behavior of $G_q(t)$ is linear in t, characteristic of diffusive, exponential relaxation. In contrast, the real space analogue, the mean-square width $w^2(t-t') = \langle|x(t) - x(t')|^2\rangle$ is not linear in any of the simple cases; instead, $w^2(t) \propto t^{1/2}, t^{1/3}, t^{1/4}$, for EC, TD, PD, respectively. (Cf. Note 1.) Fluctuations of positions along the step edge—in contrast to those of the q-modes—are interdependent. This problem of coupled Brownian oscillators has a rich history (Wax, 1954).

There are other ways to obtain many of these results. Decades ago Mullins (1957, 1959, 1963) showed the fruitfulness of formulating the problem in terms of a step chemical potential. Bales and Zangwill (1990) used the linear kinetic approximation that the step velocity is proportional to the difference between the adatom concentration near the step edge and its equilibrium value. Pimpinelli *et al.* (1993) trisect each system into a fluctuating step, a reservoir of atoms enabling the fluctuations, and a pipe connecting the two, through which the exchange of atoms occurs. From this incisive perspective, they can quickly account for a large number of limiting cases, including multi-step situations, but sacrifice the factors of π and the like appearing in more precise derivations. Elsewhere in this volume, B. Blagojević and P.M. Duxbury formulate the problem in terms of the probability P(y) that atoms leaving the step at one point return to this step a distance *y* away. Not only do they retrieve the early-time growth of the mean-square width in the three limiting cases EC, TD, and PD, but they can achieve intermediate values $t^{1/(\alpha+1)}$ if $P(y) \propto y^{-\alpha}$. It is not immediately clear how the form of P(y) relates to the physical nature in our formulation. We (Khare and Einstein, 1996,1997) have also been able to produce cross-over behavior between limiting cases by considering a unified formulation that considers all three mechanisms simultaneously, as well as reproducing and extending the multistep behavior of Pimpinelli *et al.*

As an illustration of the application to actual data of the analysis procedure developed at Maryland, we consider the case of an isolated step on Ag(110) directed 30° from the close-packed $[1\bar{1}0]$ direction, measured by STM at room temperature by Reutt-Robey's group at Maryland (Ozcomert *et al.*, 1994; Pai *et al.*, 1996; Reutt-Robey and Pai, 1997). A best fit of the early-time measurements of the autocorrelation function is $w^2(t) = 33.7Å^2 \cdot t^{0.49}$, consistent with EC. In the capillary wave analysis, the lowest value of q was 2.1×10^{-3} Å^{-1}, corresponding to a wavelength 3000Å. Some half-dozen values of q up to eight times that lowest value were analyzed for up to 1000 sec. From the fits of $G_q(t)$, A (and thence $\tilde{\beta} \approx 18$ meV/Å) and τ_q were obtained. In the plot of τ_q^{-1} vs. q, the fit to q^2 was much better than the alternatives, supporting the view that the fluctuations are EC limited. From the prefactor of this fit and the deduced $\tilde{\beta}$, we find a mobility $\Gamma_a = 1.8 \times 10^2$ Å^{3}s^{-1}, leading to $\tau_a \approx 350$ msec. For an isolated step along the $[1\bar{1}0]$ direction, the stiffness is over 8 times as large, but $\tau_a \approx$ 400 msec, indicating that the ability of steps to supply Ag atoms to the terrace ($\tau_a^{-1} \approx 3$ [events] per second [per step site]) has little dependence on step orientation (and so kink density) (Pai *et al.*, 1996). For Si surfaces at much higher temperatures, also examples of $\tau_q^{-1} \propto q^2$, the mobilities are much higher: for Si (111) at 900°C, $\tau_a^{-1} \approx 10^6$ atoms/sec (Alfonso *et al.*, 1992; Bartelt *et al.*, 1993); for Si (100) at 700–1200°C, $\tau_a^{-1} \approx 10^3 - 10^6$ dimers/sec (Bartelt *et al.*, 1994b; Bartelt and Tromp, 1996). Kuipers *et al.* (1993, 1995) had found similar fluctuations on vicinal Au(110) and Pb(111). Cases of vicinals on which fluctuations with $\tau_q^{-1} \propto q^4$ have been observed are: Ag(111) (Poensgen *et al.*, 1992), Cu(100) (Giesen, 1992; Poensgen *et al.*, 1992; Giesen-Seibert *et al.*, 1993, 1995; Masson *et al.*, 1994; Barbier *et al.*, 1996) and Ag(100) (Hoogeman *et al.*, 1996; Wang *et al.*, 1996). To date there have been no observations of $\tau_q^{-1} \propto |q|^3$; as discussed in the penultimate section, there may be reasons for this related to the isolated-step approximation.

APPLICATIONS TO SYSTEMS AWAY FROM EQUILIBRIUM

Our hope in finding mobilities from well-defined experiments in equilibrium is to be able to use them in situations away from equilibrium. As an example, we consider the problem of step unbunching. When vicinal Ag(110) is oxidized, it is energetically favorable for the oxygen to form chains on the terraces in the [001] (next-nearest neighbor) direction. Since it is energetically favorable to have a few long chains rather than several short ones, the surface phase separates into wide flat terraces covered with (1×n) chains and closely-spaced bunches of steps with temperature-dependent average misorientation. The oxygen can be rapidly removed by dosing with CO. The surface is left in an unstable configuration, and the steps relax back to the original uniform vicinality. Using formalism developed by Rettori and Villain (1988), we seek to account quantitatively for this behavior in an essentially 1D picture (viz. average position of each step as a function of time or CO exposure).

To start this discussion we need an expression for the free energy as a function of the local misorientation from the terrace [or facet] plane, or more precisely, the free energy per projected area, namely

$$f(\phi,T) = f_0(T) + \beta / \ell + g / \ell^3, \tag{6}$$

where $f_0(T)$ is the free energy per area of the terrace, and ℓ = (step height h)/tan(ϕ) is the distance between steps. Thus, β was defined after eqn. (3), and g/ℓ^2 is the repulsion between steps. Such an interaction always exists because of entropic repulsion due to the non-crossing of steps and is usually enhanced considerably by an elastic repulsion. The formula for g is given in Note 2; when the repulsion is purely entropic, g reduces to $(\pi k_B T)^2/6\tilde{\beta}$ (Williams *et al.* 1994). The repulsion produces what amounts to a 2D pressure defined as the negative derivative of the surface free energy with respect to surface area, with the number of steps held fixed (Ozcomert *et al.*, 1993). Since the width can be taken as constant,

$$p(\ell) = -\partial(\ell f(\ell)) / \partial\ell\big|_{\#\ steps} = -f_0 + 2g / \ell^3. \tag{7}$$

Thus, the pressure difference on the two sides of the step is proportional to the difference of the inverse cubes of the terrace widths (neglecting possible intereactions with more distant steps). Again in the overdamped limit, the step velocity $\partial x/\partial t$ is proportional to the pressure from the terrace behind the step minus the pressure from the terrace ahead of the step. Since the motion is again step diffusion, the prefactor ought to contain the same transport coefficient as that for equilibrium fluctuations, Γ_a for EC or $D_{su}c_{su}$ for TD, in either case divided by k_BT. Alternatively, this can be described as a current produced by the gradient of a chemical potential associated with each step (Rettori and Villain, 1988).

As a theoretical check of these ideas, Bartelt *et al.* (1994a) created in an SOS model a step bunch of 5 steps by initially confining them to half the lattice [in the $\hat{x}$ direction], then watching them evolve to nearly uniform spacing. There was no energetic interaction between

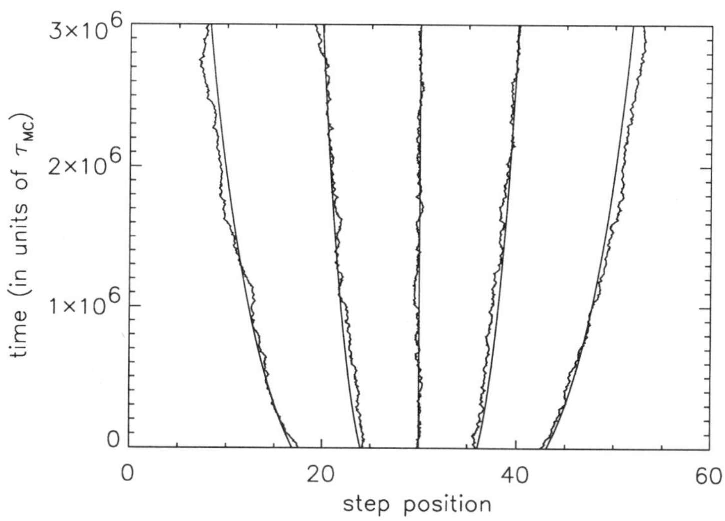

Fig. 2. Time evolution of the average position of steps in a step bunch relaxing back to their equilibrium distribution. The fluctuating lines are generated by Monte Carlo simulation, while the smooth curves come from the theory of Rettori and Villain (1988). From Bartelt *et al.* (1994a), with permission.

steps, just entropic. In Fig. 2 we show the average positions $\langle x_i(y,t)\rangle_y$ of the steps (i = 1,...5) as functions of time. Coplotted with these wiggly curves are smooth curves obtained by integrating the first-order equations of motion of each of the steps. The value of $D_{su}c_{su}$ in these curves is similar to that obtained from an analysis of the fluctuations of an isolated single step as well as from an explicit calculation of both D_{su} and c_{su} for a flat surface.

The experimental evidence for this scenario, mentioned at the outset of this section, is less compelling since it is harder to control this sort of dosage-sensitive STM experiment than a Monte Carlo simulation. (Specifically, it problematic to convert from CO dosage to the evolution time from initial instability; the analysis would be better if the surface could be instantaneously de-oxidized.) From an earlier examination of the terrace-width distribution for Ag(110)2°→[001], Ozcomert *et al.* (1993) concluded that to a good approximation the step-step interactions were purely entropic repulsions (by finding a good fit to a free-fermion form (Joós *et al.*, 1991)). (But see also Pai *et al.* (1994) for remarkable behavior under different conditions.) From the relationship (Bartelt *et al.*, 1992)

$$\tilde{\beta} = k_B T(y-y') / \left\langle (x(y)-x(y'))^2 \right\rangle \tag{8}$$

(rather than capillary-wave theory), they also determined the stiffness to be $\tilde{\beta} = k_BT/(5Å) \approx 5$ meV/Å for steps in low-symmetry directions. In Fig. 3 we show a fit of the experimentally measured average (over y) of each step position as a function of time (or CO exposure), along with smooth curves indicating a best fit to integrated Rettori-Villain equations for EC (Ozcomert *et al.*, 1995; Reutt-Robey and Pai, 1997). The best fit is obtained with $\tau_a \approx 200$ msec, similar to (about half) the value deduced from capillary-wave analysis of equilibrium fluctuations.

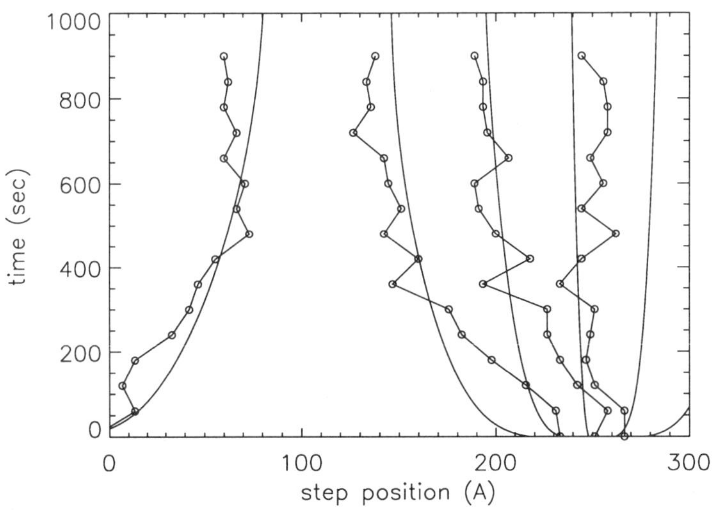

Fig. 3. Time evolution of the positions of the steps on vicinal Ag(110) after removal of the surface oxygen by reaction with CO. As in Fig. 2, the smooth curves are from the theory of Rettori and Villain (1988). Replotted from Reutt-Robey and Pai (1997), with permission.

APPLICATIONS TO DIFFUSION OF LARGE SINGLE-LAYER CLUSTERS

It is by now well established (Kellogg, 1994; Wang and Ehrlich, 1990) that field ion microscopy (FIM) can be used to monitor the diffusion of small clusters of atoms or vacancies on metallic surfaces. In recent years (room-temperature) STM has allowed for quantitative measurements of the Brownian-like motion of large single-layer clusters of 100's of atoms or vacancies on surfaces. As the cluster engages in this diffusive process, its center of mass can be tracked in a succession of images. One then expects the mean-square displacement of this center of mass to be proportional to elapsed time, the proportionality constant being four times the cluster diffusion constant D_C. Such behavior was indeed observed by Trevor and Cidsey (1990) for Au(111), by de la Figuera *et al.* (1994) for vacancies on Cu(111) dosed with Co, by Wen *et al.* (1994, 1996, 1997) for Ag islands on Ag(100), by Wendelken *et al.* (1997) for Cu islands on Cu(100), and by Morgenstern *et al.* (1995) for vacancies on Ag(111). In each case, again as expected, D_C decreases as the number N of atoms or vacancies in the cluster (or, equivalently, its average mean-square radius R^2) increases. Long ago, Binder and Kalos (1980) argued that D_C should decrease by some integer power of R: $D_C = D_{Co}R^{-\alpha}$, where the integer α depends on the atomistic process governing the diffusion. Morgenstern *et al.* (1995) found that $\alpha = 1.97 \pm 0.39$ for R between 20 and 150 atomic spacings (viz. 2.9Å), while Wen *et al.* (1994) reported $1/2 \leq \alpha \leq 1$. Issues of interest include: what these values of α indicate about the atomistic processes underlying the diffusion, and what the prefactor D_{Co} reveals about activation energies.

While there have been several other different direct approaches (Morgenstern *et al.*, 1995; Van Siclen, 1995; Sholl and Skodje, 1995; Soler, 1994,1996) to this problem, we here pursue the perspective that the fluctuations of the cluster can be viewed as fluctuations of its boundary, which is a closed single-height step (Khare *et al.*, 1995; Khare and Einstein, 1996). Our approach is to adapt the formalism for open nearly straight steps to closed nearly circular steps. Denoting by $r(\theta,t)$ the radial distance of the edge from the center of mass, we define a normalized deviation $g(\theta,t)$ from a perfect circle and do the equivalent of capillary-wave decomposition:

$$g(\theta,t) \equiv (r(\theta,t) - R)/R = \sum_n g_n(t) e^{in\theta}. \qquad (9)$$

The Langevin equation for $g_n(t)$ is essentially the same as eqn. (2) for $x_q(t)$, with τ_n^{-1} replacing τ_q^{-1} and $R^{-1}\eta_n$ replacing η_q. Our previous calculations for straight steps can be carried over to circular steps by making the replacement $q \rightarrow n/R$. Since the displacement of the center of mass at time t is given by $r_{CM}^2 = x_{CM}^2 + y_{CM}^2$, we find

$$D_c \equiv \langle r_{CM}^2 \rangle / 4t = R^2 \langle |g_1|^2 \rangle / t = R^2 f_n \tau_n (1 - e^{2t/\tau_n}) / 2t \sim k_B T R \tau_1^{-1}(R) / \pi\tilde{\beta} = D_{co} R^{-\alpha} \qquad (10)$$

where we have again used the result that in equilibrium $\langle |g_n(t)|^2 \rangle = k_B T / 2\pi R \tilde{\beta} n^2$ (Nozières 1992). [See also Note 3.] Thus, again the relation between the microscopic and macroscopic perspectives occurs thruough the characteristic time. For the cases EC, TD, and PD, since $\tau_q^{-1} \propto q^2, q^3, q^4$, we now have $\tau_1^{-1} \propto R^{-2}, R^{-3}, R^{-4}$, and so $\alpha = 1, 2, 3$, respectively, or equivalently, $D_C \propto N^{-1/2}, N^{-1}, N^{-3/2}$. (In late-stage coarsening by cluster coalescence—in contrast to the near-constant size regime treated here—Sholl and Skodje (1996) show that the average cluster radius increases like t^{β}, where $\beta = 1/(\alpha+2) = 1/3, 1/4, 1/5$, respectively, rather than the $t^{1/3}$ behavior for all 3 cases in the limit of Ostwald ripening. Furthermore, they find the dynamic scaling law for n_s, the density of islands of area s: $n_s(t) \sim t^{-2\beta} f(s/t^{2\beta})$.)

To check whether this behavior, based on a continuum viewpoint, is applicable to vacancy clusters on the scale of the experiments, we performed Monte Carlo simulations using the standard Metropolis algorithm. Since the goal was not to replicate any experiment, we invoked several simplications and "tricks" to bring out the central physics with minimal complications. We used a square lattice with just an (attractive) nearest-neighbor (NN) energy $-\varepsilon$. We worked at $k_B T/\varepsilon = 0.6$ (0.5 for TD), well below the roughening temperature of the corresponding SOS model but high enough so that the equilibrium shape was nearly circular. For EC, we used straightforward Glauber (atom hops to/from "reservoir") dynamics, adjusting the chemical potential to keep the number of vacancies about constant. Alternatively, after removing an atom at some random value of θ, we could simply immediately reattach it at some other random position along the periphery, avoiding the nuisance of adjusting a chemical potential for the reservoir. (In this approach it is important when scaling the Monte Carlo data to include the fact that the chance of such a move per unit increases proportional to the circumference, contributing a factor of R to $D_c(R)$.) For TD, we used Kawasaki (atom hops to [NN] vacancy) dynamics. To prevent the vacancy cluster from dissolving, we forbade vacancy diffusion from the cluster boundary into the surrounding atomic lattice. To enhance motion across the terrace, we reduced the energy of an isolated atom in the interior of the monolayer pit to ε [from 4ε]; this had the added benefit of suppressing atom-cluster formation in the pit. For PD we again used Kawasaki dynamics, but with the modification that only NNN (next nearest neighbor), not NN hops were allowed. This "trick" enhanced the probability of creation, along a straight edge, of atom-vacancy (notch) pairs and prevents diffusing atoms from being trapped in corners. We considered clusters of size 100, 400, 1600, and 6400 vacancies [embedded in a much bigger lattice] and found from log-log plots of D vs. R the best-fit exponents $\alpha = 0.97$, 2.03, and 3.1, respectively, in excellent agreement with the predictions of the continuum theory.

In this framework, the Morgenstern *et al.* (1995) experiment for Ag(111) pits is an example of TD, as they themselves concluded from an argument following the approach of Pimpinelli *et al.* (1993). Microscopically, the picture is that Ag atoms cannot surmount the barrier, so that they are trapped inside the pit. Thus, there are not the particle fluctuations associated with EC. On the other hand, Wen *et al.* (1994, 1996, 1997) find behavior more similar to EC. There are considerable particle fluctuations: Wen *et al.* (1994) remark that they exclude islands which decrease in area by more than 20% during the course of the observations. Moreover, the islands are more nearly square than circular (Wen *et al.*, 1996, 1997); much of the evaporation may occur by an edge-peeling mechanism (Van Siclen, 1995; Evans *et al.*, 1997) which is rate-limited by the detachment of a corner atom and so virtually independent of island size. In that case, the experimental exponent α could be more like 1/2 than 1. On the other hand, Wendelken *et al.* (1997) have just reported PD behavior for Cu islands on Cu(100). They also considered Ag islands on Ag(100), and there are preliminary indications that α is much greater than 1 and close to 3 (Pai 1997), consistent with measurements of vicinal Ag(100) (Wang *et al.*, 1996).

Our Langevin analysis also produces exact expressions for the prefactors D_{co} for the three cases: Γ_a/π, $D_{su}c_{su}\Omega^2/\pi$, and $D_{st}c_{st}\Omega^2/\pi$. [See Note 4.] The last of these is the 2D analogue of the 3D expression derived by Gruber (1967). To check the numerical values of D_{co} obtained from the y-intercepts of the log-log plots, we also computed the diffusion constant directly by applying a weak potential gradient **F** to straight steps (or to adatoms on a flat terrace for TD) and seeking the resulting average velocity $\bar{\mathbf{v}}$. Thence, the carrier diffusion constant can be calculated from the Einstein-Nernst relation $D = k_B T|\bar{\mathbf{v}}|/|\mathbf{F}|$. The resulting values agree to within 25% with those from the log-log plots. It is tempting to extract activation energies from the prefactors, an activity in which we have participated (Khare and Einstein, 1996). While the numbers obtained are semiquantitatively sensible, the level of correspondence to the real physical numbers depends on the accuracy of the presumptions made by the investigator about the microscopic Hamiltonian and how the macroscopic parameters depend on these energies.

UNIFIED FORMALISM, CROSSOVER, AND STEP REPULSIONS

Most of the preceding has been couched in terms of three separate, distinct cases. For the cluster problem we have presented a unified formalism encompassing all 3 limits and permitting the examination of crossover between them. We sketch the derivation, referring the interested reader desiring more details to Khare and Einstein (1996). The starting point is to define a chemical potential for the cluster edge analogous to that for a step (Mullins, 1957, 1959, 1963; Bonzel and Mullins, 1996), having the form $\mu_s = -\Omega R^{-1}\tilde{\beta}\partial^2 g/\partial\theta^2$. We next apply the radiation boundary condition in the adiabatic or quasistatic approximation (Bales and Zangwill, 1990; Cahn and Taylor, 1994). With the assumption of a steady-state concentration of carriers on both the interior and the exterior terraces, the diffusion equation for the concentration reduces to a Laplace's equation. At any point along the edge, the net flux is determined by the normal component of the flux from the interior and exterior terraces—assumed to be *linear* in the difference between the concentrations on the two terraces—and by the motion along the island periphery. On the outside (inside) of the nearly-circular step this flux is equated to the mobility $\Gamma_{+(-)}$ times the difference of the carrier concentration just outside (inside) the step minus $(1 + (\mu_s/k_BT))$ times the concentration far from the edge. This provides the boundary condition needed to solve Laplace's equation for the concentration. Furthermore, the sum of the net attachments on the two sides determines the motion of the step (and hence of the island as a whole) $R\partial g/\partial t$. Decomposing into circular modes as in eqn. (9) and inserting into the equivalent of eqn. (2), we find a complicated expression for the characteristic times τ_n of the modes. For simplicity we assume an atom island, with all the atomic motion on the exterior. We define two characteristic lengths: 1) $R_{su} = \Omega D_{su}/\Gamma$ is essentially the ratio of the tracer (atomic) diffusion constant of the terrace D_{su} to the mobility. When it is large, diffusion over the terrace is much greater than attachment or detachment, so the motion is limited by the latter and so more likely to be EC than TD. 2) $R_{st} = (a_\perp \Omega D_{st}/\Gamma)^{1/2}$ is a similar ratio of the tracer diffusion constant along the step to the mobility. To determine the cluster diffusion constant D_c, we again need only τ_1^{-1}. We find

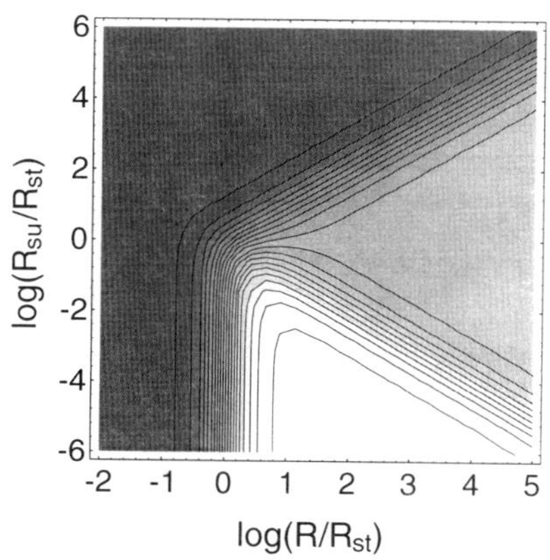

Fig. 4. Countour plot, with gray-scale shading of the effective exponent α_{eff} as a function of the common logarithms of the dimensionless ratios R/R_{st} and R_{su}/R_{st}. The large regions of gray, light gray, and white represent α = 1, 2, and 3, indicative of EC, TD, and PD, respectively. The crossover regions are relatively narrow. From Khare and Einstein (1996), with permission.

$$D_c = \frac{c_{su}\Omega\Gamma}{\pi R}\left[1+\frac{(R/R_{st})^2}{1+(R/R_{st})(R_{su}/R_{st})}\right]^{-1} \tag{11}$$

From eqn. (11) it is straightforward to compute the exponent $\alpha_{eff} = -\partial \log D_c/\partial \log R$ that one extracts from log-log plots of data. Fig. 4 gives a contour plot of this effective exponent in terms of two dimensionless ratios of the three lengths. For very small clusters, EC eventually dominates (although the continuum approximation may well fail before this limit is reached). For very large clusters (perhaps unphysically large, depending on the size of R_{su} and R_{st}, TD eventually is reached. The most important feature is that the crossover regimes are relatively narrow, little over a decade in R. This suggests that α_{eff} should attain a constant value if the data contains a large range of sizes and that one should not find values of α_{eff} other than 1, 2, or 3 for this constant. Contrary findings indicate either problems with the experiment or significant physics missing from the theoretical analysis (e.g. the edge-peeling of Ag(100) islands).

Some clarification of the names of the three regimes should be made. In the extreme case $D_{su} = 0$, so $R_{su} = 0$, eqn. (11) reduces to $D_c^{-1} \propto R[1 + (R/R_{st})^2]$, and we find smooth crossover directly from $\alpha = 3$ to $\alpha = 1$, as suggested by the bottom of Fig. 4. In this limit, atoms cannot escape from the step to the terrace, even if they can detach. Physically, when the atomic motion along the periphery is very long range, the local mass flow is effectively driven by the curvature rather than its second derivative, and we find EC-like behavior. Thus, EC denotes only that attachment/detachment limits the rate, and not that there is a finite carrier concentration on the terrace. What we label TD was called "correlated EC" by Soler (1996) and by Van Siclen (1995). The essential physical mechanism characterizing this regime is evolution by single-atom jumps from one site on the island edge to another, mediated by a concentration field on the nearby terrace region.

Bonzel and Mullins (1996) have carried out a similar analysis for an isolated straight step, and we (Khare, 1996; Khare and Einstein, 1997) have extended our formalism to treat both such steps and a vicinal surface, i.e. an infinite array of steps. As for islands, one can examine the crossover between the three limiting regimes for isolated steps. The main result, again, is that the crossover regions comprise a rather narrow portion of phase space. We also recover the important cases of transport between steps when the q-dependence of τ_q^{-1} and the early-time dependence of the mean-square width do not correspond to the corresponding behavior of an isolated step, cases D and E in Pimpinelli *et al.* (1993). In case D, there is no diffusion along the step edge ($D_{st}= 0$). Since $D_{su}|q|/\Gamma << 1$, the transport is terrace-diffusion limited, but now $|q|\ell << 1$. As a result, $w^2 \sim t^{1/2}$ and $\tau_q^{-1} \propto \ell^{-1}q^2$: the latter inequality leads to a factor of q being replaced by ℓ^{-1}. For case E, atomic motion along the step edge is again forbidden, $D_{su}|q|/\Gamma << 1$, and $|q|\ell << 1$, but now there is also the condition of a perfect (infinite) Schwoebel barrier: atoms approaching a step from the upper side are reflected back rather than crossing over the step and possibly attaching to it. Following through the algebraic reductions from taking the appropriate formal limits, we find that $\tau_q^{-1} \propto \ell\, q^4$ rather than q^3, and $w^2 \sim t^{1/4}$. Thus, it is important to measure step fluctuations for different vicinalities to be certain of the correct assignment of transport mode. On the other hand, if one does find q^3 or $t^{1/3}$ behavior, it most likely is due to the TD mechanism. We are in the process of characterizing the crossover between TD and case D or case E behavior.

Other crossover behavior can arise when one moves to a regime where the continuum picture is not valid. For examples, Giesen-Seibert *et al.* (1995) show that for PD, at very early times w^2 behaves like $t^{1/2}$ rather than $t^{1/4}$ because the dynamics are dominated by random walks of kinks. In their simulations the effective exponent decreases smoothly with increasing temperature, with no evident crossover in any of the fixed–T log–log plots of w^2 vs. t. They also show how to take into account "fast events," viz. rapid, inconsequential back-and-forth motion of atoms ("blinkers"). This work builds on an earlier analysis (Giesen-Seibert and Ibach, 1994) in which they examine the structure of the probability distribution of the time between jumps as a function of the number of scans and the time of each scan, showing that the result does not depend simply on the product of these two arguments and that this sort of analysis can be used to filter out blinker events. Masson *et al.* (1994) propose a way to scale the step-step correlation function in terms of the STM scanning speed to allow the separation of diffusive behavior at fast scanning speeds from rapid temporal fluctuations at slow speeds.

A final issue of importance is the role of energetic repulsions between the steps. From eqn. (6), expanding about the average position of the step, we find the leading correction to the step free energy discussed while treating eqn. (1) is cx^2, where $c = 6g\ell^{-4}$. Then in the EC case, $\tilde{\beta}\partial^2 x/\partial y^2$ (implicitly) in eqn. (1) is replaced by $\tilde{\beta}\partial^2 x/\partial y^2 - 2cx$, leading ultimately to the replacement of $\tilde{\beta}q^2$ in A and $\tau(q)$ by $\tilde{\beta}q^2 + 2c$. This effect should only be noticeable for $q \leq 2\ell^{-2}\sqrt{(3g\tilde{\beta})}$. Nonetheless, it can mask crossover behavior expected in the long-wavelength limit. For the PD case, Masson *et al.* (1994) provide an expression for $\langle|x(y,t) - x(0,0)|^2\rangle$ in terms of a generalization of the equation in Note 1.

CONCLUSIONS

We have shown how capillary-wave analysis in a Langevin framework is a conceptually enlightening and calculationally fruitful way to explore the quantitative data on step fluctuations that is now becoming available. The same perspective that is used for conventional steps near equilibrium can be applied to the closed, nearly circular steps defining a monolayer island and can assist the study of step bunches far from equilibrium. Most of phase space is dominated by one of three distinct physical mechanisms of atomic motion (EC, TD, and PD), the entire problem can be treated in a unified way within a single Langevin equation. It is consequently possible to examine the crossover between these limiting cases.

As a continuum approximation, this approach should break down by the atomistic level. For islands it is presumably inappropriate for the small clusters imaged with FIM. More importantly, in many cases the stiffness may not be nearly anisotropic, as we have assumed it to be in our analysis. Then, as perhaps for Ag(100) islands, new mechanisms may play a role. For vacancy clusters, there can be trapping in corners in systems that might seem to be cases of PD from consideration of vicinal surfaces.

In future work we plan to extend this approach to consider the effect of external fields due to applied potentials or adsorbed species, as well as the modifications when surface islands can change their mean size (ripening or decaying). Correspondingly, there is noteworthy current work on the effect of sublimation or deposition on the step fluctuations of a vicinal surface (E.g. Pierre-Louis and Misbah, 1996). It would also be interesting to consider the effects of weak pinning potentials.

ACKNOWLEDGEMENTS

This work was supported by NSF grants MRG-DMR-9103031 and MRSEC-DMR-96-32521. TLE also benefited from a Humboldt U.S. Senior Scientist Award and the hospitality of the IGV/KFA Jülich. SVK thanks DOE for support while at Ohio State. We gratefully acknowledge extensive collaboration with N.C. Bartelt, as well as ongoing collaboration with E.D. Williams and J.E. Reutt-Robey and their experimental groups, and helpful conversations with S. Kodiyalam.

NOTES

1. To obtain the early-time behavior of the mean-square width, we take the continuum limit of the sum over q, note that the integrand is even, convert to dimensionless variables, and integrate by parts (Khare, 1996):

$$w^2(t) = \frac{k_B T}{2\pi\tilde{\beta}} \int_{-\infty}^{\infty} \left[1 - \exp(-2tA_n|q|^n)\right] q^{-2} dq = \frac{k_B T}{\pi\tilde{\beta}} (2tA_n)^{1/n} \Gamma\left(1 - \frac{1}{n}\right),$$

where A_j is the prefactor defined by $\tau_q^{-1} \equiv A_n|q|^n$, and $\Gamma(z)$ here is the gamma function. Here $n = 2, 3, 4$, for EC, TD, and PD, respectively, and $\Gamma(1/2) = \sqrt{\pi} = 1.77245...$, $\Gamma(2/3) = 1.3541...$, and $\Gamma(3/4) = 1.2254...$ In this expression one of the time indices in eqn. (3) is taken as 0, implying the step is initially straight. If both time variables are taken as large, but with a fixed difference *t*, then step fluctuates initially. Then in the intermediate expression above, the prefactor doubles and the exponent halves, so that $2^{1/n}$ is replaced by 2 in the final expression, as noted for the PD case by Masson *et al.* (1994). See also Appendix A of the chapter by Blagojević and Duxbury. (In our later section on island diffusion, α corresponds to n-1.)

2. The full, correct formula for the amplitude $g(\theta,T)$ in the expansion of the free energy per projected area, eqn. (6), is (Williams *et al.* 1993, 1994):

$$g(\theta,T) = \frac{\pi^2 k_B T b^2(\theta,T)}{24 a_{\|} h^3} \left\{1 + \left[1 + \frac{4A_{el} a_{\|}}{k_B T b^2(\theta,T)}\right]^{1/2}\right\}^2,$$

where h is the step height, A_{el} is the amplitude of the energetic ℓ^{-2} decay, due typically to elastic repulsions, and $b^2(\theta,T)$ is the diffusivity of the step, i.e. $\langle|x(y) - x(y')|^2\rangle = b^2(y-y')/a_{\|}$ for small $|y-y'|$; in terms of the stiffness, $b^2(\theta,T) = k_B T a_{\|} / \tilde{\beta}(\theta,T) \cos^3\theta$.

3. In the middle of eqn. (10) we have made an early time approximation which requires some justification since the observation time is much longer than τ_1. However, τ_1 in turn is much greater than the time for macroscopic events: e.g. in the EC case, $\tau_1 = (k_B T/\tilde{\beta} a)(R/a)^2 \tau_a$, where the first term is somewhat smaller than unity and the second factor considerably greater. Next we can consider the position of the center of mass after each of M atomic events:

$$\mathbf{r}_{CM}(t) \equiv \mathbf{r}_{CM}(t_M) = \sum_{i=1}^{M} \left[\mathbf{r}_{CM}(t_i) - \mathbf{r}_{CM}(t_{i-1})\right]$$

In then computing $\langle \mathbf{r}_{CM}^2(t)\rangle$ to find the cluster diffusion constant, we note that these differences are all in the early-time regime. The diagonal terms each contribute $4D_c(t_i - t_{i-1})$ while the off-diagonal terms have random sign; hence the diagonal term is of order M while the off-diagonal sum is of order unity. If we were to compute shape fluctuations, as discussed in Khare and Einstein (1996), the diagonal term is also limited in size rather than increasing linearly with total time, confounding application of the method to experimental data.

4. From eqn. (4) we can identify $\Gamma_{st} \equiv D_{st} c_{st} \Omega^2$, i.e. only the combination $D_{st} c_{st}$ enters the transport coefficient. From this perspective the separation of D_{st} and c_{st} is somewhat arbitrary, leading to a subtle difference in notation between Khare *et al.* (1995) and Khare and Einstein (1996). In the former, c_{st} is the actual carrier density along the step, and D_{st} is then the actual diffusion constant along the edge. In the latter reference, $c_{st}^{[eff]}$ is defined as $c_{su} a_\perp$ in conjunction with eqn. (23), leading to an effective D_{st} [e.g. in eqns. (9) and (23)] equalling $D_{st}^{[actual]} c_{st}/(c_{su} a_\perp)$.

REFERENCES

Alfonso, C., Bermond, J.M., Heyraud, J.C., and Métois, J.J., 1992, *Surface Sci.* 262:37.

Bales, G.S., and Zangwill, A., 1990, *Phys. Rev. B* 41:5500.

Barbier, L., Masson, L., Cousty, J., and Salanon, B., 1996, *Surface Sci.* 345:197.

Bartelt, N.C., Goldberg, J.L., Einstein, T.L., and Williams, E.D., 1992, *Surface Sci.* 273:252.

Bartelt, N.C., Goldberg, J.L., Einstein, T.L., Williams, E.D., Heyraud, J.C., and Métois, J.J., 1993, *Phys. Rev. B* 48:15453.

Bartelt, N.C., Einstein, T.L., and Williams, E.D., 1992, *Surface Sci.* 276:308.

Bartelt, N.C., Einstein, T.L., and Williams, E.D., 1994a, *Surface Sci.* 312:411.

Bartelt, N.C., Tromp, R.M., and Williams, E.D., 1994b, *Phys. Rev. Lett.* 73:1656.

Bartelt, N.C., and Tromp, R.M., 1996, *Phys. Rev. B* 54:11731.

Binder, K., and Kalos, M.H., 1980, *J. Stat. Phys.* 22:363.

Bonzel, H.P., and Mullins, W.W., 1996, *Surface Sci.* 350:285.

Cahn, J.W., and Taylor, J.E., 1994, *Acta Metall. Mater.* 42:1045.

de la Figuera, J., Prieto, J.E., Ocal, C., and Miranda, R., 1994, *Solid State Comm.* 89:815.

Einstein, T.L., in *Physical Structure (Handbook of Surface Science, vol . 1)*, W.N. Unertl, ed., Elsevier, Amsterdam (1996), chap. 11.

Evans, J.W., Thiel, P.A., and Wang, R., 1997, *Phys. Rev. B* 55:xxx.

Giesen, M., Frohn, J., Poensgen, M., Wolf, J.F., and Ibach, H., 1992, *J. Vac. Sci. Technol. A* 10:2597.

Giesen-Seibert, M., Jentjens, R., Poensgen, M., and Ibach, H., 1992, *Phys. Rev. Lett.* 71:3521; 73:E911 (1994).

Giesen-Seibert, M., and Ibach, H., 1994, *Surface Sci.* 316:205.

Giesen-Seibert, M., Schmitz, F., Jentjens, R., and Ibach, H., 1995, *Surface Sci.* 329:47.

Gruber, E.E., 1967, *J. Appl. Phys.* 38:243.

Hohenberg, P.C., and Halperin, B.I., 1977, *Rev. Mod. Phys.* 49:435.

Hoogeman, M.S., Schlosser, D.C., and Frenken, J.W.M., 1996, *Phys. Rev. B* 53:R13299.

Joós, B., Einstein, T.L., and Bartelt, N.C., *Phys. Rev. B* 43:8153.

Kardar, M., 1994, *Turkish J. of Phys.* 18:221.

Kellogg, G.L., 1994, *Phys. Rev. Lett.* 73:1833; *Surface Sci. Rep.* 21:1.

Khare, S.V., 1996, Ph.D. thesis, University of Maryland, College Park (unpublished).

Khare, S.V., Bartelt, N.C., and Einstein, T.L., 1995, *Phys. Rev. Lett.* 75:2148.

Khare, S.V. and Einstein, T.L., 1996, *Phys. Rev. B* 54:11752.

Khare, S.V. and Einstein, T.L., 1997, *Bull. Am. Phys. Soc.* 42:619 and preprint.

Kuipers, L., Hoogeman, M.S., and Frenken, J.W.M., 1993, *Phys. Rev. Lett.* 71:3517.

Kuipers, L., Hoogeman, M.S., Frenken, J.W.M., and van Beijeren, H., 1995, *Phys. Rev. B* 52:11387.

Masson, L., Barbier, L., Cousty, J., and Salanon, B., 1994, *Surface Sci.* 317:L1115.

Morgenstern, K., Rosenfeld, G., Poelsema, B., and Comsa, G., 1995, *Phys. Rev. Lett.* 74:2058.

Morgenstern, K., Rosenfeld, G., and Comsa, G., 1996, *Phys. Rev. Lett.* 76:2113.

Mullins, W.W., 1957, *J. Appl Phys.* 28:333.

Mullins, W.W., 1959, *J. Appl Phys.* 36:77.

Mullins, W.W., in: *Metal Surfaces: Structure, Energetics and Kinetics*, N.A. Gjostein and R.W. Roberts, eds., Am. Soc. Metals, Metals Park, Ohio, 1963), P. 17.

Nozières, P., in: *Solids Far From Equilibrium*, C. Godrèche, ed., Cambridge Univ. Press, Cambridge (1992), chap. 1.

Ozcomert, J.S., Pai, W.W., Bartelt, N.C., and Reutt-Robey, J.E., 1993, *Surface Sci.* 293:183.

Ozcomert, J.S., Pai, W.W., Bartelt, N.C., Einstein, T.L., and Reutt-Robey, J.E., 1994, *J. Vac. Sci. Technol. A* 12:2224.

Ozcomert, J.S., Pai, W.W., Bartelt, N.C., and Reutt-Robey, J.E., 1995, *Mat. Res. Soc. Symp. Proc.* 355:115.

Pai, W.W., 1997, priv. comm.

Pai, W.W., Ozcomert, J.S., Bartelt, N.C., Einstein, T.L., and Reutt-Robey, J.E., 1994, *Surface Sci.* 307-309:747.

Pai, W.W., Bartelt, N.C., and Reutt-Robey, J.E., 1996, *Phys. Rev. B* 53:15991.

Pai, W.W., and Reutt-Robey, J.E., 1996, *Phys. Rev. B* 53:15997.

Pierre-Louis, O., and Misbah, C., 1996, *Phys. Rev. Lett.* 76:4761, and preprints.

Pimpinelli, A., Villain, J., Wolf, D.E., Métois, J.J., Heyraud, J.C., Elkinani, I., and Uimin, G., 1993, *Surface Sci.* 295:143.

Poensgen, M., Wolf, J.F., Frohn, J., Giesen, M., and Ibach, H., 1992, *Surface Sci.* 274:430.

Rettori, A. and Villain, J., 1988, *J. Phys. France* 49:257.

Reutt-Robey, J.E., and Pai, W. W., in: *Surface Diffusion: Atomistic and Collective Processes*, M.C. Tringides and M. Scheffler, eds., NATO-ASI Series, Plenum Press, New York (1997).

Rosenfeld, G., Morgenstern, K., and Comsa, G., in: *Surface Diffusion: Atomistic and Collective Processes*, M.C. Tringides and M. Scheffler, eds., NATO-ASI Series, Plenum Press, New York (1997).

Sholl, D.S., and Skodje, R.T., 1995, *Phys. Rev. Lett.* 75:3158.

Sholl, D.S., and Skodje, R.T., 1996, *Physica A* 231:631.

Siggia, E.D., 1996, priv. comm.

Soler, J.M., 1994, *Phys. Rev. B* 50:5578.

Soler, J.M., 1996, *Phys. Rev. B* 53:R10540.

Sutherland, B. 1971, *Phys. Rev. A* 4:2019.

Van Siclen, C.DeW., 1995, *Phys. Rev. Lett.* 75:1574.

Wang, P., Pfnür, H., Khare, S.V. Einstein, T.L., Williams, E.D., Pai, W. W., and Reutt-Robey, J.E., 1996, *Bull. Am. Phys. Soc.* 41:189, and preprint.

Wang, S.C., and Ehrlich, G., 1990, *Surface Sci.* 239:301.

Wax, N., ed., *Selected Papers on Noise and Stochastic Processes*, Dover, New York (1954).

Wen, J.M., Chang, S.-L., Burnett, J.W., Evans, J.W., and Thiel, P.A., 1994, *Phys. Rev. Lett.* 73:2591.

Wen, J.M., Evans, J.W., Bartelt, M.C., Burnett, J.W., and Thiel, P.A., 1996, *Phys. Rev. Lett.* 76:652.

Wen, J.M., Evans, J.W., Bartelt, M.C., Zhang, C.M., Burnett, J.W., and Thiel, P.A., 1997, *J. Vac. Sci. Technol. A* 15:xxx.

Wendelken, J.F., Pai, W.W., Swan, A.K., and Zhang, Z., 1997, *Bull. Am. Phys. Soc.* 42:687.

Williams, E.D., Phaneuf, R.J., Wei, J., Bartelt, N.C., and Einstein, T.L., 1993, *Surface Sci.* 294:219.

Williams, E.D., Phaneuf, R.J., Wei, J., Bartelt, N.C., and Einstein, T.L., 1994, *Surface Sci.* 310:E451.

Williams, E.D., and Bartelt, N.C., in *Physical Structure (Handbook of Surface Science, vol. 1)*, W.N. Unertl, ed., Elsevier, Amsterdam (1996), chap. 2.

KINETIC RATE LAW ISSUES IN THE MORPHOLOGICAL RELAXATION OF RIPPLED CRYSTAL SURFACES

Jonah D. Erlebacher

Division of Engineering and Applied Sciences
Harvard University
9 Oxford St.
Cambridge, MA 02138

INTRODUCTION

The relaxation of a sinusoidally rippled crystal surface below the roughing transition temperature, T_R, can exhibit significantly different behavior than predicted by the classical theory of surface relaxation (continuum/isotropic surface energy/valid above T_R) due to Mullins[1]. Computing power has begun to allow attack of this problem using sufficiently large system sizes, and for this reason the topic is receiving a great deal of attention and it is the subject of a number of papers in this symposium. The primary quantity one measures in this type of study is the scaling behavior of the amplitude decay curve, $h(t)$, with respect to the sinusoid wavelength, λ, and temperature, T. A number of theories predict that $h(t)$ should scale as λ^{α}, (i.e., $h(t,\lambda) = f\left(t/\lambda^{\alpha}\right)$, where the value of α depends on the particular geometry of the relaxing surface. We show here that the kinetic rate laws that control atomic motion also affect the value of α.

Certain aspects of the problem have been generally agreed upon. Particularly, the kinetics of relaxation of unidirectional modulations (i.e., one-dimensional sinusoids) is seen to proceed according to the pinch-off mechanism, originally due to Rettori and Villain[2]. In this picture, the surface is primarily static. There is an uppermost terrace at the top of the sinusoid, and a corresponding terrace at the bottom; a series of steps separates the two terraces. These steps are on average straight but are free to fluctuate from their average position. We are interested in the time, τ_P, for sides of the upper- or lower-most terrace to fluctuate into each other. This is the pinch-off event. Once a pinch-off has been nucleated, dissolution of the top and bottoms layers of the surface proceeds quickly as atoms are ejected from the highly curved steps crossing the terrace that has pinched off.

In the simulations, it is evident after each layer of the sinusoid has peeled off, the steps of the newly exposed terrace on the next layer down have not reached their equilibrium "width" (rms deviation from its average position); rather, they are still essentially straight.

Dynamics of Crystal Surfaces and Interfaces
Edited by Duxbury and Pence, Plenum Press, New York, 1997

Thus, the relaxation time and its scaling behavior is dominated by the time for the pair of straight steps flanking each other across the extremal terraces to generate fluctuations large enough that a statistical ensemble of pinch-off events will be formed. This view is subtly different than the kind of calculation made by Duport, et al[3], who find the time for a terrace to evaporate once the statistical distribution of pinch-off events has been formed. In loose analogy to nucleation and growth (dissolution), our simulations suggest that at low temperatures and long wavelengths, the nucleation mechanism (pinch-off) is rate limiting, and dominates the scaling behavior of the amplitude decay curve.

Previous Monte Carlo work simulating surface relaxation has employed the Metropolis algorithm[4]. We have presented work elsewhere describing surface relaxation below T_R in which the *Kinetic* Monte Carlo (KMC) algorithm was used instead[5]. There is a fair amount of discussion as to the validity of comparing results given by each simulation technique. One aspect of the problem is simply the comparison between techniques. KMC gives a physically relevant time step, the Metropolis algorithm does not necessarily do so, etc.

A more interesting problem is that the Metropolis Monte Carlo studies used a different (physically simplified) kinetic rate law for atomic motion than the KMC work. That is, the rules governing the rate at which atoms jump from one configuration to the next were fundamentally different. This can have serious implications for such dynamic phenomena as step fluctuations, adatom mobility, etc. In this paper, we describe the physical differences between the rate laws used in the previous work, and then present results using just one of the simulation techniques, namely KMC, but comparing both kinds of rate laws.

It is interesting to note in passing that comparison of results using KMC vs. Metropolis are few, and suggest that correspondence between the two methods is weak at best[6]. However, we have found that the results of Murty, et al[4], who used the Metropolis algorithm, can be essentially duplicated using KMC. Murty has also seen this correspondence[7].

Some definitions used in this paper:

D_s = adatom diffusivity on a terrace
$C(T)$ = equilibrium adatom concentration on a terrace
$\tilde{\beta}(T)$ = step stiffness,
a = lattice parameter

COMPARISON OF MONTE CARLO SIMULATION TECHNIQUES

In this section, the mechanics of the kinetic Monte Carlo algorithm (KMC)[6] are compared to the Metropolis algorithm[8].

In any discrete system, there will be only a limited number of types of atomic transitions that can occur, $\{k_i\}$. For the surface relaxation case considered here, these correspond to diffusion events across a terrace, off kinks, etc. In KMC, the state of the simulation system at any time t is defined by the number of atoms that are able to undergo each of these transition types, $\{n_i\}$. All of the transitions are operating in parallel, so the average time between transitions is

$$\Delta t = \frac{1}{\sum n_i k_i} . \tag{1}$$

The probability of a transition of type i occurring in this time interval is

$$P_i = \frac{n_i k_i}{\sum n_j k_j} \quad . \tag{2}$$

The KMC algorithm works by (i) tabulating $\{n_i\}$, (ii) calculating the P_i's and picking the kind of transition to occur, (iii) randomly picking one of the atoms allowed to undergo that transition and moving it, (iv) incrementing the time step according to the current value of Δt.

The Metropolis algorithm is fundamentally thermodynamic, and is guaranteed to move a system into a lowest energy configuration via a Markov chain. Finding low energy states has been its primary use, but the dynamics of the relaxation to equilibrium are often of interest. To implement the Metropolis algorithm, one chooses an atom at random from the entire simulation field. By calculating the energy change $\Delta U = U_{initial} - U_{final}$ between its current position and a neighboring position, one finds the probability of the atom to undergo that transition according to

$$P = \begin{cases} \exp(-\Delta U / k_B T) & \Delta U > 0 \\ 1 & \Delta U < 0 \end{cases} . \tag{3}$$

By calculating P for that atom, one finds whether or not it jumps. One iteration of this procedure is called a Monte Carlo time step (MCS), but there is no intrinsic prescription for Δt. Usually, particular measures of the state of a system are simply plotted against MCS/site.

For the simulation of surface relaxation, KMC has two advantages over Metropolis. Firstly, since an atom moves at every iteration regardless of temperature, lower temperatures can be studied. Secondly, the dynamic (vs. thermodynamic) nature of the algorithm yields a proper time step, whereas is debated whether or not Metropolis does so[6].

KINETIC RATE LAWS USED TO STUDY SURFACE RELAXATION

The results shown in this study are limited to the KMC algorithm. In principle, to model a realistic system the set $\{k_i\}$ can be found using molecular dynamics simulations, or other similar techniques; these can then be used as input into KMC. Here, our purposes are more general. A square lattice is examined, and there are two simplified rate laws of interest. The first is often used in KMC simulations of deposition[9], and is termed *i*-kinetics. A second set represents a kinetics that is analogous to the hopping probabilities used in the Metropolis simulations. We term the latter Δ*i*-kinetics.

i-kinetics

This rate law corresponds to an atom jumping out of a potential well by completely breaking i lateral bonds. The transition rate is independent of the energy of the final configuration, i.e, contains no information about the site to which the atom hops, see Figure 1(a). One sets

$$k_i = \nu_0 \exp\left(\frac{-(\varepsilon_B + i\varepsilon)}{k_B T}\right), \tag{4}$$

where ν_0 is a hopping attempt frequency, ε_B is a substrate bonding energy, and ε is a lateral bond energy.

Δi-kinetics

Δi-kinetics is contrasted with i-kinetics in that the activation barrier contains information about both the initial and final state when the atom hops. One way to implement this is shown in Figure 1(b). In this case, we can write the kinetic law in a form analogous to hopping probabilities in the Metropolis algorithm:

$$k_{\Delta i} = \begin{cases} \nu_0 \exp\left(-\dfrac{\varepsilon_B}{k_B T}\right) & \Delta i < 0 \\ \nu_0 \exp\left(-\dfrac{\varepsilon_B}{k_B T}\right) \exp\left(-\dfrac{\Delta i \cdot \varepsilon}{k_B T}\right) & \Delta i > 0 \end{cases} \tag{5}$$

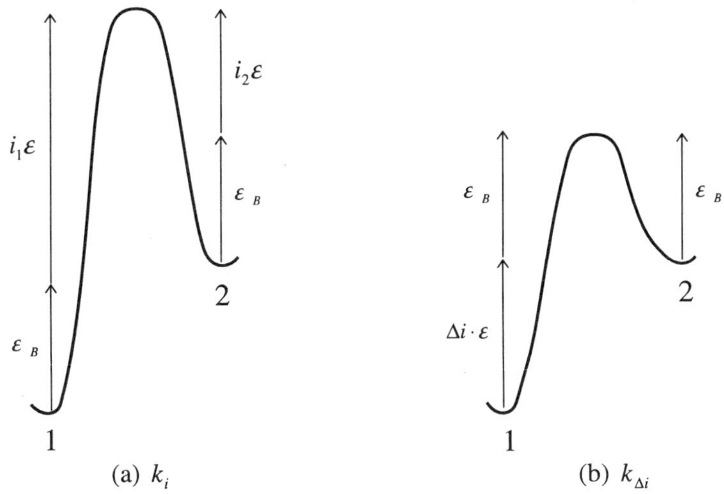

Figure 1. Activation barriers for an atom jumping from site 1 to site 2 through a transition state: (a) i-kinetics, and (b) Δi-kinetics.

Effects of Kinetic Law on Step Fluctuations

Both i-kinetics and Δi-kinetics have the same thermodynamic creation energies for all configurations. For example, on a simple cubic lattice, the adatom formation energy will be $\Delta\varepsilon_{adatom} = 2\varepsilon$, using bond counting arguments, and thus the equilibrium adatom concentration on an infinite terrace will be $C_{eq}(T) \propto \exp(-2\varepsilon/k_B T)$. However, we expect some qualitative differences in the dynamics of surface relaxation behavior when using the two approaches, and these differences are seen in the simulation results shown below. In particular, there is no relative difference in activation barrier height for adatom diffusion either on a step or on a terrace, that is, $\Delta i = 0$ in both cases. Thus, adatoms on steps move as fast as adatoms on terraces. In contrast, for i-kinetics, atoms on steps ($i = 1$) move significantly slower than atoms on terraces ($i = 0$). Thus, an adatom obeying i-kinetics which has newly evaporated from a kink site has more of a chance to move away from the step than an equivalent atom obeying Δi-kinetics.

The implication of this behavior suggests that there will be a quantitative difference in the kind of step fluctuation dynamics observed for each kinetic law system. For i-kinetics, we expect steps to fluctuate by variations in the flux of adatoms hitting the step from a uniform, quickly moving sea of equilibrated adatoms on the adjoining terrace. In this case, the time to create a step fluctuation of amplitude y (perpendicular to both the surface normal and the average step direction) will be given by

$$\tau_i \propto \frac{\tilde{\beta}(T)}{D_s C k_B T} y^4 \tag{6}$$

according to arguments given previously[5]. Assuming that the structure relaxes by peeling off layers one at a time, then $y_{\text{crit}} \propto \lambda$ always, and this expression gives the overall temperature and wavelength scaling behavior. Values for $\tilde{\beta}(T)$ using i-kinetics can be found in Reference 5.

For Δi-kinetics, on the other hand, adatoms diffuse rapidly along steps, and we can expect a tendency to "wash out" the variation of the terrace adatom flux onto the step. Thus, at the very least, we can say that step fluctuations in this system should form more slowly than for i-kinetics, i.e.,

$$\tau_{\Delta i} \propto y^{\alpha}, \tag{7}$$

where $\alpha > 4$. This situation does not correspond to any of the fluctuating step cases enumerated by Pimpinelli, et al.[10], and we will not attempt a derivation here to predict the scaling exponent.

SUMMARY OF RESULTS USING i-KINETICS

One potential shortcoming of our simulation results presented in Ref. 5 was the geometry of the simulation system. In that work, multiple wavelengths were placed on a square field. It was thought that the size of the system parallel to the direction of the sinusoid steps was not big enough to get good statistics. The results here have been repeated in the "strip" geometry, the dimensions of which are one wavelength in the direction perpendicular to the step, and ranged from 2048-16536 lattice units in the parallel direction, depending on the computing power needed to model the particular wavelength. A careful comparison shows that our results using the strip geometry are indistinguishable from the results presented in Ref. 5, but we show the new results here for completeness.

It is found that for data at each temperature, data collapse occurs for time scaling by λ^4, as shown in Figure 2. This is consistent both with the classical model of Mullins[1], as well as the pinch-off model. The difference between these two models is the temperature dependence of the relaxation rate. For classical behavior, one expects the characteristic time to scale according to

$$\tau_M^{-1} \propto \frac{D_s C}{k_B T} \tag{8}$$

whereas one expects the pinch-off characteristic time to scale according to

$$\tau_P^{-1} \propto D_s C k_B T \quad , \tag{9}$$

since the line stiffness[11], $\tilde{\beta}(T) \propto (k_B T)^2$, and τ_P takes the form shown in Eq. (6).

In the simulations, $D_s(T) = \nu_0 \exp(-\varepsilon_B / k_B T)$ and $C(T) \propto \exp(-2\varepsilon / k_B T)$ are predetermined by the assigned bond energies, so we can test Eqs. (8) and (9) for data collapse *with no adjustable parameters*. These data collapses are shown in Figure 2, and show that data collapse is more consistent with the classical model at temperatures near T_R, and more consistent with the pinch-off model at temperatures well below T_R.

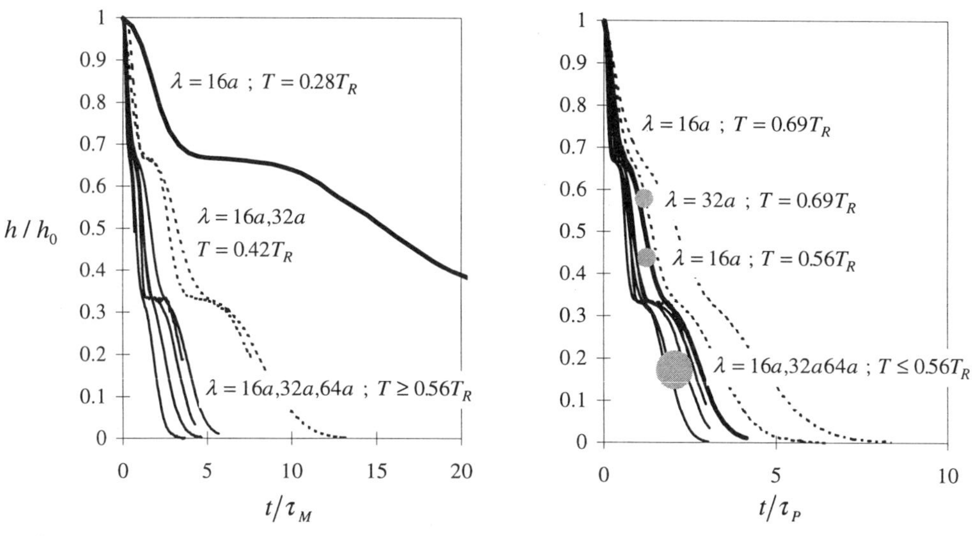

Figure 2. Data collapses for relaxation of one-dimensional sinusoidal modulations, $0.28T_R \leq T \leq 0.69T_R$, $16a \leq \lambda \leq 64a$: (a) classical temperature dependence, $\tau_M^{-1} \propto \frac{\exp(-\varepsilon^*/k_B T)}{k_B T} \lambda^{-4}$, (b) pinch-off model, $\tau_P^{-1} \propto \frac{k_B T \exp(-\varepsilon^*/k_B T)}{\tilde{\beta}(T)} \lambda^{-4}$. In both cases, $\varepsilon^* = \varepsilon_B + 2\varepsilon$ and $h_0 = 3a$.

Figure 3 shows some topography of surfaces illustrating the pinch-off mechanism; here $T = 0.56T_R$, and $\lambda = 64a$. At time t_0, a layer of the sinusoid has relaxed away exposing a pair of terraces with relatively straight steps. At time t_1, a single pinch-off event has occurred and at time t_2 all relevant pinch-offs have been nucleated. By time t_3, the layer has relaxed away.

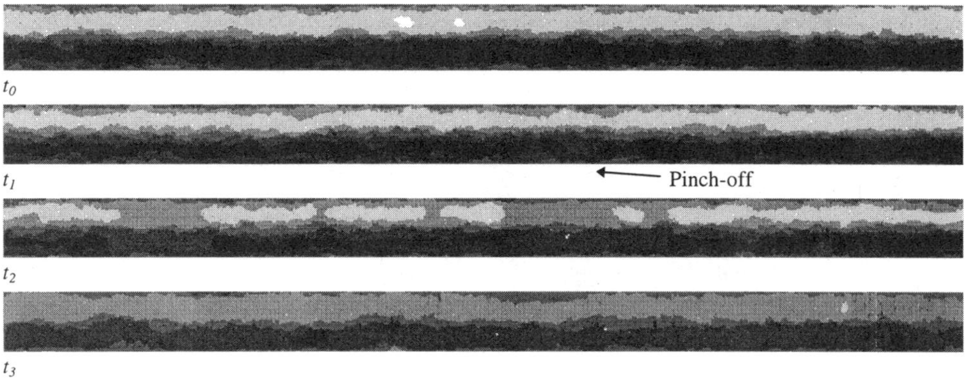

Figure 3. Snapshots of a surface obeying *i*-kinetics illustrating the pinch-off mechanism. $T = 0.56T_R$, $\lambda = 64a$. $t_0\nu_0 = 1.75\text{x}10^5$ (no pinch-off), $t_1\nu_0 = 2.62\text{x}10^5$ (first pinch-off), $t_2\nu_0 = 3.78\text{x}10^5$ (steady-state pinch-off), $t_3\nu_0 = 5.24\text{x}10^5$ (terrace dissolution virtually complete).

It is interesting to note as well, that for large wavelength one sees an asymmetry in the initial behavior of the topmost and bottommost terraces reflected in each's adatom concentration. At short times, the concentration of adatoms on the top terrace is slightly higher than the equilibrium concentration, and it is slightly lower on the bottom one, as shown in Figure 4(a). The reason for this is that creation of an adatom on the top terrace causes the top step to move away from its immediate neighbor, reducing the repulsive step-step interaction energy. On the other hand, creation of an adatom on the bottom terrace causes the bottom step to move toward its immediate neighbor, increasing the repulsive step-step interaction energy. This is illustrated in Figure 4(b). At late time, in the fully relaxed state, the adatom concentration is of course at its equilibrium value.

SUMMARY OF RESULTS USING Δ*i*-KINETICS

Figure 5(a) shows the amplitude decay curves using Δi-kinetics for $T = 0.56T_R$. It almost exactly reproduces a figure given by Murty and Cooper, who used the Metropolis algorithm[4], but is extended to the long time behavior. Amplitude decay curves for $h_0 = 3a, 4a$ are shown. The scaling behavior at $T = 0.56T_R$ seems to follow λ^5 scaling, agreeing with a model of Ozdemir and Zangwill[12] in which the driving force for surface relaxation is entropic step repulsion. Entropic repulsion, however, is a different mechanism than the step fluctuation picture. Nonetheless, whichever picture one chooses, one expects to find a scaling exponent greater than 4.

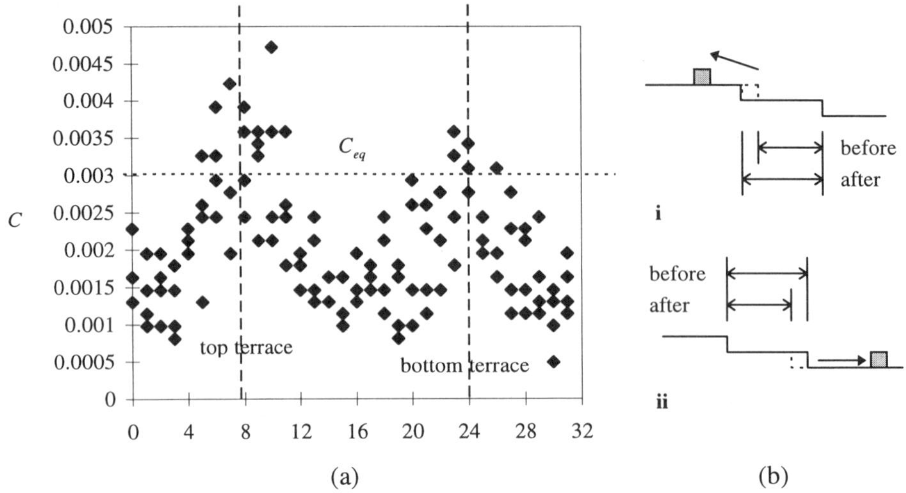

Figure 4. Early time adatom concentration profile for $T = 0.56T_R$, $\lambda = 32a$: (a) concentration, C vs. position, (b) schematic illustrating lower step interaction energy upon creation of adatoms on the top terrace (i) and higher step interaction energy upon creation of adatoms on the bottom terrace (ii). $C_{eq} = \exp(-2\varepsilon/k_B T) = 0.003$. The minima in adatom concentration are due to an artifact in the counting statistics on the small terraces on the slopes of the sinusoid.

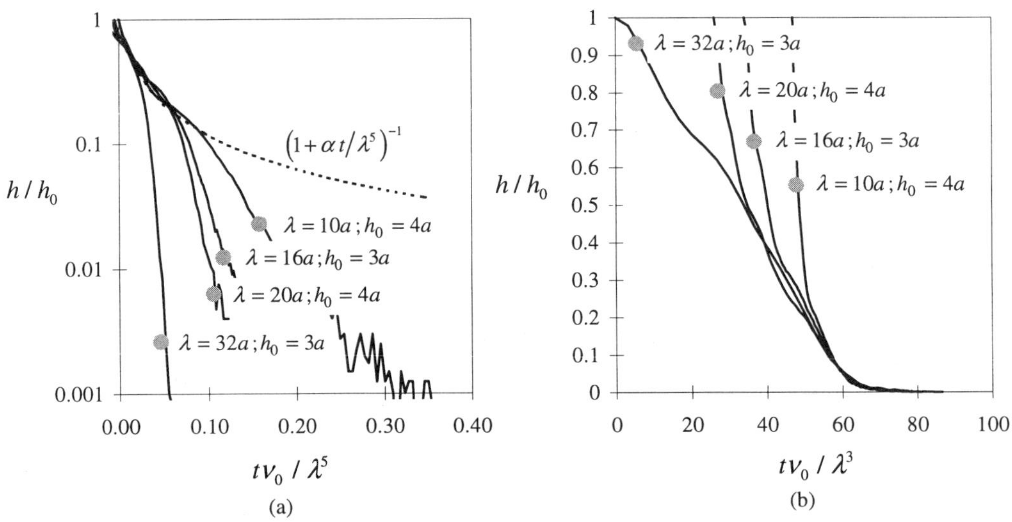

Figure 5. Data collapses for Δi-kinetics, $T = 0.56T_R$, $\lambda = 10a, 16a, 20a, 32a$: (a) λ^5 scaling at short times; $(1+\alpha t/\lambda^5)^{-1}$ is the form predicted for the collapsed amplitude decay curve in Ref. 12. The data deviate from "data collapse" as λ increases. (b) λ^3 scaling at long times. Note that figure (a) has a log scale, while (b) is linear in h/h_0.

This is not to imply one cannot have both steps fluctuating and entropic repulsion simultaneously. However, one can play some games with the data which will suggest that fluctuation phenomena is the primary mechanism. Rather than collapse the amplitude decay curves around $t = 0$, the simulation start time, we collapse the amplitude decay around some late time t_f. We define t_f to be that time at which all of the curves have hit the same small but finite amplitude; here, this amplitude was chosen to be $h/h_0 = 0.001$. Once t_f is found for each decay curve, each curve is translated along the time axis so that all the t_f overlay. This procedure is done in accordance to the analogy between nucleation (of pinch-off events) and growth (evaporation of pinched-off islands); we eliminate the transient effects of nucleation by "starting the clock" once nucleation of pinch-off events has ceased. This is done in Figure 5(b), and it can be seen that data collapse occurs nicely for λ^3 scaling at long times. λ^3 scaling at very long times is not in itself surprising. After all, when we are confronted with an amplitude $h = 1$ sinusoid that has pinched off, essentially we are left with an array of bumps, a two-dimensional sinusoidal modulation. This kind of modulation decays according to a linear λ^3 scaling behavior according to a well-accepted theory and is supported by simulation[2, 5].

What is surprising in Figure 5 is that for longer wavelengths, the λ^3 scaling behavior seems to persist to amplitudes greater than one layer height. The implication of this observation is that bumps are beginning to form at amplitudes greater than $h = 1$, earlier than when there is just the lowermost terrace remaining. In fact, striking examples of this behavior were observed for all temperatures $T < 0.56T_R$, $\lambda \geq 32a$, temperatures and wavelengths not accessible to simulation using the Metropolis algorithm. Figure 6 shows snapshots of $T = 0.42T_R$, $\lambda = 64a$, $h_0 = 3a$, at illustrative times. The entire sinusoid has pinched off over multiple layers creating a well-formed array of bumps. We repeated the simulations in the square (multiple wavelengths) geometry and the effect persisted.

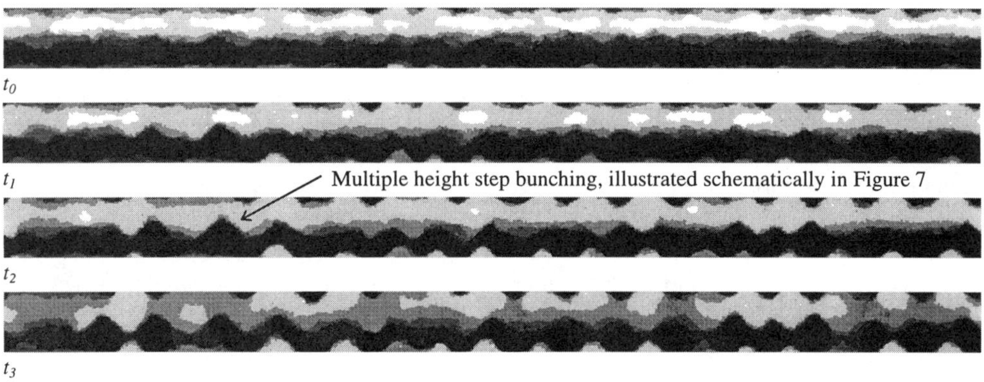

Figure 6. Snapshots of a surface obeying Δi-kinetics illustrating spontaneous islanding at low temperatures. $T = 0.42T_R$, $\lambda = 64a$. $t_0\nu_0 = 4.31\times10^6$ (pinch-off mechanism operative), $t_1\nu_0 = 6.7\times10^6$, $t_2\nu_0 = 9.09\times10^6$, $t_3\nu_0 = 1.87\times10^7$ (islanding apparent).

A mechanism that may qualitatively explain the effect is illustrated in Figure 7, showing a kinetic pathway in which an atom jumps from kink to kink site from the uppermost to the lowermost terrace. Since $\Delta i = 0$ in all these jumps, there is essentially no kinetic barrier to migration. By bunching steps close to each other while simultaneously creating a large amount of kink sites per steps, many $\Delta i = 0$ paths from terrace to terrace are created. The supposition here is that this is effect is kinetically favorable, leading the surface to spontaneously island and then relax as if it were two-dimensionally modulated. So, at long wavelength (small h_0/λ) or low temperature, a system obeying Δi-kinetics has a low energy path to equilibrium, an alternate to pinch-off. No such alternate energy path was seen in any of our simulations using i-kinetics.

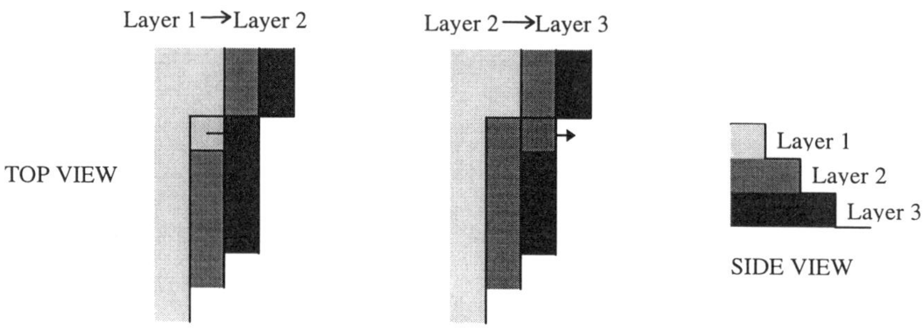

Figure 7. Schematic illustrating a $\Delta i = 0$ pathway for an adatom moving from the topmost terrace to the bottommost one.

SUMMARY

A comparison study of surface relaxation below roughening using two simplified rate laws for adatom motion has been made. Both rate laws, termed i-kinetics and Δi-kinetics, show kinetic pathways to equilibrium whose scaling behavior is consistent with theory. For i-kinetics, pinch-off is always observed to be the operative mechanism in surface relaxation (over the temperature and wavelength range accessible to simulation), and is consistent with the observed λ^4 scaling behavior. The temperature dependence of the amplitude decay is also consistent with pinch-off, but very different from the temperature scaling expected classically (and uniformly used in experiment). For Δi-kinetics, pinch-off is observed to be operative at higher temperatures and short times, leading to $\lambda^{\alpha>4}$ scaling behavior, again consistent with theory. However, at low temperatures and large wavelengths, the surface is observed to spontaneously island, leading to λ^3 scaling behavior at long times, consistent with the relaxation of surfaces with bidirectional modulations.

ACKNOWLEDGMENTS

This research was supported by DE-FG02-89ER45401. The author acknowledges the NDSEG fellowship program for support, and thanks M. J. Aziz for many helpful discussions.

REFERENCES

1. W. W. Mullins, J. Appl. Phys. **28** (1957), 333. W. W. Mullins, J. Appl. Phys. **30** (1959), 77.
2. A. Rettori, J. Villain, J. Phys. (Paris) **49** (1988), 257. W. Selke, T. Bieker, Surf. Sci. **281** (1993), 163.
3. C. Duport, A. Chame, W. W. Mullins, J. Villain, J. Phys. (Paris), in press.
4. Z. Jiang, C. Ebner, Phys. Rev. B **40** (1989), 316. P. C. Searson, R. Li, K. Sieradzki, Phys. Rev. Lett. **74** (1995), 1395. M. V. Ramana Murty, B. H. Cooper, in press.
5. J. D. Erlebacher, M. J. Aziz., accepted to Surface Science, 1996.
6. H. C. Kang, W. H. Weinberg, J. Chem. Phys. **90** (1989), 2824. A. M. Bowler, E. S. Hood, J. Chem. Phys. **94** (1991), 5162.
7. M. V. Ramana Murty, private communication.
8. K. Binder, D. W. Heermann, Monte Carlo Simulation in Statistical Physics, Springer-Verlag, 1992.
9. G. S. Bales, D. C. Chrzan, Phys. Rev. B **50** (1994), 6057.
10. A. Pimpinelli, J. Villain, D. E. Wolf, J. J. Metois, J. C. Heyraud, I. Elkinani, G. Uimin, Surf. Sci. **295** (1993), 143.
11. E. D. Williams, R. J. Phaneuf, J. Wei, N. C. Bartelt, T. L. Einstein, Surf. Sci. **294** (1993), 219. E. D. Williams, R. J. Phaneuf, J. Wei, N. C. Bartelt, T. L. Einstein, Surf. Sci. **310** (1994), 451.
12. M. Ozdemir, A. Zangwill, Phys. Rev. B, **42** (1990), 5013.

GRAIN BOUNDARY MOTION IN ALUMINIUM BICRYSTALS

G. Gottstein[1], L. S. Shvindlerman[2], D. A. Molodov[1], U. Czubayko[1]

[1] Institut für Metallkunde und Metallphysik, RWTH Aachen, Kopernikusstr. 14, 52074 Aachen, Germany.

[2] Institute of Solid State Physics, Russian Academy of Sciences, Chernogolovka, Moscow Distr., 142432 Russia.

INTRODUCTION

Grain boundary motion is the fundamental process in recrystallization and grain growth, which determine and control microstructure formation during annealing of cold worked materials. Grain boundaries in polycrystal are not alike in their properties, especially in their kinetics. On the contrary grain boundary properties depend on grain boundary structure, i.e. on its crystallography and chemistry, as evident from the formation of pronounced crystallographic textures produced during recrystallization. Thus, for microstructure control during recrystallization and grain growth it is indispensible to understand the mechanisms that underlie grain boundary kinetics. Experiments on polycrystals can provide information only on average grain boundary behaviour but not on the relationship between grain boundary structure and mobility. The structural dependence of boundary mobility can only be retrieved from investigations of defined grain boundaries in materials with controlled purity. The current study will focus on the orientation dependence of grain boundary mobility in aluminium bicrystals and on the effect of solute atoms on grain boundary motion in Al.

EXPERIMENTAL

The experiments were carried out on high purity aluminium from different producers (Table 1) and Al-Ga alloys. The impurity concentration in the used materials was determined by glow discharge mass-spectrometry. The total impurity content in pure aluminium (0.4 - 7.7 ppm) was defined as the sum of the concentration of all found elements.The residual resistivity ratio $RRR = \rho(273K)/\rho(4.2K)$ of the materials was measured by the method prescribed by the U.S. National Bureau of Standards.

The investigation of boundary mobility was confined to <111> tilt boundaries with misorientation angles in the vicinity of the special misorientation $\Sigma 7$ ($\varphi = 38.2°$). These

Table 1.. Materials notation and purity.

Material	RRR	Total impurity content, ppm
Al I	21400	0.4
Al II	15000	1.0
Al III	5100	3.6
Al IV	6500	4.9
Al V	11800	7.7

boundaries are known to be play a dominant role in recrystallization and grain growth owing to their high mobility[1]. To measure the grain boundary mobility, the motion of grain boundaries under the action of a constant driving force was recorded (Fig. 1). The driving force p was provided by the surface tension of a curved grain boundary: $p = \sigma/a$, where σ is the grain boundary surface tension, and a the width of the shrinking grain. It is particularly stressed that the boundary remains self-semilar during its motion.

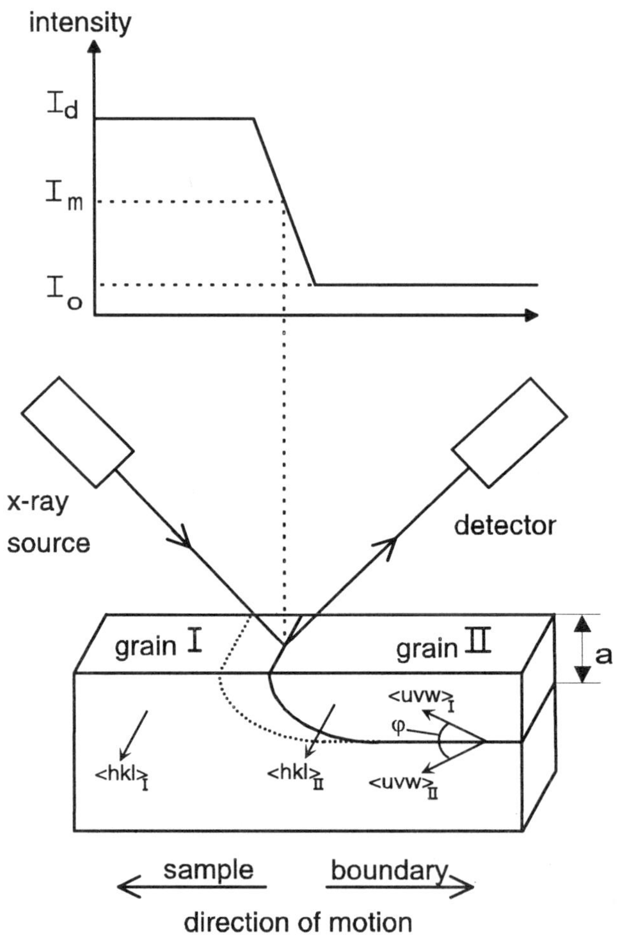

Fig. 1: Geometry of used bicrystals and measuring principle of the XICTD[4]

Details of bicrystal growth and sample preparation are given elsewhere[2,3]. The velocity of grain boundary motion was measured using a specially designed X-ray device for continuous tracking of a moving grain boundary (XICTD) with an accuracy better than 2%[4]. For the measurement of grain boundary motion the bicrystal sample is mounted under the incident X-ray beam such that one crystal is in Bragg-position while the other is not. A scan of the X-ray beam along the sample surface would result in an intensity profile as shown in Fig. 1. The maximum intensity I_d is recorded as long as the X-ray spot is entirely located on the surface of crystal I. When the X-ray spot has completely moved over to crystal II the detected intensity attains the value I_o. The boundary position can be associated with the position where the intermediate intensity $I_m = (I_o + I_d)/2$ is detected. When the boundary moves, the sample is concurrently displaced such that the reflected X-ray intensity remains constant during the measurement. Thus, the velocity of the moving grain boundary is equal to the speed of sample movement at any moment during the experiment.

Fig. 2 shows a schematic sketch of the XICTD. The computer records the X-ray intensity from the detector and activates the stepping motor of the sample holder such that the recording intensity remains constant. Because of the size of the X-ray beam the absolute accuracy of locating the grain boundary position is approximately 15 μm. The device can measure a grain boundary velocity in a wide range between 1 μm/s to 1000 μm/s and allows up to 4 measurements of the boundary position per second. The inaccuracy of velocity determination depends on the frequency of measurements and amounts to 0.5-2.5%. The hot stage of the device allows a sample temperature between 20°C and 1300°C. During the measurement of grain boundary motion the temperature is

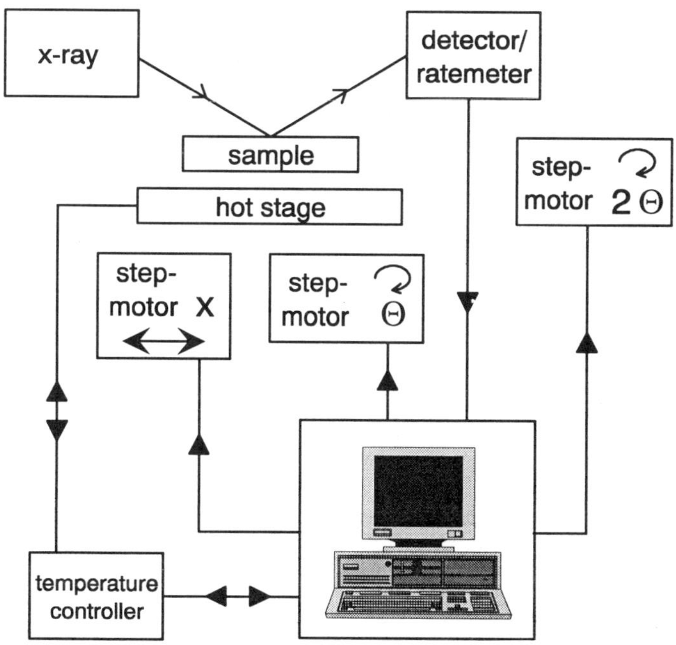

Fig. 2: Schematic arrangement of the X-ray Interface Continuous Tracking Device (XICTD) and its controls.

kept constant within ±0.3°. To account for thermal expansion of the sample during temperature changes the Bragg-angle is accordingly adjusted. To avoid surface oxidation the sample and the hot stage were exposed to a nitrogen gas atmosphere.

During the experiment the boundary displacement is recorded. Its derivative with regard to time is the velocity v of grain boundary motion, which is related to the driving force p by the boundary mobility $m = v / p$. For convenience we use the reduced boundary mobility

$$A \equiv v \cdot a = A_0 \exp\left(-\frac{H}{kT}\right) = m\sigma, \qquad (1)$$

where H is the activation enthalpy of migration and A_0 the pre-exponential mobility factor. In the following we refer to A as mobility for brevity.

RESULTS

Orientation Dependence of Grain Boundary Motion

As was already shown in the past, for instance by Aust and Rutter[5] or Shvindlerman et al.[6-8] the mobility of tilt grain boundaries depends on axis <*hkl*> and angle φ of misorientation. Studies of the grain boundary mobility in Al bicrystals[6] have shown that tilt grain boundaries with <111> rotation axis and rotation angle of about 40° have the highest mobility. This is commonly understood such that grain boundaries with highly periodic coincidence structure (so called low Σ or special boundaries) move faster than off-coincidence (random) boundaries, i.e. the special Σ7 (38.2°<111>) tilt boundary was identified as the fastest boundary in Al.

However, from growth selection experiments[1,9,10] it was known that the rotation angle of the fastest boundary was distinctly larger than 38.2° (Fig. 3).

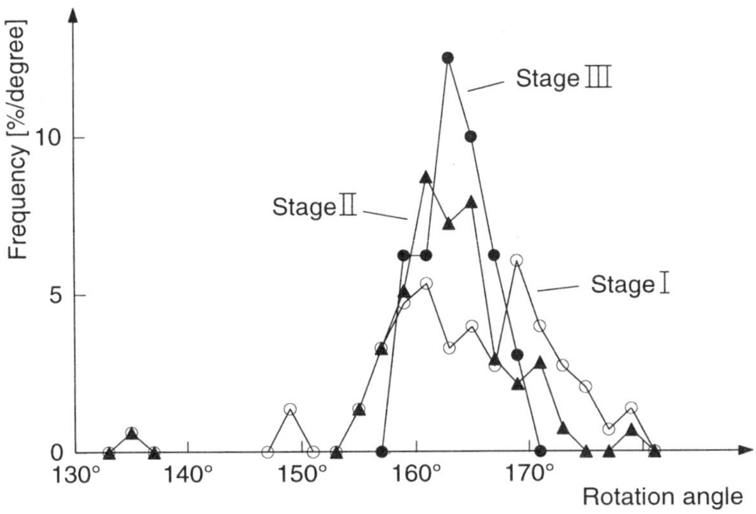

Fig. 3: *Growth selection in 20% rolled aluminum single crystals as observed at three consecutive stages. Frequency of the rotation angles around the best fitting <111> rotation axes*[9].

We investigated the misorientation dependence of grain boundary mobility on a fine scale in the angular interval 37°-43°<111> with angular spacing 0.3°-0.6°. The experiments reveal that both the activation enthalpy and the preexponential factor are at maximum for a misorientation angle φ=40.5° and at minimum for the exact Σ7 orientation (Fig. 4). Therefore, one is tempted to conclude that the Σ7 boundary has the highest mobility. However, the mobility of boundaries with different misorientation angles do have a different temperature dependence, and there is a temperature, the so-called compensation temperature T_c, where the mobilities of all investigated boundaries of differently misoriented grains are the same. As a result, for $T>T_c$, the mobility is higher for grain boundaries with higher activation energy, in particular it is at maximum for φ=40.5°, while for $T<T_c$ the exact Σ7 boundary moves fastest (Fig. 5).

This result explains the apparent contradiction between growth selection experiments and recrystallization experiments. The problem resulted only from the wrong tacit assumption that the preexponential factor is essentially independent of misorientation so that only the activation enthalpy controls mobility. Growth selection experiments have to

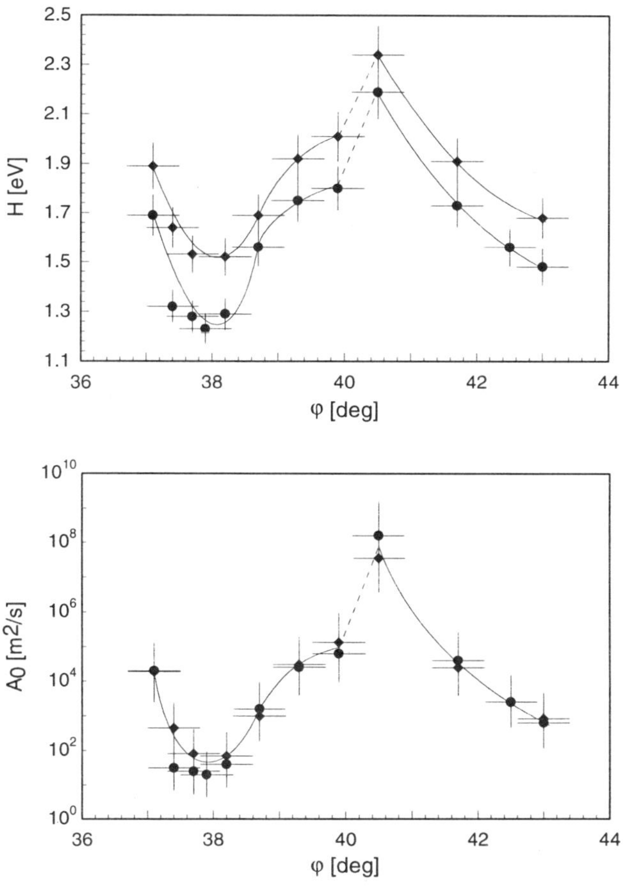

***Fig. 4**: Activation enthalpy H and preexponential factor A_o for <111> tilt boundaries in pure Al of different origin (● - Al I; ◆ - Al II)*

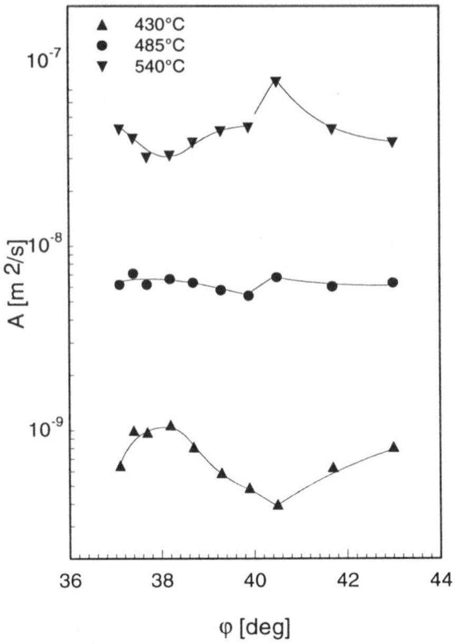

***Fig. 5**: Mobility dependemce of <111> tilt grain boundaries on rotation angle in pure Al (Al II)*

be conducted at very high temperatures (above 600°C), i. e. in the temperature regime, where, according to results of the current study, the mobility of the 40.5° <111> boundary is the highest due to its high preexponential factor. The reason for the changing maximum mobility orientation in different temperature regimes is obviously the orientation dependence of both, the activation enthalpy and the preexponential factor. In fact, both are related to each other in a linear fashion (Fig. 6), i.e.

$$H = \alpha \ln A_o + \beta \tag{2}$$

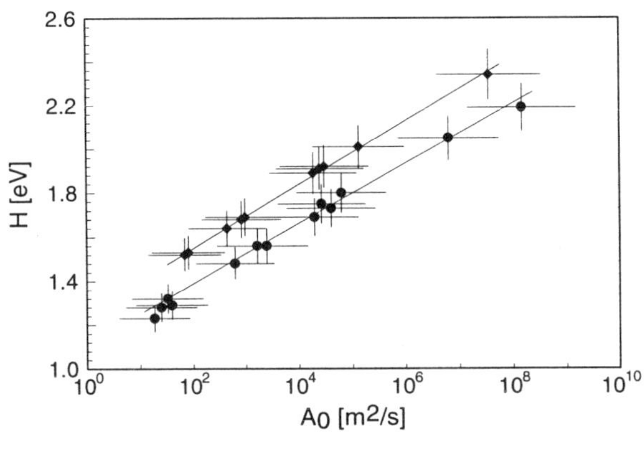

***Fig. 6**: Dependence of migration activation enthalpy on preexponential mobility factor for <111> tilt grain boundaries in Al I (●) and Al II (◆).*

where α and β are constants. This correlation is referred to as the compensation effect[11]. The value of $\alpha = kT_c$ defines the compensation temperature T_c. Effectively the compensation effect states that at high temperatures ($T>T_c$), i.e. above the compensation temperature T_c, the process with the highest activation energy proceeds fastest, while at low temperatures ($T<T_c$), grain boundaries with the lowest activation energy exhibit the highest mobility and at $T = T_c$ all boundaries move at the same rate. It is important to keep this in mind, when discussing the temperature dependence of grain boundary mobility.

Effect of Impurities on Grain Boundary Motion

It is well known that grain boundary mobility is strongly affected by dissolved impurities. Grain boundary motion is almost always slowed down by solute impurity atoms. However, little is known how solutes affect the activation enthalpy of grain boundary migration and the preexponential mobility factor and how much this effect depends on the grain boundary crystallography. The effect of impurities on grain boundary motion was addressed by the impurity drag theories of Lücke and coworkers[12] and Cahn[13]. These theoretical approaches are based on the assumption that there is an interaction between impurities and the grain boundary such that the impurities prefer to stay with the grain boundary and, therefore, on grain boundary migration move along with the boundary. Accordingly, the boundary becomes loaded with impurities and will move more slowly than the free boundary. This manifests itself in a high activation energy and a concentration dependent preexponential factor for the loaded boundary. The theories predict that the activation energy is independent of impurity concentration and that the preexponential factor decreases with increasing impurity content in a hyperbolic fashion. This is at variance, however, with experimental results. As obvious from Fig. 7 the activation energy changes with concentration actually more strongly than the preexponential mobility factor does. This experimental result can only be understood in the conceptual framework of the impurity drag theory, if an interaction among the impurities in the grain boundary is taken into account, i.e. by treating the chemistry in the boundary as a real solution rather than an ideal solution. Assuming thermal equilibrium in the bulk and in the boundary the chemical potential μ_i of the alloy constituents must be equal throughout. For a binary alloy with concentrations c_1 and c_2

$$\mu_1^b(\sigma, T, c_1^b) = \mu_1^v(p, T, c_1) \tag{3a}$$

$$\mu_2^b(\sigma, T, c_2^b) = \mu_2^v(p, T, c_2), \tag{3b}$$

where the index b refers to the grain boundary and v denotes bulk properties, σ is the grain boundary surface tension. The activities a_i are related by

$$\frac{a_1^b}{a_1} = \left(\frac{a_2^b}{a_2}\right)^{\frac{\omega_1}{\omega_2}} \cdot e^{\frac{\omega_1(\sigma_2 - \sigma_1)}{kT}}, \tag{4}$$

where σ_i are the grain boundary surface tensions of the pure constituents and

$\omega_1 = -\left(\frac{\partial \mu_1^b}{\partial \sigma}\right)\Big|_{p,T,\sigma_1}$ is the partial area of the respective component in the boundary.

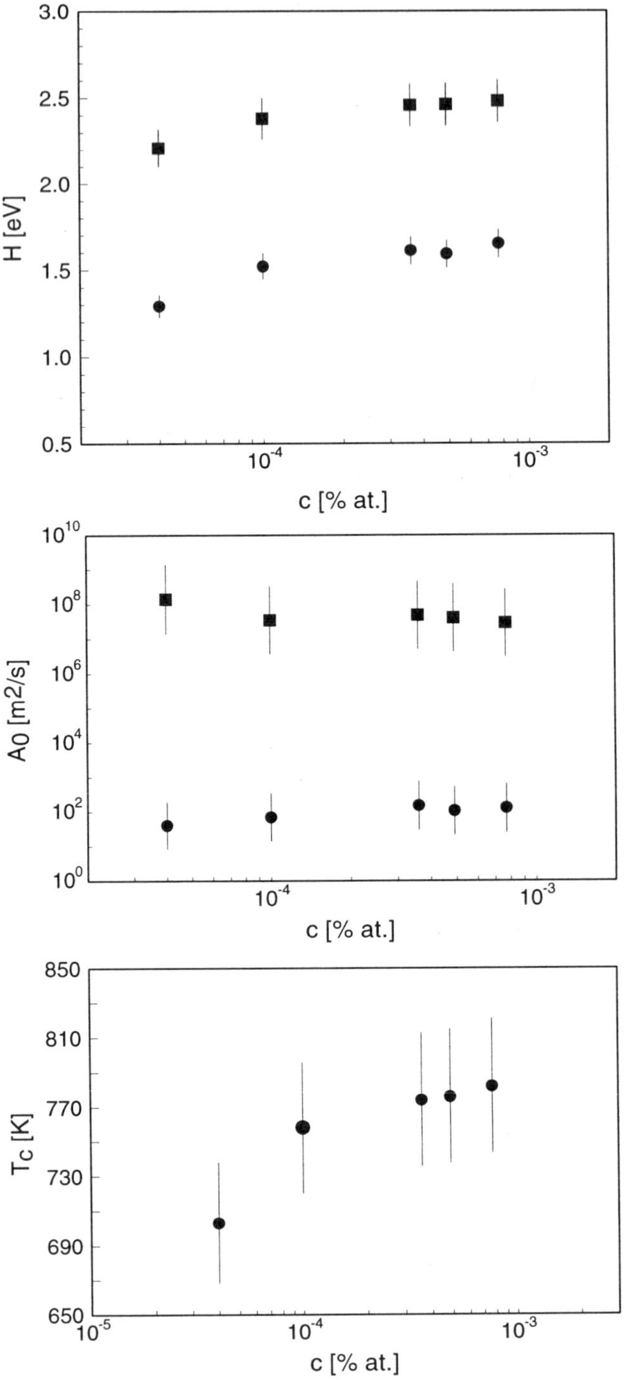

***Fig. 7**: Dependence of activation enthalpy H, preexponential factor A_0 and compensation temperature T_c in pure Al for 38.2° (●) and 40.5° (■) <111>-tilt grain boundaries.*

For a regular solution the activities read

$$a_1 = c_1 \exp\left(\frac{z \cdot \varepsilon^2 (c_2)}{kT}\right), \quad a_2 = c_2 \exp\left(\frac{z \cdot \varepsilon^2 (c_1)}{kT}\right), \tag{5}$$

where z is the coordination number and $\varepsilon = \varepsilon_{12} - 1/2(\varepsilon_{11} + \varepsilon_{22})$ is the heat of mixing. For an ideal solution in the bulk and a regular solution in the boundary, i.e. $\varepsilon = 0$ $\varepsilon^b \neq 0$, $\omega_1 \neq \omega_2$ and $c = c_1$, $B = B_o \exp^{H_i/kT}$, H_i - interaction enthalpy of impurity atoms of sort i with the boundary

$$c_1^b = c_1 \exp\left(-\frac{z^b \varepsilon^b \left(c_2^b\right)^2}{kT}\right) \exp\left[\frac{\omega_1(\sigma_2 - \sigma_1)}{kT}\right] \cdot \left\{\frac{c_2^b}{c_2} \exp\left[\frac{z^b \varepsilon^b \left(c_1^b\right)^2}{kT}\right]\right\}^{\frac{\omega_1}{\omega_2}} \tag{6}$$

and the boundary mobility

$$m_b = \frac{m_{im}}{c^b - c} \cong \frac{m_{im}}{c^b} = \frac{m_o}{B_o c} \cdot \frac{\exp\left[-\frac{H^* + H_i + (\beta - 1) z \varepsilon (1 - c^b)^2}{kT}\right]}{\left(\frac{1 - c^b}{1 - c}\right)^{\beta}} \tag{7}$$

H^* is the activation energy for volume diffusion of the impurity atoms, $\beta = \omega_1/\omega_2$.

Fig. 8 reveals that under the assumption of reasonable values for the adjustable parameters in Eq. (7) the theoretical predictions compare well to the experimental data. The very different behaviour of special and non-special boundaries reflects an influence of grain boundary structure on grain boundary migration mechanism. In particular, impurities not only may have an effect on migration by impurity drag, but also by changing grain boundary structure itself. This was shown recently by Udler and Seidman in a Monte Carlo simulation study[14]. It is further noted that obviously the compensation temperature, which essentially corresponds to the slope of the compensation plot, H versus ln A_o, changes with composition (Fig. 6-8).

The experimental results reveal that the migration activation enthalpy is strongly affected by both, the boundary crystallography and material purity. However, in the former case the preexponential factor A_o rises with increasing H by several orders of magnitude, while in the latter case A_o remains at the same level. Therefore, the preexponential factor A_o in the investigated impurity concentration interval was found to be much less sensitive to the material purity than to a change of the misorientation angle. This result allows to conclude that the observed orientation dependence of mobility (Fig. 5), determined by both H and A_o, does not only reflect the different segregation behavior of coincidence and random boundaries, as frequently proposed[5], rather it provides evidence for an intrinsic dependence of grain boundary mobility on grain boundary structure.

$$H = H^{*} + H_i + (\beta - 1)z\varepsilon(1 - c^{b})^{2} \qquad A_o = \frac{m_o\sigma}{b_o c}\left(\frac{1 - c^{b}}{1 - c}\right)^{-\beta}$$

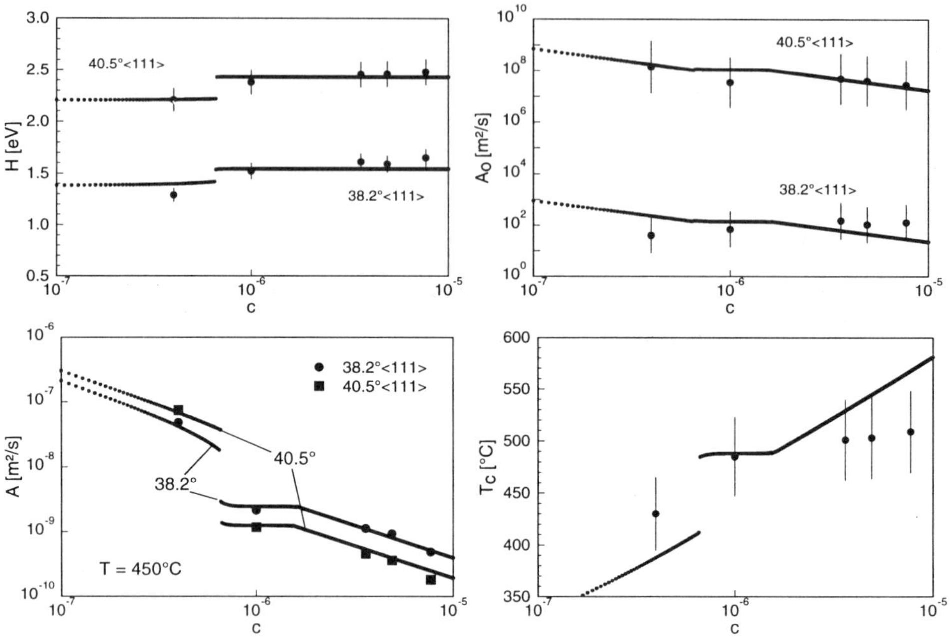

Fig. 8: *Experimental data (symbols) and results of calculations (lines) for two grain boundaries. Dependence of activation enthalpy H, preexponential factor A_0, boundary mobility A, compensation temperature T_c on the bulk impurity content. Fit parameters for 38.2°<111> boundary: $H^*=0.68$ eV, $H_i=0.86$ eV, $(z\varepsilon)=0.17$ eV, $(m_0\sigma)=3\cdot10^{-4}$ m^2/sec, and for 40.5°<111>: $H^*=1.57$ eV, $H_i=0.86$ eV, $(z\varepsilon)=0.24$ eV, $(m_0\sigma)=350$ m^2/sec.*

The experiments also reveal that the grain boundary mobility can be utilized as a measure for the total impurity content of a metal. We have investigated the grain boundary mobility of high purity Al of different producers, i. e. of different origin and, therefore, contaminated with trace elements in different concentrations. In contrast to the residual resistivity ratio (RRR) (Table 1), which is commonly used to characterize the material purity, the mobility decreases monotonously with increasing impurity content (Fig. 9). Moreover, the mobility is much more sensitive to the solute content than the electrical resistivity: for an impurity content rising by a factor of about 20, the RRR varies by a factor of 4 while the mobility (compared at 450°C) decreases by more than two orders of magnitude.

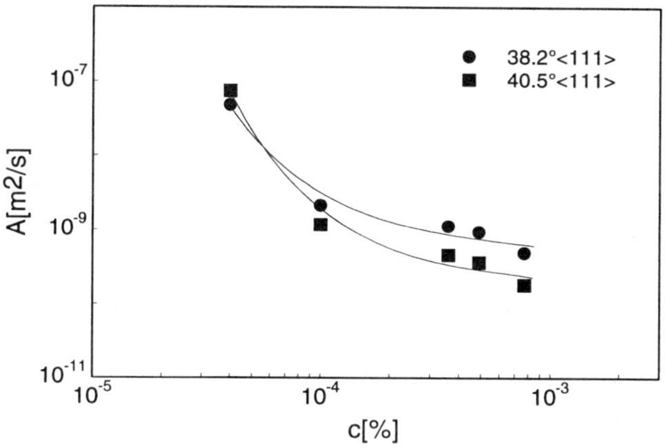

Fig. 9: Dependence of grain boundary mobility on total impurity content in pure Al.

Effect of Ga on Grain Boundary Motion

All known experiments on bicrystals and polycrystals confirm that solute atoms reduce the rate of boundary motion. However, it is important to realize, that solute atoms not necessarily hinder grain boundary motion as evident from the addition of minor amounts of gallium to aluminum (Fig. 10). Our experiments were carried out on bicrystals of both pure Al (Al III) and the same Al doped with 10 ppm Ga. Irrespective of the type of boundary, whether special or nonspecial, 10 ppm gallium in aluminum substantially increases grain boundary mobility, which means that it substantially speeds up recrystallization kinetics. Addition of 10 ppm Ga effectively increases the mobility of both investigated 38.2° and 40.5°<111> tilt boundaries but modifies the activation parameters differently. For the 38.2° (Σ7) boundary H and A_0 increase, while they decrease for the 40.5° boundary. The orientation dependence of grain boundary mobility is strongly reduced but not entirely removed. We propose to interpret these results as a change of mechanism of grain boundary migration owing to a change of boundary structure, such that a prewetting phase transition occurs and a thin layer of a Ga-rich phase forms in the boundary.

More comprehensive investigations of grain boundary motion in Al-Ga alloys have shown that enhanced grain boundary mobility is observed only in a narrow concentration interval of Ga (Fig. 11). At Ga concentrations of 50ppm and above, the grain boundary mobility decreases compared to the alloy doped with 10ppm. Such behaviour may be explained by a change of boundary interlayer thickness, which is likely to grow with increasing Ga concentration. As apparent from the concentration dependence of mobility activation parameters (Fig. 12), the boundary mobility in Ga-doped Al decreases with rising concentration due to a decreasing preexponential factor, while the migration activation enthalpy remains virtually constant over the investigated concentration range for all Al-Ga alloys. The orientation dependence of activation enthalpy and pre-exponential factor decreases upon Ga addition but remains constant at higher Ga concentration. A constant migration activation enthalpy with increasing Ga concentration is in agreement with the hypothesis that a change of the boundary migration mechanism upon Ga addition is associated with the formation of an interlayer of a Ga-rich wetting phase on the grain boundary. The activation enthalpy of boundary motion is determined by its slowest atomistic process, which, however, can not be the mass transport across the wetting phase

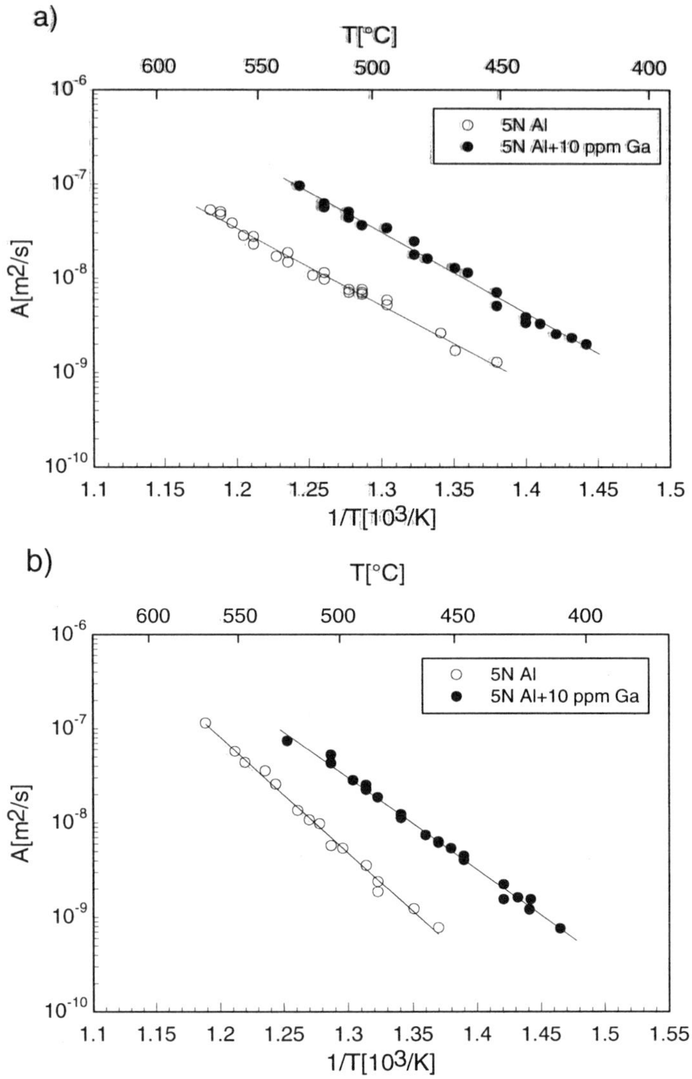

Fig. 10: *Arrhenius plot of mobility of (a) 38.2° and (b) 40.5°<111> tilt grain boundaries in pure Al and pure Al doped with 10 ppm Ga.*

interlayer, because in such case the activation enthalpy should be in the order of 0.1 eV (activation enthalpy for diffusion in liquids), i.e. much smaller than obtained by experiment. Thus, the grain boundary motion is apparently controlled by the process of detachment of transfered atoms from the shrinking grain or attachment to the growing grain, which occurs at the crystal/interlayer interface. Hence, the activation enthalpy for boundary migration must depend on structure and properties of the crystal/interlayer interface, but should obviously be independent of the bulk impurity concentration.

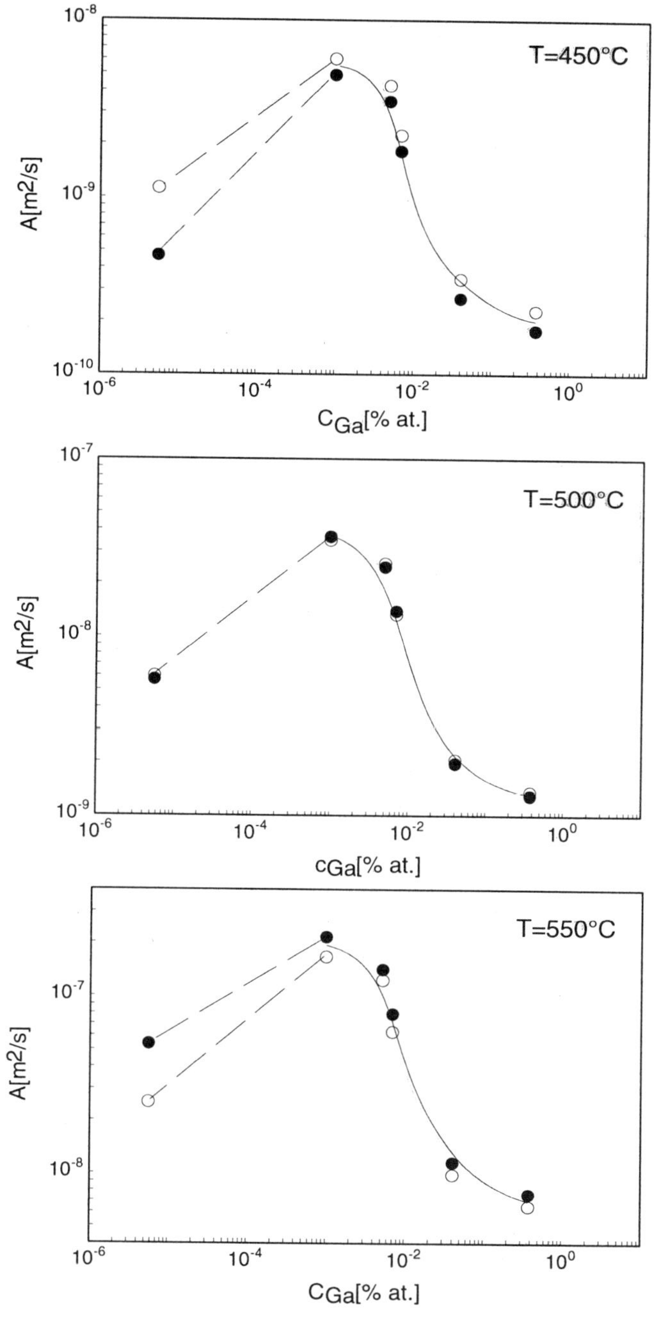

***Fig. 11**. Concentration dependence of mobility for 38.2°(○) and 40.5°(●)<111> tilt grain boundaries at different temperatures.*

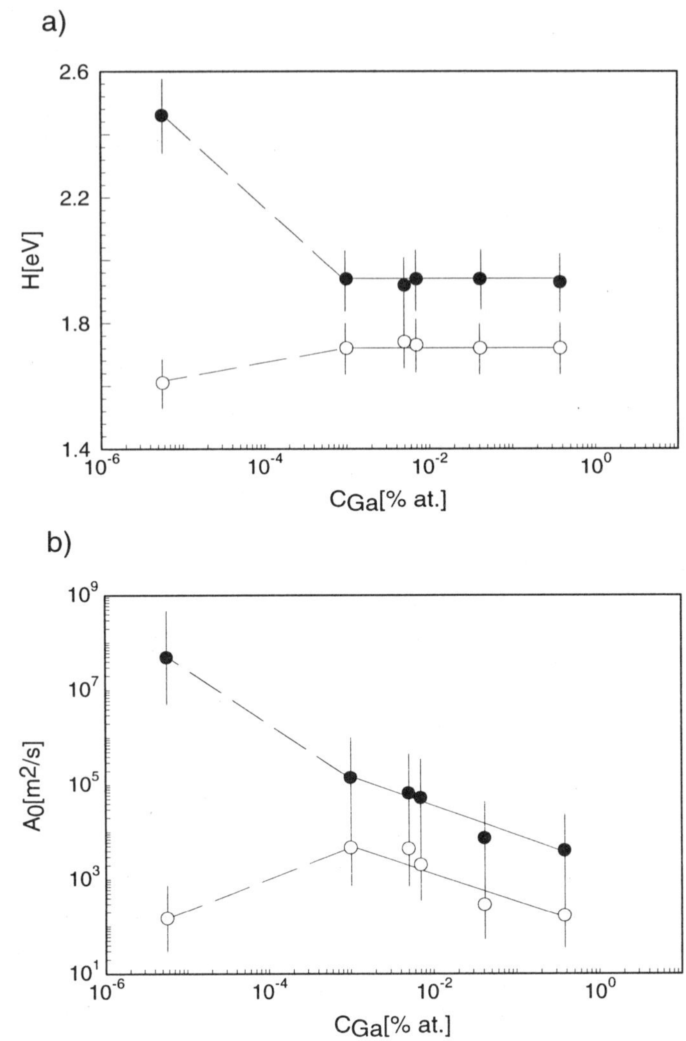

***Fig. 12**: Migration activation enthalpy H (a) and mobility pre-exponential factor* A_0 *(b) for 38.2° (○) and 40.5°(●) <111> tilt grain boundaries as a function of Ga concentration.*

CONCLUSIONS

1. Owing to the compensation effect, the maximum growth rate misorientation during recrystallization can depend on the annealing temperature. At high temperatures, grain boundaries with high activation enthalpy for migration are favored, while at low temperatures low Σ coincidence boundaries dominate.
2. The effect of solute atoms on grain boundary migration cannot adequately be described by standard impurity drag theories. A more satisfactory agreement is obtained by taking an interaction of the impurities in the boundary into account.
3. Solute atoms do not always reduce grain boundary mobility. Small amounts of Ga in Al will enhance grain boundary migration.

ACKNOWLEDGEMENT

The authors express their gratitude to the Deutsche Forschungsgemeinschaft for financial support of the project.

REFERENCES

1. K. Lücke, The orientation dependence of grain boundary motion and the formation of recrystallization textures, *Can. Met. Quart.* 13:261 (1974).
2. D. A. Molodov, J. Swiderski, G. Gottstein, W. Lojkowski and L. S. Shvindlerman, Effect of pressure on grain boundary migration in aluminium bicrystals, *Acta metall. mater.* 42:3397 (1994).
3. D. A. Molodov, U. Czubayko, G. Gottstein and L. S. Shvindlerman, Mobility of <111> tilt grain boundaries in the vicinity of the special misorientation $\Sigma7$ in bicrystals of pure aluminium, *Scripta metall. mater.* 32:529 (1995).
4. U. Czubayko, D. A. Molodov, B.-C. Petersen, G. Gottstein and L. S. Shvindlerman, An X-Ray Device for Continuous Tracking of Moving Interfaces in Crystalline Solids, *Meas. Sci. Technol.* 6:947 (1995).
5. J. W. Rutter and K. T. Aust, Migration of <111> tilt grain boundaries in high purity lead, *Acta Metall.* 13:181 (1965).
6. V. Yu. Aristov, V. L. Mirochnik, L. S. Shvindlerman, Mobility of <111> tilt grain boundaries in Al, *Sov. Phys. Solid State* 18:137 (1976).
7. V. G. Sursaeva, A. V. Andreeva, Ch. V. Kopezki and L. S. Shvindlerman, Mobility of <1010> tilt grain boundaries in Zn, *Phys. met. metall.* 41:98 (1976).
8. D. A. Molodov, B. B. Straumal and L. S. Shvindlerman, The effect of pressure on migration of <001> tilt grain boundaries in tin bicrystals, *Scripta metall.* 18:207 (1984).
9. G. Ibe, W. Dietz, A. C. Fraker and K. Lücke, Vorzugsorientierungen bei der Rekristallisation gedehnter Einkristalle aus Reinst-Aluminium, *Z. Metallk.* 61:498 (1970).
10. G. Ibe, K. Lücke, Growth selection during recrystallization of single crystals, in: *Recrystallization, Grain Growth and Textures*, 1966, p. 434.
11. L. S. Shvindlerman, G. Gottstein, The compensation effect in thermally activated interface processes, to be published.
12. K. Lücke and H. Stüwe, The theory of grain boundary motion, in: *Recovery and Recrystallization of Metals,* L. Himmel, ed., Interscience, New York (1963).
13. J. W. Cahn,The impurity-drag effect in grain boundary motion, *Acta Metall.* 10:789 (1962).
14. D. Udler, D. N. Seidman, Solute-Atom Segregation/Structure Relations at High.Angle (002) Twist Boundaries in Dilute Ni-Pt Alloys, *Interface Sci.*, 3:41 (1995).

AN INSTABILITY ANALYSIS OF HETEROEPITAXIAL INTERFACES VIA A DISCRETE ATOM METHOD

Jong K. Lee

Department of Metallurgical and Materials Engineering
Michigan Technological University
Houghton, MI 49931

INTRODUCTION

The microstructural development of elastically strained two-phase systems has been a subject of great interest. Two well-known cases are the formation of islands during thin film processing [1-3] and the development of spatial correlations during coarsening of coherent precipitates in nickel-based superalloys [4,5]. Our understanding of this subject, however, has been limited due to the lack of a computational technique through which one can analyze the elastic state associated with *arbitrarily-shaped* inclusions whose elastic constants are *different* from those of the matrix phase. Eshelby [6] was the pioneer in the field of coherency strain who devised the seminal equivalency method and thus brought much understanding to the coherency strain problem; however, the method is limited to a special geometry - a single ellipsoidal particle [7-9]. Since his work, several numerical techniques have been developed, but most involve either computations of an *elastically homogeneous* state, or approximate solutions for integro-differential equations when faced with an *inhomogeneous* system [10-11]. Recently, statistical mechanics has been combined with continuum elasticity to develop a computer simulation, termed the discrete atom method (DAM), of the evolution of coherent precipitate morphologies and phase equilibria [12-15]. In this work, the DAM is applied to address the morphological evolution of the heteroepitaxial thin films composed of two-dimensional, *dislocation-free* crystals. Through a linear stability perturbation technique or other continuum elasticity approach, there has been a great deal of theoretical investigations on the heteroepitaxy instability in the past [16-18], but most of these works are concerned with the early stages of the instability. The main thrust of this work is to examine the complete path of such evolution processes.

DISCRETE ATOM METHOD

As the DAM was described in detail elsewhere [12,13], only a brief review is accounted here. According to classical statistical mechanics, the configurational free energy is given by $A_q = -k_BT \ln Z_q$, where k_B is Boltzmann's constant and T is the absolute temper-

Dynamics of Crystal Surfaces and Interfaces
Edited by Duxbury and Pence, Plenum Press, New York, 1997

ature. Z_q is the configurational partition function, and is equal to:

$$Z_q = \int exp(-\Phi/k_B T)\, d\vec{q} \qquad (1)$$

In the DAM, the Hamiltonian, Φ, is given by:

$$\Phi = \frac{1}{4}\sum_{i=1}^{N}\sum_{j=1}^{m} k_{ij}(r_{ij}-a_{ij})^2 + \sum_{l} E_l^S \qquad (2)$$

where N is the total number of atoms, m is the number of interacting atoms with the i-th atom, k_{ij} is the spring constant between i-th and j-th atom, r_{ij} is the distance, a_{ij} is r_{ij} at stress-free state, and E_l^S is the interfacial energy of l-th interface atom.

When all k_{ij} are equal to k and only nearest neighbor interactions ($m = 6$) are considered, a two-dimensional triangular lattice becomes elastically isotropic with $\lambda = \mu = 0.433k$ [19]. *For a heteroepitaxial system,* k and $a(1 + \varepsilon)$ are assigned to the film phase as the spring constant and the lattice parameter, respectively, whereas k^* and a are assigned to the substrate phase. ε represents a misfit strain. Since the main concern is stress effects, an isotropic interface energy is assumed. An interface atom is defined as the atom having unlike or broken nearest neighbor bonds, and has a specific interfacial energy depending on the number of unlike or broken bonds. Morphological evolution is then examined through a Monte Carlo process [20], which, by exchanging different species of atoms, generates a Boltzmann-weighted chain of the configurations of a given system. All the computations are performed under the conditions of *a constant* number of atoms of a given species, a pure dilatational misfit, and a plane strain. The diffusion temperature is kept low enough so that surface roughening is negligible and yet atoms have sufficient mobility for shape change. With $k = 1.38\text{x}10^{-18}$ J/a^2 and isotropic interfacial energy, $\gamma_o = 2.5\text{x}10^{-21}$ J/a, a typical temperature of 30 K is used for a system in which the melting point of the matrix phase is about 1000 K.

RESULTS AND DISCUSSION

In an *unstressed* system, if a thin film undergoes shape change by minimizing interfacial energy, the final shape should be consistent with the equilibrium surface dynamics, and is to confirm the well-known Young's equation at a triple junction [21,22]. This is demonstrated in Fig. 1, where three different substrate-vapor surface energies, $\gamma_{sv,}$ are employed to examine the morphological evolution. A thin film region, initially made of four monolayers (4 ML), are allowed to diffuse at a low temperature, 30 K. A periodic boundary length of 1000 a is used, and the film-vapor surface energy, γ_{fv}, and film-substrate interface energy, $\gamma_{fs,}$ are equal to γ_o. On the left side of Fig. 1, the early stages of islanding is displayed, where M indicates the number of Monte Carlo steps in units of one million. On the right, the final equilibrium shapes are pictured, each confirming Young's relationship, $\gamma_{sv} = \gamma_{fs} + \gamma_{fv}\cos\theta$. Surface perturbations and thus the island formation are instigated by the thermal fluctuations whose amplitude is comparable to the thin film thickness. Indeed, when the thickness is over 16 ML, island formation is suppressed, consistent with the stable state of an unstressed planar surface.

A solid interface accompanies a torque, which is a measure of the resistance to a change in orientation [21]. In the previous case of Fig. 1, the torque terms of the substrate-vapor and film-substrate interface are assumed to be infinite, while that of the film-vapor interface is zero. Thus both the substrate-vapor and film-substrate interface remain flat during the entire simulation. When all the interfaces are allowed to freely pucker with zero torques, the system can further reduce its energy. The zero-torque condition is investigated in Fig. 2, where thin films of 4 ML evolve into lense-like equilibrium shapes at the end. For comparison, the corresponding Wulff constructions [21,22] are also portrayed.

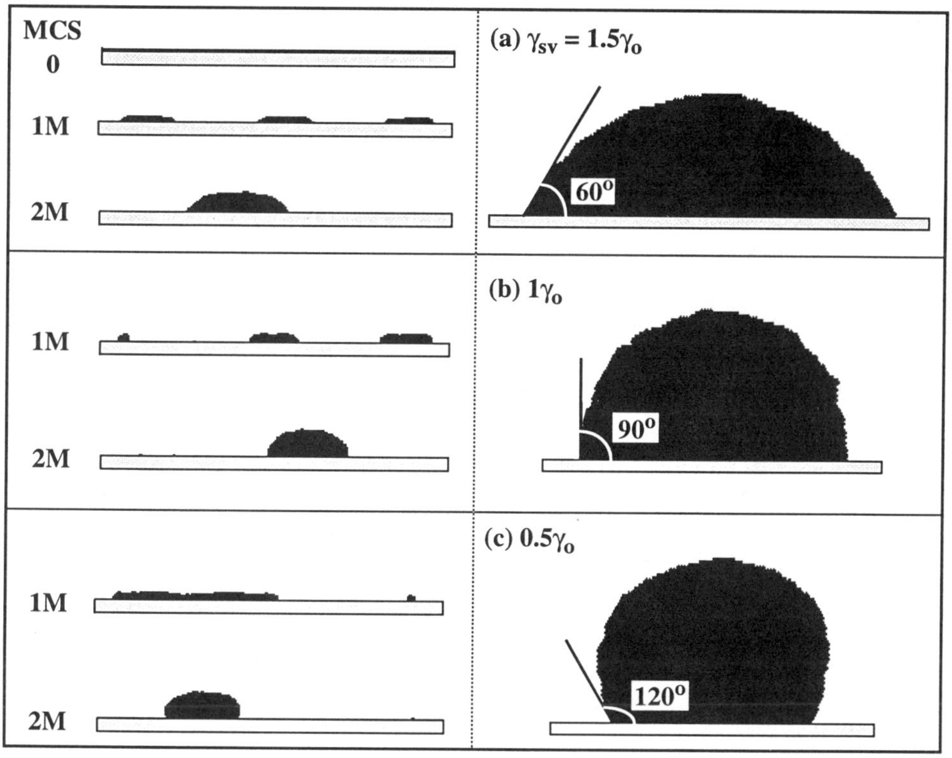

Figure 1. Morphological evolution of *unstressed* thin-film regions, made of initially four monolayers, through thermal fluctuations. The early stages of islanding are examined on the left side, while the final equilibrium shapes are shown on the right for three different cases of substrate-vapor surface energy, γ_{sv}.

Let us examine the instability of *strained* thin films. In Fig. 3, thin films of 30 ML are coherently bonded to the hard substrates. The film phase has a misfit strain, $\varepsilon = 0.01$, relative to the substrate phase, and the periodic length is equal to 200 a. The three interface energies are identical to each other: $\gamma_{sv} = \gamma_{fv} = \gamma_{fs} = \gamma_o$. Both phases are elastically isotropic, but the shear modulus of the substrate is twice that of the film ($\mu^* = 2\mu$). On the left-hand side, an infinite-torque condition is imposed to the substrate-vapor and film-substrate interfaces, whereas torque terms are equal to zero on the right. In the absence of the coherency strain, these films are stable as their thickness is well over 16 ML. With a coherency strain, surface undulations induced by thermal fluctuations become growing waves. By the time of 2M, six waves are definitely seen to have established, and these numbers are in agreement with the continuum linear elasticity prediction [16].

The waves then coarsen into a fewer ones, and their valleys or troughs become 'cracktips', which advance to the film-substrate interface. The number of the advancing cracktips depends on the misfit strain. In the infinite-toque condition, the tips end at the substrate, but become wider at the triple junction of film-substrate-vapor. A typical island features an 'acorn' shape. In the zero-torque condition, the tips advance much deeper with the puckering ability of the substrate interface. As the film tries to minimize its contact area with the substrate, the tips create crevices between the film and substrate.

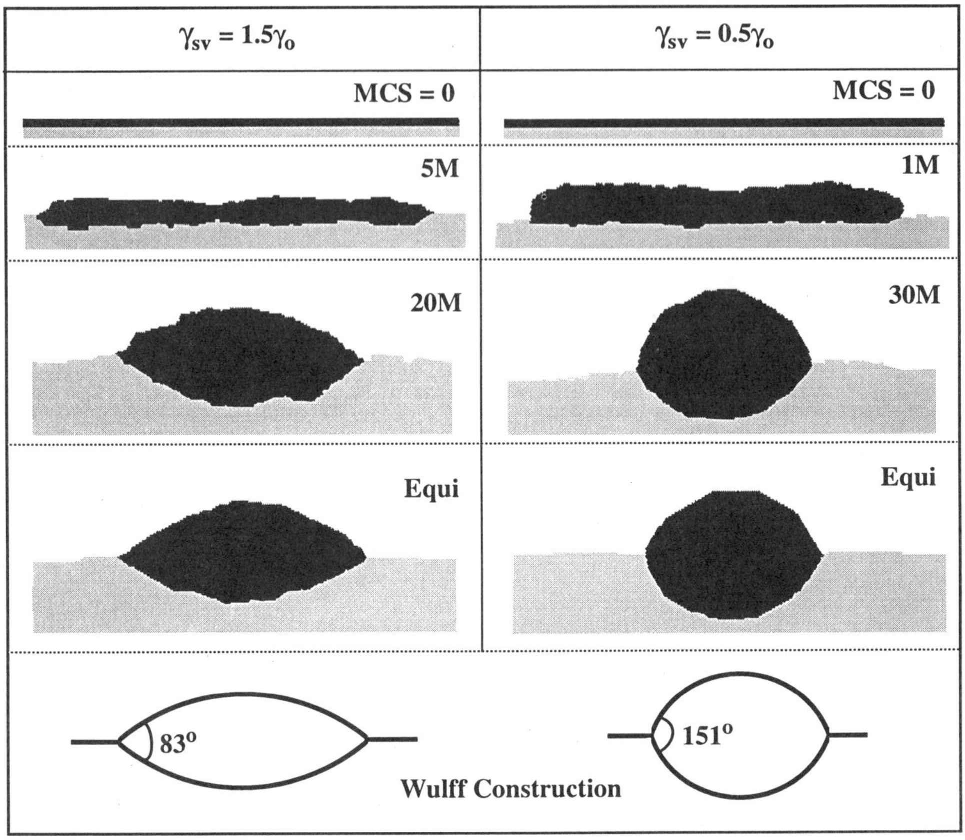

Figure 2. Morphological evolution of *unstressed* thin-films under the zero-torque condition, in which all the interfaces are allowed to pucker freely. The corresponding Wulff constructions are also pictured for comparison.

If the misfit strain is less than a critical value, the undulations cannot mount cracktips, as demonstrated in Fig. 4, where a periodic length is equal to 100 a and film thickness is 30 ML. With the same physical parameters employed for Fig. 3, no islands are created if the misfit strain is less than 0.006. When the misfit strain is less than but close to the critical value, a permanent wave structure sets in the film as in the case of $\varepsilon = 0.005$. If the misfit strain is further reduced, coherency-induced undulations are swept away by thermal fluctuations.

In search for true equilibrium shapes, a smaller system with a periodic length equal to 100 a is used to study isolated islands. In Fig. 5, such isolated islands are examined as a function of misfit strain, ε. All other input conditions are similar to those of Fig. 3. When $\varepsilon = 0$, the island becomes a semi-circle in the infinite-torque condition, whereas a lense shape with a dihedral angle of 120° should be established in the zero-torque condition. Albeit some facetting due to the discrete nature, the shapes are consistent with the Wulff construction. For $\varepsilon \neq 0$, the shapes of $\varepsilon = 0$ are used as the initial configurations. As ε increases, the triple junction of film-vapor-substrate (marked by open arrow heads) shifts toward the center of the original film-substrate interface. Thus, when the coherency strain is large, as in the case of $\varepsilon = 0.02$, much of the original film-substrate interface region is replaced by a thin layer of a vapor phase, creating crevice cracks. This, of course, is a consequence of strain energy reduction by lessening the contact between the film and substrate phase. Eventually, there comes a time when the increase in the interface energy can not be negated by the decrease in the strain energy, and an equilibrium reaches. It is interesting to notice that a strain energy relief often accompanies 'island on island' phenomena as shown in the infinite-torque case. Clearly, some of these small islands (marked by solid arrow heads) are aberrations due to statistical fluctuations, but others are quite persistent.

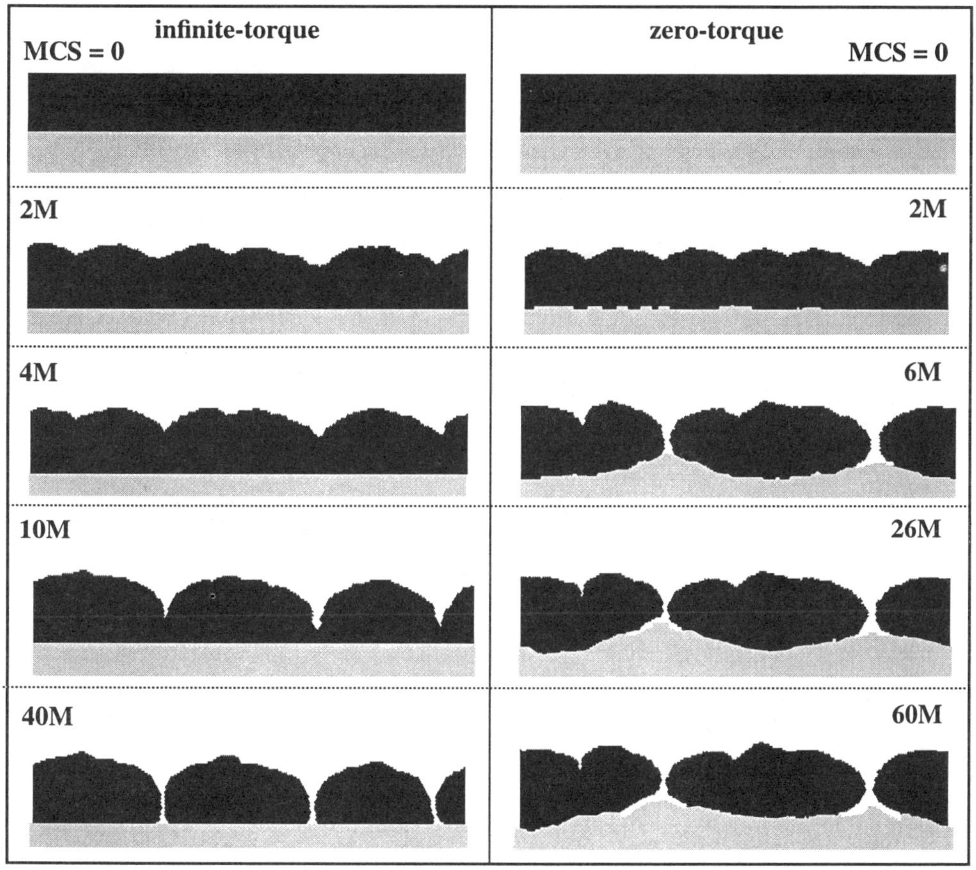

Figure 3. Morphological evolution of *strained* thin-films under the infinite- and zero-torque conditions. Thin films of 30 ML and are coherently bonded to the hard substrates whose shear modulus is twice that of the film ($\mu^* = 2\mu$). The film phase has a misfit strain, $\varepsilon = 0.01$, relative to the substrate phase.

Stress distributions are examined for the 'acorn'-shaped island with $\varepsilon = 0.01$ of Fig. 5. In Fig. 6b, the normal stress components along the y-axis passing through the island center (marked as 'b' in Fig. 6a) are plotted as a function of the distance from the substrate base. With a positive misfit, both normal stress components are positive within the substrate region ($y < 43\ a$), indicating the substrate is in tension as expected. Entering into the island region, the σ_{11} ($= \sigma_{xx}$) component changes its sign, but the σ_{22} component remains positive for a while. Most of the island is in a compressive state, which is no surprise. As the free surface is neared, however, both stress components switch their signs, and the area becomes a state of a tension. In (c), the two normal stress components and the shear stress, σ_{12}, of the substrate are plotted along the substrate surface. As a periodic boundary condition is imposed along the x-axis, the σ_{11} component shows a tension when the substrate is in contact with the film, a compression otherwise. Finally, Fig. 6d displays the substrate stress field along the vertical line passing through the midpoint between two islands. Note that the σ_{11} changes its state from a tension at the base region to a compression at the proximity of the free surface. Both the island and substrate have regions of tensional as well as compressive stress fields.

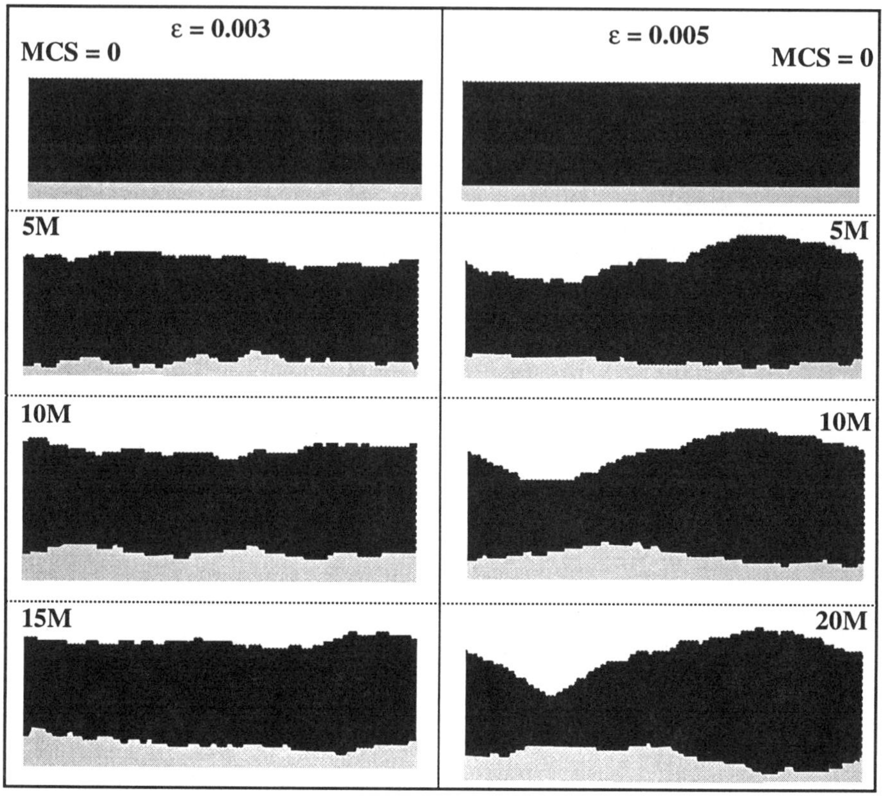

Figure 4. Thin films with small misfit strains under the zero-torque condition. As the misfit strains are less than the critical value, 0.006, no islands are created out of the undulations. The periodic length is equal to 100 *a* and the film thickness is 30 ML.

Let us consider a three-layer lamellar structure. In Fig. 7, a hard phase ($\mu^* = 2\mu$) is sandwiched between two soft phases with shear moduli, μ. The soft phase has a misfit strain of 0.01 relative to the hard phase. The periodic boundary length is equal to 200 *a*, and each layer initially has 50 ML. Again, all the interfacial energies are taken to be equal to γ_o, and the zero-torque condition is imposed. Once formed, the islands merge together by migrating through the hard phase region. Eventually, a phase separation develops along the direction perpendicular to the original layered structure. When a soft phase is sandwiched between two hard phases, however, the three-layer is quite stable as shown in Fig. 8.

The instability of the two lamellar structures may be understood in terms of Eshelby's inclusion theory [6,7]. According to the theory, a hard coherent precipitate with a dilatational misfit strain is elastically stable when it takes on a spherical shape in an infinite matrix. A soft coherent precipitate, on the other hand, takes on a plate-like shape as the minimum strain energy shape. Thus, the soft-hard-soft layered structure of Fig. 7 is simply a wrong configuration for the hard phase as its initial geometry is a plate, and morphological evolution to a new, stable configuration should be expected. For the hard-soft-hard layered structure of Fig. 8, however, the soft phase surrounded by the hard phase has a right geometry - a plate, thus there shouldn't be any instability. Alternatively, the instability of such lamellar structures, at least for the early stage of evolution, can be studied with a linear stability perturbation technique [23].

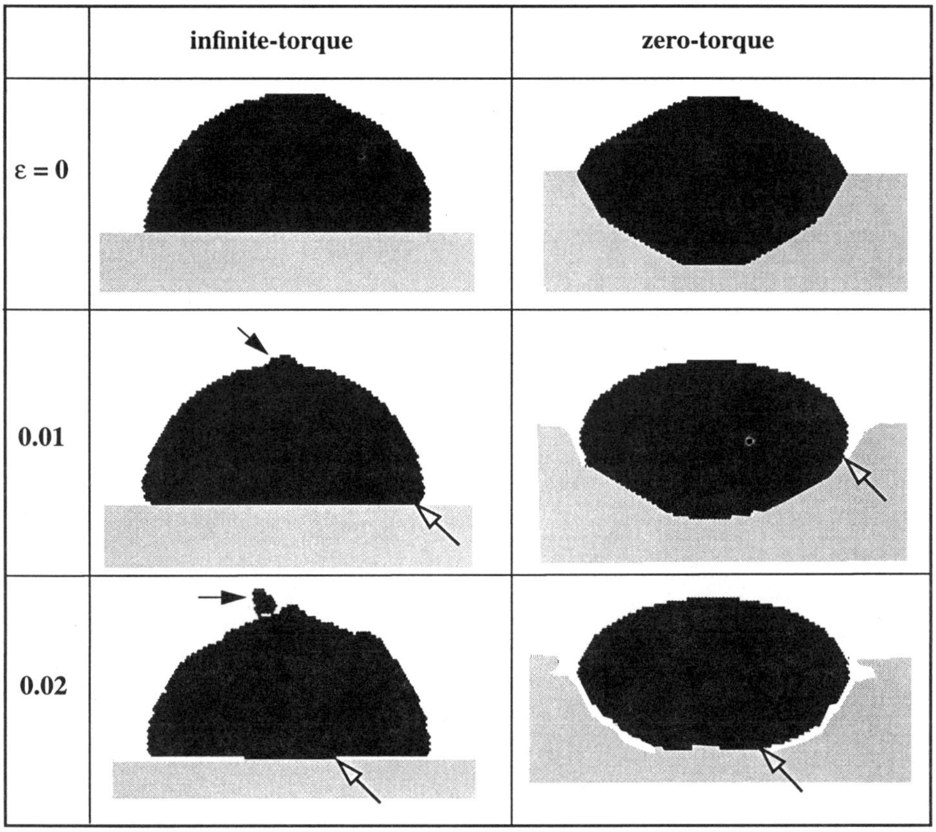

Figure 5. Isolate islands with different ε under the infinite- and zero-torque conditions. The shear modulus of the substrate phase is twice that of the film ($\mu^* = 2\mu$), and all the interface energies are identical to each other. Large open arrow heads mark triple junctions, and small solid ones indicate islands on the top of a main island.

SUMMARY

Morphological evolution of strained heteroepitaxial thin films are examined via a discrete atom method under a dislocation-free plane strain condition. When the misfit strain is over a critical value, surface undulations create growing waves, whose troughs become cracktips. The cracktips advance to the original film-substrate interface, converting a film into islands. A stress analysis reveals that both the island and substrate phase consist of the regions of a compressive as well as a tensional stress state in order to accommodate the strain energy. A three-layer lamellar structure, made of soft-hard-soft phases, is intrinsically unstable if the misfit strain exceeds a critical value. On the other hand, a hard-soft-hard lamellar structure is stable against surface perturbations.

There are several aspects not addressed in this study. First, the role of both threading and misfit dislocations is neglected as the lattice is assumed to be dislocation-free. In the DAM, anharmonic atomic interactions can be introduced into a lattice such that the elastic interaction between a threading dislocation and a coherency strain can be analyzed, or a strain relief through nucleation of misfit dislocations can be examined [14,15]. Clearly, if misfit dislocations can be easily produced, the extent of the film-substrate interface separation should be reduced. Secondly, only isotropic elasticity is considered in the examples, but anisotropic elasticity can be treated, without additional difficulty, by introducing directional spring constants, k_{ij}. With a triangular lattice, the atomic interactions should be extended to the second nearest neighbors if a cubic anisotropy is desired [13]. Since the harmonic potentials introduced in the Hamiltonian (Eq. (2)) represent a central-pairwise interaction, the elastic constants follow the Cauchy relationship, i.e., $C_{12} = C_{44}$ in cubic anisotropy. If $C_{12} \neq C_{44}$ is desired, a volume-dependent energy term should be introduced into the Hamiltonian [24]. Finally, the current study is limited to a two-dimensional case, but its extension to a three-dimensional model should be possible by constructing an appropriate lattice such as a face centered cubic or diamond structure. Some of these aspects have been already explored for the study of morphological evolution of coherent precipitates embedded in an infinite matrix, and are being investigated for the heteroepitaxial instability problem.

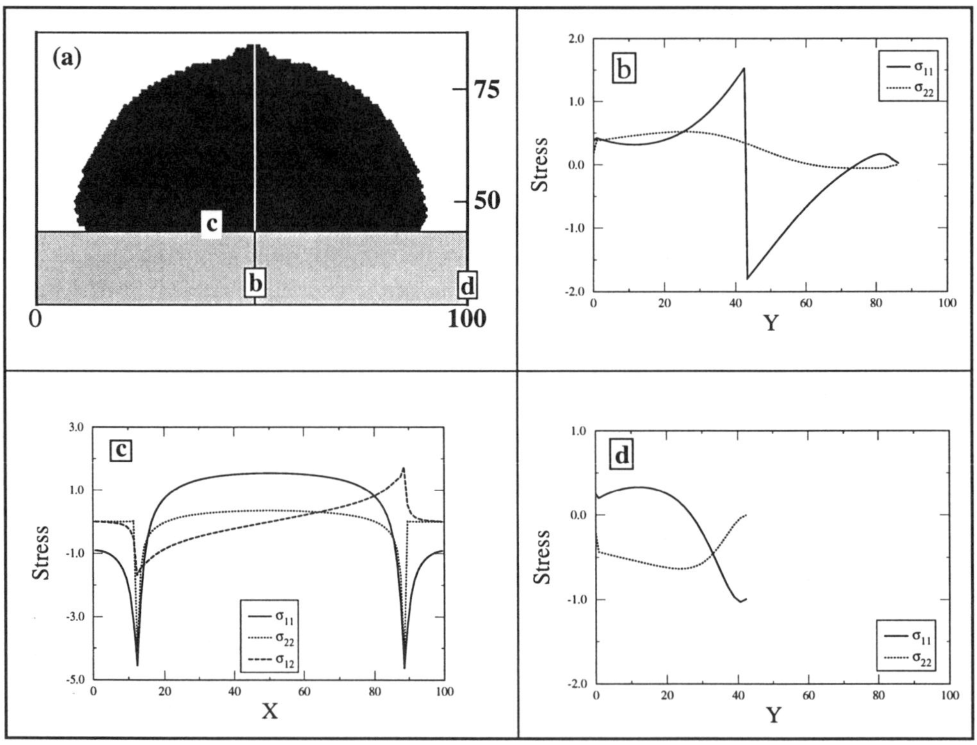

Figure 6. Stress analyses for an 'acorn-shaped' island with $\varepsilon = 0.01$ under the infinite-torque condition: (b) normal stress components, σ_{11} and σ_{22}, along the y-axis passing through the island center as a function of the distance from the substrate base, (c) the stress components of the substrate along the substrate surface, and (d) the substrate stress field along the vertical line passing through the midpoint between two islands.

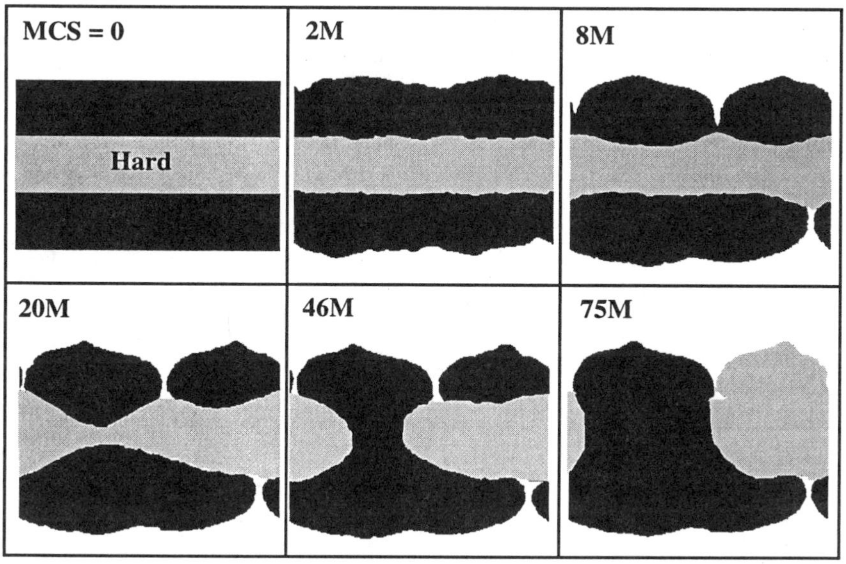

Figure 7. Morphological evolution of a soft-hard-soft lamellar structure. A hard phase is sandwiched between two soft phases with ε = 0.01. Each layer has 50 ML, and the periodic length is equal to 200 *a*.

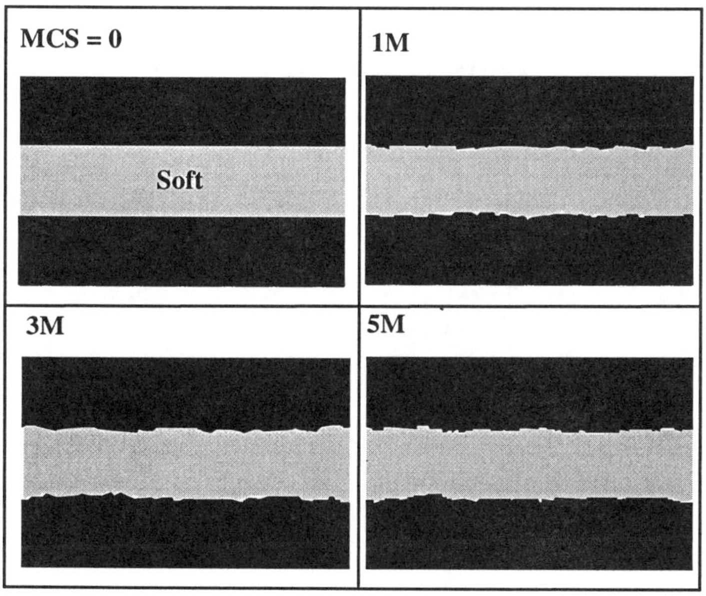

Figure 8. Stable state of a hard-soft-hard lamellar structure. A soft phase with ε = 0.01 is sandwiched between two hard phases. Each layer has 50 ML, and the periodic length is equal to 200 *a*.

ACKNOWLEDGMENTS

The author wishes to thank Professor Mohan Krishnamurthy for the enlightening discussions he shared with during the course of this work. The research was supported by the U.S. Dept. of Energy under Grant DE-FG02-87ER45315, for which much appreciation is expressed.

REFERENCES

1. A. G. Cullis, *MRS Bulletin* **21**, 21 (1996).
2. D. E. Jesson, K. M. Chen, and S. J. Pennycook, *MRS Bulletin* **21**, 31 (1996).
3. M. Krishnamurthy, J. S. Drucker, and J. A. Venables, *J. Appl. Phys.* **69**, 6461 (1991).
4. M. Meshkinpour and A. J. Ardell, *Mater. Sci. Eng.* **A185**, 153 (1994).
5. Y. S. Yoo, D. Y. Yoon, and M. F. Henry, *Metals and Materials* **1**, 47 (1995).
6. J. D. Eshelby, *Prog. Solid Mech.* **2**, 89 (1961).
7. J. K. Lee, D. M. Barnett, and H. I. Aaronson, *Metall. Trans.* **8A**, 963 (1977).
8. T. Mura, *Micromechanics of Defects in Solids*, 2nd ed., p. 177, Martinus Nijhoff, Dordrecht (1987).
9. A. G. Khachaturyan, *Theory of Structural Transformations in Solids*, p. 213, Wiley & Son, New York (1983).
10. M. E. Thompson, C. S. Su and P. W. Voorhees, *Acta Metall.* **42**, 2107 (1994).
11. S. Satoh and W. C. Johnson, *Metall. Trans.* **23A,** 2761 (1992).
12. J. K. Lee, *Scripta Metall.* **32**, 559 (1995).
13. J. K. Lee, *Metall. Trans.* **27A**, 1449 (1996).
14. J. K. Lee, in *Micromechanics of Advanced Materials* (eds., S. N. G. Chu et al.) p. 41, TMS, Warrendale, PA (1995).
15. J. K. Lee, in *Superallys 1996* (eds., R. D. Kissinger et al.) p. 211, TMS, Warrendale, PA (1996).
16. D. J. Srolovitz, *Acta Metall.* **37**, 621 (1989)
17. L. B. Freund and F. Jonsdottir, *J. Mech. Phys. Solids* **41**, 1245 (1993).
18. B. J. Spencer and D. I. Meiron, *Acta Metall.* **42**, 3629 (1994).
19. W. G. Hoover, W. T. Ashurst, and R. J. Olness, *J. Chem. Phys.* **60**, 4043 (1974).
20. K. Binder, *Monte Carlo Methods*, p. 1, Springer-Verlag, New York (1979).
21. J. W. Cahn and D. W. Hoffman, *Acta Metall.* 22, 1205 (1974).
22. J. K. Lee and H. I. Aaronson, *Acta Metall.* **23**, 799 (1975).
23. N. Sridhar, J. M. Rickman and D. J. Srolovitz, *Acta Metall.* to be published.
24. M. I. Baskes and C. F. Melius, *Phys. Rev.* **B20**, 3197 (1979).

AB INITIO SIMULATIONS OF THE SI (100) SURFACE: STEPS AND MELTING

C.M. Roland, M.G. Wensell, Q.M. Zhang, P. Boguslawski, Z. Zhang and J. Bernholc

Department of Physics
North Carolina State University
Raleigh, NC 27695

INTRODUCTION

Thin film growth is one of the most important aspects of materials science. Over the past few decades, methods such as molecular-beam epitaxy and chemical vapor deposition have made possible the fabrication of semiconductor devices with outstanding electronic properties. These device applications have fueled a continuing effort aiming to understand and ultimately control the physics of growth on smaller and smaller length scales. At present, our understanding of growth on semiconductor surfaces is far from complete. The field has, however, benefitted enormously from the recent advent of scanning tunneling microscopy (STM), low energy electron microscopy (LEEM) and atomic force microscopy (AFM), all of which have enabled the visualization of the semiconductor surface in unprecendented detail. These experiments have given a new impetus to theoretical studies of growth on semiconductor surfaces, so that a coherent understanding at the atomistic level is beginning to emerge.[1]

In this article, we summarize our recent *ab initio* studies of the Si (100) surface,[2,3,4] the most important semiconductor surface for device applications. The focus will on be the structure and growth properties of the Si (100) steps, and the melting of the (100) surface. That steps are a central component of the Si (100) surface has long been recognized. They arise not only because of the natural atomic-level roughness that any surface possesses over large length scales, but also because of the anisotropy of the surface stress created by the surface reconstruction.[5] The latter is particularly important, because it organizes the distribution of the steps on the surface,[6] which may in turn be manipulated by changing the strain in the thin film.[7,8] Steps also play an important role during growth. At high temperatures and low adatom concentrations, growth takes place primarily via a step flow mechanism. Adatoms make their way over the flat terraces via diffusion, until they encounter a step edge where they are more readily incorporated into the crystal. This is the so-called "step-flow" regime.[9] At lower temperatures, this process is in competition with the nucleation, growth and coalescence of islands, referred to as the "nucleation-dominated" regime. At very low temperatures, limited epitaxy or an amorphous deposit may form.[10] Issues related to the melting of the Si (100) surface arise naturally when characterizing the high-temperature behavior of the surface.

A brief outline of this paper is as follows. We begin with a brief review of the methodology, followed by a discussion of the atomic and electronic structure of the flat Si (100) surface and the single-height steps. The growth properties of these these structures for adatoms is then presented, followed by a description of initial simulations of the melting of the surface. We end with a short summary.

Dynamics of Crystal Surfaces and Interfaces
Edited by Duxbury and Pence, Plenum Press, New York, 1997

METHODOLOGY

The main tool of investigation was quantum molecular dynamics (QMD) simulations using both a plane-wave and real-space multigrid implementation. The plane wave Car-Parrinello (CP) scheme[11] was used to calculate the electronic structure and the growth properties of the Si (100) surface and steps, while the finite-temperature simulations of the melting were carried out with the real-space multigrid method.[12] Some relevant details are as follows. The CP method combines local density theory with nonlocal pseudopotentials to compute the electronic structure and the forces on the atoms. Typically, a slab geometry consisted of seven atomic layers of Si, with the two surfaces separated by an equivalent amount of vacuum. The dangling bonds of the bottom layer were saturated with H atoms, and the atoms of the lowest two Si layers were fixed at the ideal bulk lattice sites. The in-plane unit cell period was 6x4, so that each structure consisted of about 160 Si and 48 H atoms. Because of the large cell size, summations over the Brillouin zone were approximated by the Γ point. A kinetic energy cutoff of 8 Ry was used. The codes were highly parallel, and run at speeds of 2.3 Gflops on four Cray C90 processors. The "adiabatic trajectory" method[13] was used to calculate the diffusion barriers of the Si (100) surface and steps. In this method, an initially relaxed adatom is pushed over the surface with a small, constant speed in a given direction, while monitoring the total energy. Because the adatom is constrained in one direction only, it is free to move in any direction perpendicular to the applied force, thereby enabling the adatom to find its optimum path. All other atoms are allowed to relax continuously in response to the motion of the adatom. We have found that this method represents a viable and realistic alternative to the costly point-by-point determination of the potential energy surface. Tests show that the estimated error of this procedure is less than 0.1 eV. Because of the high costs associated with these simulations, the paths investigated were restricted to the most important ones for growth, as previously identified with classical molecular dynamics simulations.

The finite-temperature simulations of Si (100) melting were carried out with a real-space multigrid method.[12] Instead of using plane waves, the wavefunctions ψ_n, the densities and potentials are all represented on a global grid. The Kohn-Sham equations are given by a generalized eigenvalue form:

$$A[\psi_n] + B[V_{eff}\psi_n] = \epsilon_n \psi_n,$$

where A and B are the components of a compact discretization operator, the so-called Mehrstellen operator, which, to leading order, represent the kinetic energy and unity operators, respectively. Also, V_{eff} is an effective potential and ϵ_n the corresponding eigenvalue. When this equation is solved in combination with multigrid methods, accelerated convergence is obtained, which is particularly useful when dealing with ill-conditioned systems or atoms that require a high-energy cutoff. Another advantages of this scheme reside in the inherently local nature of the formulation, which allows for easy parallelization of the codes. Also, with the grids, one is no longer constrained to periodic boundary conditions.

For the finite-temperature simulations, the temperature of the Si ions were controlled with a chain of five, linked Nose-Hoover thermostats.[14] Because the electrons are always quenched back onto the Born-Oppenheimer surface after every timestep, no additional thermostat is needed for the electrons. Details of the configurations were similar to those with the CP scheme, except that the in-plane cells consisted of 16 atoms per layer and the basic timestep of the simulation was 100 a.u.

STRUCTURE OF SI (100) SURFACE AND STEPS

The Si (100) surface reconstructs by forming dimers of (2x1) symmetry, which arrange themselves into parallel rows. It is now well established that these dimers buckle to form the higher-order p(2x2) and c(2x4) reconstructions. This buckling is due to a transfer of an electron from the lower to the upper atom of the dimer, which opens up a gap between the occupied and unoccupied states. At room temperatures or above, the buckled dimers oscillate in time, and therefore appear symmetric under

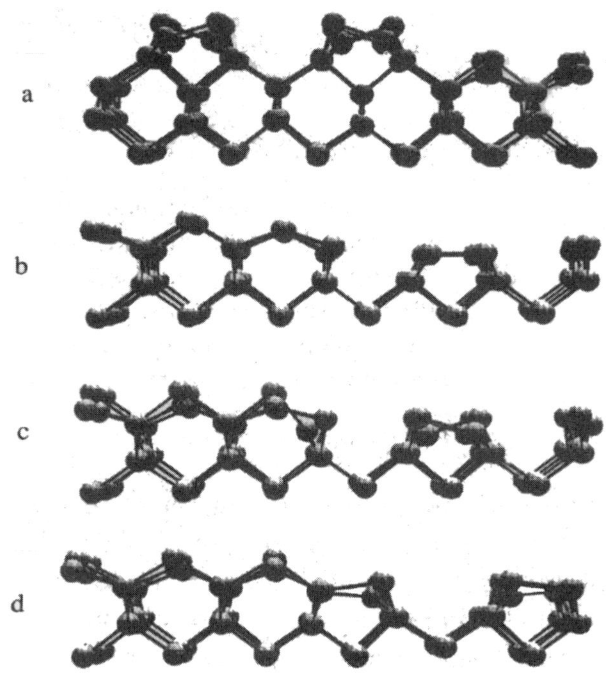

Figure 1. Calculated side-views of the equilibrium configurations of (a) S_A step; (b) non-rebonded S_B step edge with symmetric dimers; (c) non-rebonded dimers with p(2x2) reconstructions; and (d) rebonded S_B step edge with p(2x2) reconstruction.

the STM images.[15] We have calculated the energies of the different reconstrunctions on the flat surface, and find, in agreement with other studies,[16] that the c(2x4) reconstruction is the ground state structure with an energy gain of about 0.24 eV per surface dimer and a buckling of 0.72Å. The p(2x2) reconstruction has a similar buckling and is nearly degenerate with the c(2x4) structure.

The formation of dimer rows on the Si (100) surface largely determines the structure of steps and their orientation. Both single- and double-height steps are important. Because of the diamond structure of the bulk Si lattice, the orientation of the reconstruction is forced to alternate on terraces separated by single-height steps. The orientation is, however, preserved for steps separated by double-height steps. Following the notation of Chadi,[17] S_A denotes the single-height step edge with the dimer axis on the upper terrace oriented perpendicular to the step edge, while the S_B step edge has the dimer axis on the upper terrace oriented in a perpendicular direction to the step edge. There are two types of S_B step edges: one with, and one without rebonded atoms at the lower step edge. Similarly, D_B will denote the corresponding double-height step. To date, there has been considerable work on the energies of these steps as a function of terrace length.[6,18] Because of the dimer rows, the Si (100) surface is anisotropically strained, some of which may be relieved at the edges of the single-height steps. This leads to the formation of different phases on the surface, and a transition from the

(a)

(b)

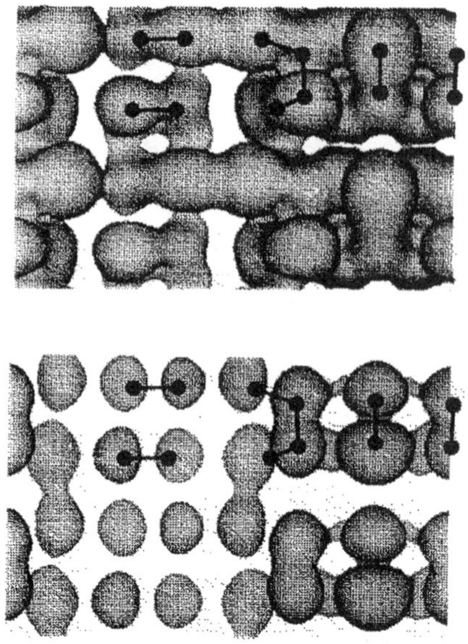

Figure 2. Simulated STM images of the non-rebonded S_B step edge. (a) the filled-state image and (b) the empty-state image, both under "high-current" conditions.

single- to the double-height steps as a function of temperature and miscut,[6] which may be controlled in some measure with bulk strain.[7,8]

The calculated atomic configurations of the single-height steps is shown in Fig. 1. For the S_A step edge, we find that the the c(2x4) reconstruction is induced spontaneously on the upper terrace, but is degenerate with respect to the formation of the c(2x4) and p(2x2) reconstruction on the lower terrace, within the accuracy of our calculations. The buckling energy is 0.14 eV per surface dimer, which is somewhat less than the value on the flat terrace. This reduced buckling energy reflects a slightly lower buckling of 0.62Åmeasured at the step edge.

As both variants of the S_B step edge (rebonded and non-rebonded) have been observed in STM experiments,[1] both of these types were considered in our calculations. The findings are as follows. The rebonded S_B step edge, the p(2x2) reconstruction is more stable than the c(2x4) by $\sim$ 80 meV per surface dimer, and indeed, this is the reconstruction seen in high-resolution STM images. This lowering of the energy is induced by the rebonding at the step edge, which decreases the buckling to 0.58Å. In the case of the non-rebonded step edge, the buckling is 0.74Å, which is slightly larger when compared to the buckling on the flat terraces.

STM images of the step edges as a function of the bias voltage and current intensity have been calculated using the approximations of Tersoff and Hamann,[19] and compared to the experimental images. In the case of a negative bias voltage, the electrons tunnel from the sample to the tip and the corresponding filled states are probed. Under a positive bias voltage, the electrons tunnel from the tip to the sample and the empty states are accessed. Figure 2 shows a sample image of the non-rebonded S_B step edge under different bias voltages. In the filled state image, note the apparent formation of a weak bond between the atoms on the upper and lower terraces, which is a result of the close proximity between the atoms. Note also the large, apparent buckling of the surface. The measured apparent buckling is 1.2Å, which is to be compared to the actual buckling of 0.70Åon the surface. This is to be contrasted with the results obtained from the empty-state image. Here, only a small apparent buckling is visible. Also, the sign of the buckling has changed: the up atoms of the dimer actually appear to be lower that the down atoms by about 0.1Å. These differences between the filled and empty state images have previously been observed in STM experiments, where the initial interpetation was that they were due to tip-induced changes of the surface reconstruction.[20] The present results indicate that the images simply reflect the electronic structure of the surface.

DIFFUSION BARRIERS

Under conditions of step flow, the ability to grow good crystalline material is related to the mobility of the adatoms on the surface. These must be able to diffuse freely and find the proper crystal lattice sites for growth, wherever these are available. In this section, we discuss our calculations of the diffusion barriers on the Si (100) surface and the single-height steps. We shall restrict our discussion to the motion of adatoms even though there is considerable evidence that mass transport via dimer diffusion plays a role at high temperatures as well.[21]

We begin our discussion with the diffusion of a Si adatom over a flat terrace. This problem has previously been addressed with *ab initio* calculations for the case of symmetric dimers.[22] The main result is that diffusion is highly anisotropic on the surface, with fast diffusion taking place over the top of the dimers with a saddle point energy of about 0.60 eV. Slow adatom diffusion is predicted to take place across the dimer rows with a barrier of $\sim$ 1.0 eV. Experiments based on a number counting of the island density are in agreement with these results.[23]

We have reexamined this problem for the case of a buckled surface, with both c(2x4) and p(2x2) reconstructions.[24] While our results are in general agreement with those of the symmetric case, there are some differences resulting from the effect of the adatom on the local dimer tilt. As in the symmetric case, the global minimum is situated between the dimers, as marked by M on Fig. 3a. Other local minima are labeled A, B and C, which have binding energies 0.5, 0.7 and 0.1 eV less than the global minimum at M. Note that the binding at sites next to a downwardly tilted dimer atom is slightly stronger than the sites next to an upwardly tilted dimer atom. This may be understood in terms of the charge transfer taking place as the buckling proceeds. Essentially, there is a charge transfer from the down to the up-tilted atom, which can therefore make a slightly stronger bond with the adatom.

Diffusion over the surface is still highly anisotropic. For motion over the top of the dimers, we obtain intervening barriers of 0.70 and 0.55 eV, respectively. To move across the dimer rows, a barrier of 0.95 eV must be surmounted. The barriers for diffusion in the channels separating the dimer rows are quite sensitive to the tilt of the dimer, with 0.75 eV being the lowest barrier encountered when the adatom moves past a down dimer.

The anisotropy in the diffusion barriers sets up natural diffusion barriers for step flow. Thus, in the absence of any additional barriers at the step edges, and under the constraint of a constant external flux, one can expect the rate with which adatoms reach the S_A step edge from the lower terrace to be higher than the rate with which they come in from the upper terrace. The opposite will hold for the S_B case.

Figure 3b shows the structure of the S_A step, with a c(2x4) reconstruction on the terraces. The geometry of this step edge resembles that of a dimer row on a flat terrace, as seen from the channel separating the dimer rows. There are three main binding sites, labelled A, B and C, at this step edge, with A and B located in front of a down- and up- atom of a tilted dimer, respectively. The binding energy is 0.35 and 0.60 eV less than that of the global minimum on the flat surface. The site B closely resembles that of the global minimum on the flat terrace, and indeed, they have virtually the same energy. Because the binding at the S_A step edge is close to that of flat terrace, we can expect the S_A step edge to be a weak sink.

Motion of an adatom along this step edge resembles the motion in the channel separating dimer rows on the flat terraces. For $C \rightarrow B$ the activation energy is 0.65 eV (1.25 eV for the reverse ho), and 0.60 eV for $B \rightarrow A$ (0.25 eV for the reverse). However, motion along the step edge should still be possible, because at high-temperatures the dimers will oscillate on a timescale of picoseconds bringing the barrier down from its high values of 1.25 eV. Adatoms can also easily escape the step edge onto the lower terrace. Relatively small barriers of 0.35 and 0.60 eV are present for jumps to positions

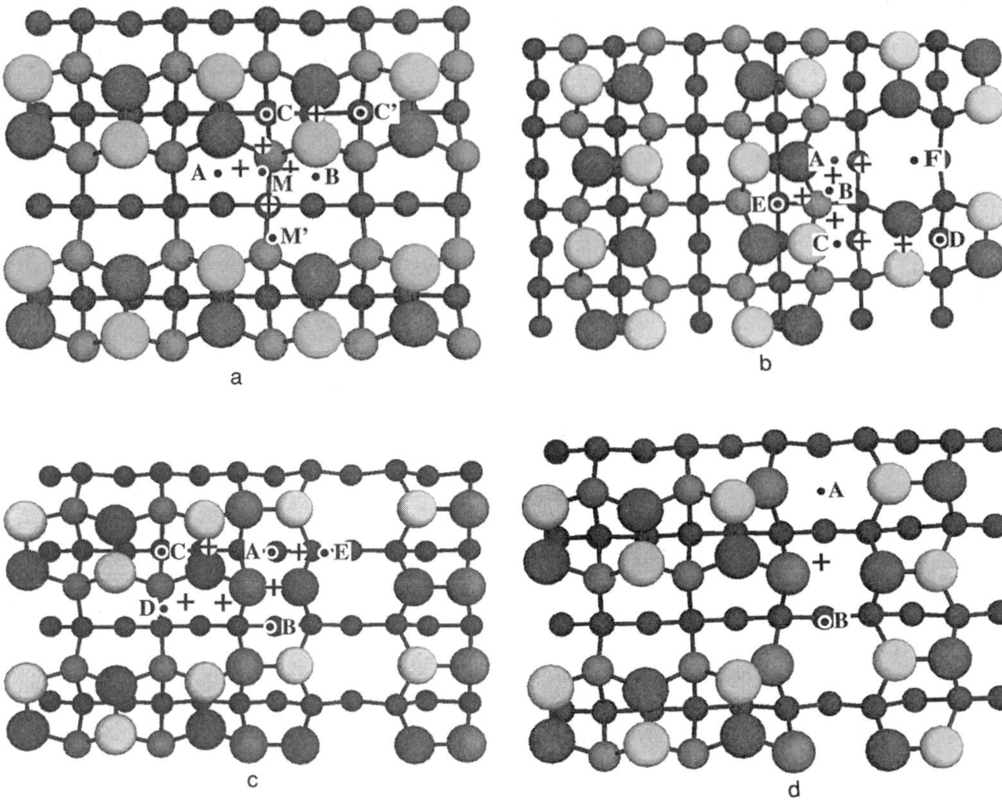

Figure 3. Top view of an (a) Si(100) surface with p(2x2) reconstruction; (b) S_A step edge; (c) rebonded S_B step edge; and (d) the non-rebonded S_B step edge. The labels (A, B, etc) indicate positions of minima discussed in the text. The shading of the lighter atoms indicates the "up" atom of the dimer, and the crosses the approximate positions of the saddle points.

between the dimer rows ($A \rightarrow F$) and over the top of the dimers ($C \rightarrow D$). The reverse motions require energies of 0.40 and 1.25 eV, respectively. Adatoms approaching over the top of the dimers therefore have a good chance of being reflected. For atoms to escape onto the upper terrace ($B \rightarrow E$), the barrier is 0.85 eV. To summarize, because the binding sites at the S_A step edge are relatively weak, and the barriers for adatoms to escape from the step edge are low, this step edge must be considered a weak sink.

These properties are to be contrasted with those of the rebonded S_B step edge, shown in Fig. 3c. At this step edge, there are several minima where the binding is significantly stronger than on the flat surface. For instance, the binding at sites A, B and E are 0.35, 0.15 and 1.0 eV deeper than the global minimum on the flat surface. Clearly, S_B should act as a strong sink.

Most of the adatoms will reach this step edge from the upper terrace. Adatoms coming in over the top of the upper terrace dimers encounter a barrier of 0.85 eV, while the barrier is only 0.70 eV if the adatoms come in from between the dimer rows. A relatively large activation barrier of more than 1.0 eV is required to come in from the lower terrace, so that this direction of approach is unlikely. To escape onto the upper terrace requires hops of 1.1 eV for $A \rightarrow C$ and 0.70 eV for hops $B \rightarrow D$. Clearly, in contast to the S_A step, because the minima are deeper and the barriers for escape are higher (all but one, which is symmetric), the rebonded B-step should be able to effectively contain the adatoms at its step edge.

Further facilitating the growth at the S_B step edge is the relatively low diffusion barrier over top of the rebonded atoms. We obtained this to be 0.50 eV. Thus, adatoms localized at a step edge will be able to find each other fairly easily and nucleate a kink site. In contrast to the S_A step edge, the presence of an adatom has little effect on the tilt of the dimers at the step edge. This is due to the fact the atoms moving along the optimum path are relatively far away from the upper step edge atoms. Growth at a rebonded step edge leads to a non-rebonded step edge, which each atom terminating at this edge having an extra dangling bond. Surprisingly, the minimum along this step edge, along $A \rightarrow B$ is only slightly deeper, by $\sim$ 0.10 eV, than the global minimum on the flat terrace. This result differs substantially from that obtained with classical molecular dynamics simulations, where a deep potential energy well is observed. The barrier for diffusion along the nonrebonded step edge is 0.65 eV.

That growth at the S_B steps is much faster than at the S_A steps has been confirmed in a number of experiments. It also helps explain the highly anisotropic shapes of islands that form on the flat Si(100) surface. Growth takes place primarily at the ends of the islands. Little growth takes place at the sides, because these edges, which resemble the S_A steps, simply cannot localize the adatoms there.

MELTING OF THE Si (100) SURFACE

We now turn to the melting of the Si (100) surface. This is a classic problem, whose microscopic details are not well understood. This is particularly true of covalent materials like Si, whose surfaces are characterized by reconstructions, steps, islands, and other surface defects – all of which are expected to play a role in the microscopic aspects of the melting process. As a first step towards this goal, we have carried out simulations of the melting process of the Si (100) surface with finite-temperature *ab initio* methods.

The melting of the Si (100) surface has previously been investigated with classical molecular dynamics simulations.[25] However, it is now known that these potentials fail to capture much of the this process. Upon melting, Si goes from a 4-fold coordinated semiconductor to a metallic liquid.[26] The density of the liquid is about ten percent higher than in the solid. The average coordination number is between 6 and 7, which is rather low for a metal. This low coordination number is indicative of persistant remnants of covalent bonding. Moreover, recent *ab initio* simulations of the liquid show that spin effects play an important role.[27]

The simulations were carried out with a real-space, multigrid code as briefly discussed in the methodology section. A relaxed slab of Si, with the bottom layer fixed, was heated up to $1500^o K$ over a 1.5 ps period of time. Over the next 0.3 ps, the temperature was raised to $1800^o K$, well above the melting temperature of bulk Si. The simulation was then allowed to proceed for another $\sim$ 6.5 ps. Sample configurations as a function of time are shown in Fig. 4. Essentially, the results are as follows. For the first 2 ps, the dimers oscillate as may be expected at high temperatures. There is the occasional breaking and reforming of dimers across a dimer row. After about 2.3 ps, we observe the formation of a dimer vacancy, which moves predominantly in the direction of the dimer rows. The first layer then melts over the next 2 ps. The melting of the second layer proceeds much more rapidly, with the layer achieving fluidity in $\sim$ 1 ps. The third layer is just beginning to melt at the end of the simulation time.

We have charactized the structural, thermodynamic and electronic properties changes that take place during the melting process. However, here we will only briefly summarize some of the main features. Reference [4] presents a more complete treatment. As the melting of the top two layers proceeds, there is a change in the density of the layers as function of the vertical distance. While initially each of the layers is well separated and defined, the density of the top two layers is essentially uniform at the end of the simulation time. During this time, the pair-correlation function of the top four layers becomes more liquid-like, with the number of nearest neighbors increasing from less than 4, to about 4.8. At the same time, the angular distribution function

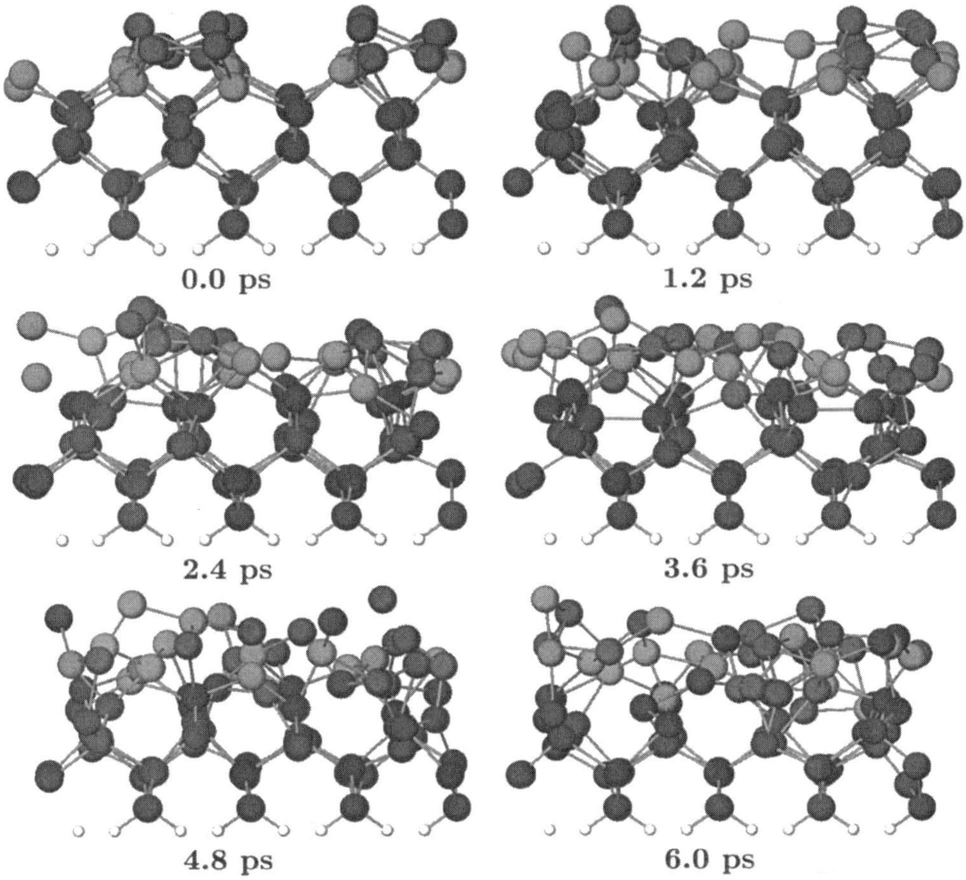

Figure 4. Side view of sample configurations of Si (100) surface at $1800^{\circ}K$ as the top two layers melt.

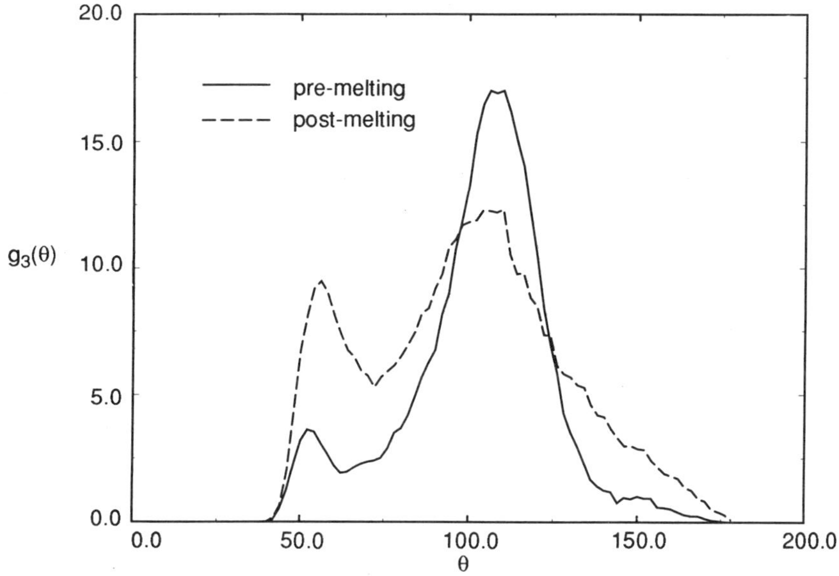

Figure 5. Angular correlation function versus time. Note the decrease in the large peak near the tetrahedral angle.

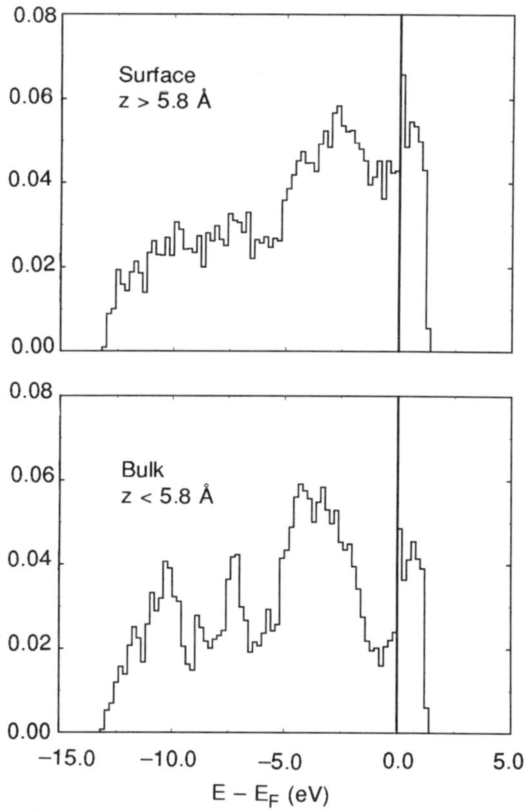

Figure 6. Local density of states in units of states per eV per atom for surface layer (top panel) and in the bulk (solid line) over the entire run.

$g_3(\theta)$ changes by decreasing its large peak about the tetrahedral angle, broadening and developing a second peak in the $50 - 60^o$ range, as shown in Fig. 5. All of this is characteristic of a melted system. As a probe of the electronic properties, we have measured the local density of states for the bulk and liquid domains. These are compared in Fig. 6. Note the relatively large buildup of states at the Fermi level for the liquid, characteristic of a metal, and the decrease in number of states for the bulk semiconductor.

SUMMARY

We have investigated the stepped Si (100) surface with *ab initio* molecular dynamics methods. In agreement with other studies, we find that the flat surface with symmetric dimers is unstable with respect to buckling, with the c(2x4) and p(2x2) reconstructions favored energetically by about 0.24 eV per dimer. The presence of the single-height steps was found to influence the reconstructions near the steps edges: at the S_A step edge c(2x4) is the spontaneously induced reconstruction, whereas the p(2x2) is induced at the S_B step edge. Simulated STM images of the step edges under different bias voltages compare well with the experimental images. The observed differences between the filled and empty-state images can be explained in terms of the electronic structure of the steps rather than any tip-induced changes. Diffusion barriers on the Si (100) surface were obtained with the "adiabatic trajectory" method. On the flat surface, fast diffusion takes place over the top of the dimer rows, and slow diffusion perpendicular to them. There is some effect of the local buckling of the dimer on the saddle point energies for diffusion. Considering the growth properties of the different steps, we find that the S_A step edge is a weak sink, as compared to the S_B step edge. In agreement with experiment, we therefore predict relatively fast growth at the S_B step edges. We have also simulated the melting of the Si (100) surface with finite temperature *ab initio* simulations. Melting of the surface was found to take place in a layer-by-layer fashion. The melting lead to the formation of a well-defined liquid-solid interface, whose structural, thermodynamic and electronic properties were characterized.

ACKNOWLEDGEMENTS

This work was supported by NSF Grants No. DMR-9500858 and DMR-9100063 and ONR Grant No. N00014-91-J-1516. The calculations were carried out at the Pittsburgh Supercomputing Center.

References

1. For a review, see: J.E. Griffith and G.P. Kochanski, CRC Rev. Solid State Mater. Sci. **16**, 255 (1990).

2. P. Boguslawski, Q.-M. Zhang, Z. Zhang and J. Bernholc, Phys. Rev. Lett. **72**, 3694 (1994).

3. Q.-M. Zhang, C. Roland, P. Boguslawski and J. Bernholc, Phys. Rev. Lett. **75**, 101 (1995).

4. M.G. Wensell, E.L. Briggs, C. Roland and J. Bernholc, to be published.

5. O.L. Alerhand, D. Vanderbilt, R. Meade and J.D. Joannopoulos, Phys. Rev. Lett. **61**, 1973 (1988).

6. E. Pehlke and J. Tersoff, Phys. Rev. Lett. **67**, 468 (1991); *ibid* 1290 (1991).

7. Y.H. Xie, G.H. Gilmer, C. Roland, P.J. Silverman, S.K. Buratto, J.Y. Cheng, E.A. Fitzgerald, A.R. Kortan, S. Schuppler, M.A. Marcus and P. Citrin, Phys. Rev. Lett. **73**, 3006 (1994); D.E. Jones, J.P. Pelz, Y.H. Xie, P.J. Silverman and G.H. Gilmer, ibid **75**, 1570 (1995); K.M. Chen, D.E. Jesson, S.J. Pennycook, M. Mostoller, T. Kaplan, T. Thundat and R.J. Warmack, *ibid* 1582 (1995).

8. D.E. Jones, J.P. Pelz, Y. Hong, E. Bauer and I.S.T. Tsong, Phys. Rev. Lett. **77**, 330 (1996).

9. W.K. Burton, N. Cabrera and F.C. Frank, Philos. Trans. R. Soc. London **243**, 299 (1951).

10. G.H. Gilmer and C. Roland, Appl. Phys. Lett. **65**, 824 (1994).

11. R. Car and M. Parrinello, Phys. Rev. Lett. **55**, 2471 (1985).

12. E.L. Briggs, D.J. Sullivan and J. Bernholc, Phys. Rev. B **52**, R5471 (1995); *ibid* **54** (1996).

13. C. Wang, Q.-M. Zhang and J. Bernholc, Phys. Rev. Lett. **69**, 3789 (1992).

14. S. Nose, J. Chem. Phys. **81**, 511 (1984); W.G. Hoover, Phys. Rev. A **31**, 1695 (1985).

15. R.A. Wolkov, Phys. Rev. Lett. **68**, 2636 (1992).

16. J. Dabrowski and M. Scheffler, Appl. Surf. Sci. **56-58**, 15 (1992).

17. J.D. Chadi, Phys. Rev. Lett. **59**, 1691 (1987).

18. T.W. Poon, S. Yip, P.S. Ho and F.F. Abraham, Phys. Rev. Lett. **65**, 2161 (1990).

19. J. Tersoff and D.R. Hamman, Phys. Rev. Lett. **50**, 1998 (1983).

20. G.P. Kochanski and J.E. Griffith, Surf. Sci. **249**, L293 (1991).

21. E. Ganz, private communication.

22. G.Brocks, P.J. Kelly and R. Car, Phys. Rev. Lett. **66**, 1729 (1991).

23. See, for example: Y.W. Mo, B.S. Schwartzentruber, K. Kariotis, M.B. Webb, M.G. Lagally, Phys. Rev. Lett. **63**, 2392 (1989).

24. A.J. Hoeven, J.M. Lenssinck, D. Dijkkamp, E.L. van Loenen and J. Dieleman, Phys. Rev. Lett. **64**, 1830 (1989).

25. F.F. Abraham and J.Q. Broughton, Phys. Rev. Lett. **56**, 734 (1986).

26. S. Sugino and R. Car, Phys. Rev. Lett. **74**, 1823 (1995).

27. I. Stich, M. Parrinello and R. Car, Phys. Rev. Lett. **63**, 2240 (1989); I. Stich, M. Parrinello and J. Hollender, *ibid* **76**, 2077 (1996).

RELAXATION OF SURFACE STEPS TOWARDS EQUILIBRIUM

Walter Selke

Institut für Theoretische Physik B
Technische Hochschule
D-52056 Aachen, Germany

ABSTRACT

The relaxation of isolated, pairs of and ensembles of steps on crystal surfaces towards equilibrium is reviewed, for systems both above and below the roughening transition temperature. Results of Monte Carlo simulations are discussed, together with analytic theories and experimental findings. Elementary dynamical processes are, below roughening, step fluctuations, step-step repulsion and annihilation of steps. Evaporation kinetics and surface diffusion are considered.

INTRODUCTION

A profile imprinted on a crystal surface will undergo morphological changes when relaxing towards equilibrium. This morphological evolution has been found, in experiments and theoretically, to be significant different above and below the roughening transition of the relevant surface.[1–7]

The thermal healing has been studied most extensively for one–dimensional gratings. *Above* roughening, the gratings acquire, for small amplitude to wavelength ratios, a sinusoidal form, as predicted by the classical continuum theory of Mullins[1] and confirmed by experiments[4,5] and Monte Carlo simulations.[6–9] The decay of the amplitude is, asymptotically, exponential in time. This is true for both evaporation dynamics and (experimentally more relevant) surface diffusion.

Below roughening, contradictory profile shapes and time laws for the decay of the profile amplitude have been suggested.[2,3,6–8,10–18] However, there is no doubt that the elementary microscopic processes are fluctuations of single steps, step–step repulsion and top step annihilation.[6,7] Each one of these processes is of much interest by itself, and will be dealt with separately in the following article, by considering single, pairs of and finally ensembles of steps. Not only gratings but also other profiles, like wires or bumps, can and will be described as ensembles of interacting steps. Of course, the

Dynamics of Crystal Surfaces and Interfaces
Edited by Duxbury and Pence, Plenum Press, New York, 1997

different types of steps may be investigated above roughening as well, using appropriate boundary conditions.

In the following, relaxation by evaporation kinetics and surface diffusion will be discussed. Results of simulations of nearest–neighbor SOS models and analytic theories are to be compared to experimental findings. A short summary concludes the article.

ISOLATED STEPS

The fluctuations of isolated steps have been studied, both theoretically[19–22], using Langevin theory, Monte Carlo simulations of SOS models, as well as exact methods, and experimentally[23,24], by scanning tunneling microscopy (caution is needed in the measurements to avoid artefacts of tip assisted motions of the steps[25]).

The various approaches suggest that different generic cases exist for the asymptotics of the time dependent step fluctuations, measured by the step width, ω. ω describes equilibrium fluctuations or the relaxation of an initially straight step, of width zero, to a thermalized step, broadened due to kinks, see Figure 1. ω grows asymptotically with time, t, as $\omega \propto t^s$. The value of the exponent s reflects the mechanism responsible for the meandering of the steps. Three generic, physically meanigful processes have been identified: (i) Step diffusion, i.e. surface atoms move along the step edge, leading to s=1/8;

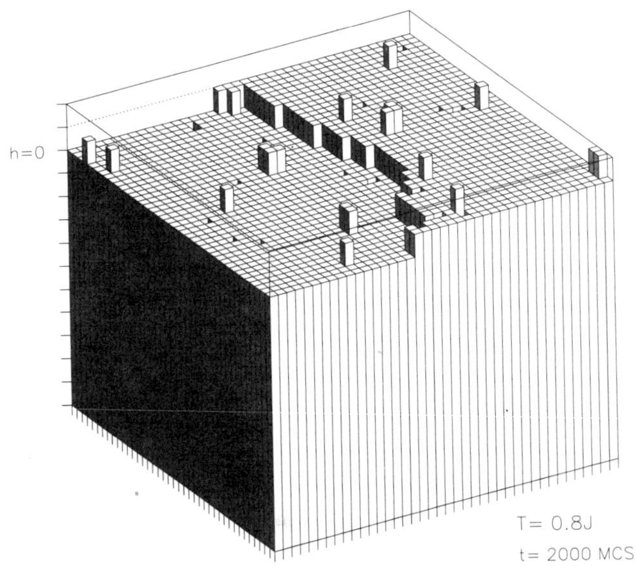

Figure 1. Monte Carlo configuration of an isolated step below the roughening transition temperature $T_R \approx$ 1.24J of the standard SOS model, at t= 2000 MCS, using evaporation kinetics.

(ii) fast terrace diffusion, i.e. the step emits an adatom which diffuses fastly on the terrace and attaches again (with a small sticking coefficient) at the step, at a position essentially uncorrelated to the initial step site of the atom. In that case, corresponding also to evaporation kinetics, one obtains s=1/4; (iii) slow terrace diffusion, where final and initial step sites of the emitted adatom are correlated, giving s=1/6. The slow diffusion is realized, e.g., by the standard Metropolis algorithm, where the transition rates for the hopping of a surface atom are determined by the Boltzmann factor of the change of energy involved in such a move (in the following, we shall use the term 'surface diffusion' for that algorithm; the Arrhenius kinetics seems to belong to the same category). Fast terrace diffusion may be realized by non-standard transition rates[19].

So far, we tacitly assumed that the upper and lower terraces next to the step are below their roughening transition temperature. By fixing the *boundary* heights of the terraces, away from the step, at, say, level 0 for the lower and level 1 (in units of the lattice spacing) for the upper terrace, one can study the time evolution of the step width ω, defined, for instance, as the second moment of the gradient of the step profile also *above* roughening. Then one obtains s=1/4 for terrace diffusion and 1/2 for evaporation kinetics, as predicted by the continuum description of Mullins[1] and confirmed by our Monte Carlo simulations.

PAIRS OF STEPS

To study the dynamics of a pair of steps, we consider two steps which are initially straight and separated by a terrace of width l_{s0}. In the SOS model on a rectangular lattice of $(L+2) \times M$ sites, the initial height configuration would be $h(x, y, t=0) = 0$ at $1 \leq x \leq L_1$, $h(x, y, 0) = 1$ at $L_1 < x \leq L_2$, and $h(x, y, 0) = 2$ at $L_2 < x \leq L+2$, with $l_{s0} = L_2 - L_1$. We shall monitor the thermal relaxation of the steps, subject to the the boundary condition $h(1, y, t) = 0$ and $h(L+2, y, t) = 2$. A typical configuration is shown in Figure 2.

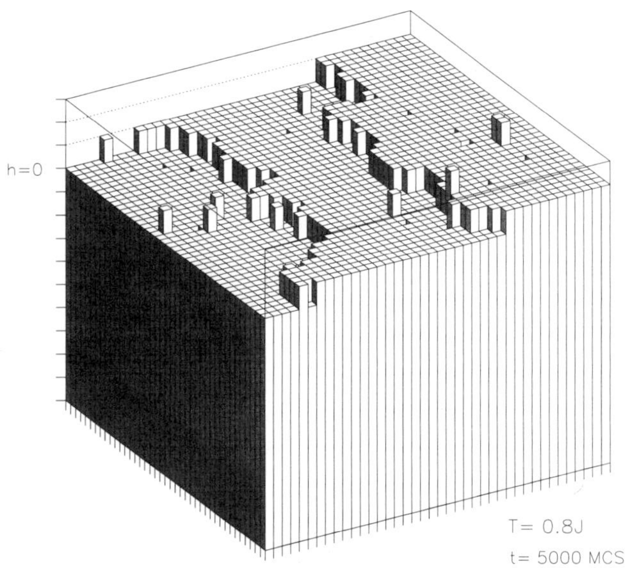

Figure 2. Monte Carlo configuration of a pair of steps below the roughening transition temperature of the standard SOS model, at t= 5000 MCS, using evaporation kinetics. The initial width l_{s0} of the center terrace is six lattice spacings.

As time goes on, *below* roughening, the steps will meander due to formation of kinks, with only rather few adatoms on the terraces. To overcome possible ambiguities in locating the step position, one may calculate the step profile $z(x,t) = [\sum_y h(x,y,t)/M]$, where the brackets, [ ], denote an average over several realizations, to smoothen the profiles (alternatively, one may increase M in a single realization).

The profiles, as observed in Monte Carlo simulations for the standard SOS model[7], display two turning points. The distance between them denotes the average step separation, $l_s(t)$. Indeed, l_s is found to increase with time, corresponding to an effective repulsion between steps, the entropic step–step repulsion (note that there is no energetic penalty for forming double steps in the standard SOS model, in contrast to the Gaussian SOS model). Of course, after some time, the boundary conditions will prevent the steps from moving further apart from each other, leading to a stationary step profile, the equilibrium profile.

In the case of surface diffusion, the Monte Carlo data[7,12] at intermediate times are consistent with the scaling behavior of the profile $z(x,t) \sim Z(xt^{-1/5})$, i.e. the step separation l_s grows with t^p, where p =1/5, in accordance with continuum theories.[2,3] Deviations occur at early and late times. They may be conveniently seen in the time–dependent effective exponent $p_e(t)$, determined from the ansatz $z = Z(xt^{-p_e})$, evaluated at a fixed height of the profile (say, z =1.5), see Ref. 6. $p_e(t)$ approaches, as time proceeds, the characteristic value 1/5 from below; at late times, it goes to 0 again, reflecting the boundary condition and the finite–size effect, leading to a stationary profile after some time. An exact calculation[26] on a simplified model for detachment and attachment of atoms at the two steps, performing a random walk in between[6,10], confirms the value 1/5 for the asymptotic exponent.

For evaporation kinetics, continuum theory predicts p= 1/4.[3]

Above roughening, the Monte Carlo data can be analysed in the same way. The turning points in the profile disappear quickly, corresponding to the fact that the step free energy vanishes in the rough phase. Nevertheless, the profile scales, for surface diffusion, with an exponent, at some intermediate times, p= 1/4, which now describes the increase of the width of the profile, in accordance with the classical theory.[27] Note that the effective exponent p_e, as computed in the simulations, displays an interesting oscillatory behavior, caused by an interference of the boundary condition with the spreading of the oscillations in the profile away from the steps (those oscillations are also described by the continuum theory of Mullins; they persist below roughening, albeit becoming much weaker as the temperature is lowered).

In the case of evaporation–condensation, continuum theory predicts p= 1/2.[27]

ENSEMBLES OF STEPS

A series or **bunch** of m initially straight and parallel steps, between heights 0 and m, may be expected to relax with the same asymptotics as a pair of steps. Modifications may occur, already for a pair of steps, when step–step interactions are present in addition to the entropic step repulsion. Here, we merely refer to recent reviews on experiments[28] and theoretical analyses[29] on the much studied phenomenon of step bunching for vicinal surfaces, which is accompanied by interesting phase transistions.

An intriguing complication arises when a series, say, of uphill steps is followed by an equal number of downhill steps, forming a one–dimensional **wire** on the surface. *Below* roughening, the thermal flattening of the wire proceeds through the *top step annihilation* mechanism[6,7]. By that, the amplitude of the wire is reduced by one lattice constant when the two steps, bordering the top terrace, meander, collide and form islands which shrink and vanish. The process may be imagined by looking at Figure 3.

The top step annihilation involves a pair of steps, an 'up' and a 'down' step. For evaporation kinetics, its dynamics may be easily simulated, or calculated in a Langevin description for a single interface fluctuating between an attractive and repulsive wall[7] (see also Ref. 30). The time needed to annihilate the steps, being initially straight and separated by a distance L_0, increases with L_0^a. The asymptotic value, for large separation L_0 of indefinitely long steps, of a is 4; however, the corresponding effective exponent reflects strong finite–size effects.[7]

The layerwise relaxation of the wires leads to two different time scales during the flattening, as readily observed in simulations.[6,7,12,13] If the amplitude is close to an integer, the decay is comparatively slow, reflecting the meandering of the top steps. At amplitudes in between, fast islanding dominates. Similarly, the profile shapes fluctuate, with a broadening near the top at integer values of the amplitude, due to an (almost) intact top terrace with only few adatoms and vacancies, see Figure 3. Of course, the detailed features may depend on various aspects of the simulations, such as, for example, the extent of the lattice along the step direction, the width and amplitude of the wire as well as the type of relaxation dynamics[7], similarly to what has been observed for the top step annihilation.

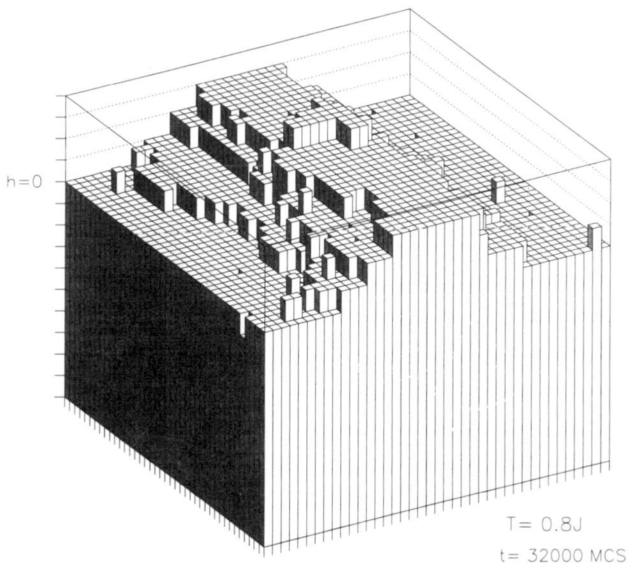

Figure 3. Monte Carlo configuration of a wire below the roughening transition temperature of the standard SOS model, at t= 32000 MCS, using surface diffusion. The initial width of the wire is twelve, and the initial height four lattice spacings.

In the case of evaporation kinetics, continuum theory[2,3] predicts, e.g., that the amplitude of the wire decays with t^{-w}, with w= 1/5. Simulations, for rather small systems, show strong deviations.[7] The simulated profile shapes also differ appreciably from the predicted ones, even when conservation of mass at the surface is taken into account, especially near the top. The differences may be traced back to the fact that the mobility at the top during the relaxation is governed by the top step annihilation, usually not included in the continuum theory; although one may incorporate it, at least in a phenomenological fashion.[3,7,18]

In marked contrast, the classical continuum theory by Mullins[1,27] describes the simulational data (profile shapes and amplitude decay) *above* roughening for wires even with small geometries surprisingly well, both for surface diffusion and evaporation–condensation[7]. The agreement may be a little bit fortuitous, because of a compensation of the *competing* effects of the anisotropic surface tension and anisotropic mobility[7], whereas continuum theory assumes isotropic quantities. In any event, the predicted decay laws with $w = 1/4$ for surface diffusion and $w = 1/2$ for evaporation kinetics are readily reproduced in the simulations.

One–dimensional **gratings** or periodic corrugations may be simulated by choosing the initial configuration $h(x, y, t = 0) = aint(A_0 sin(2\pi x/L))$, i.e. a discretised sine function, rounded to the next lower, in magnitude, integer, with wavelenght L; A_0 is the initial amplitude. SOS models of $L \times M$ sites with full periodic boundary conditions have been investigated by several authors. These studies have been motivated by pertinent experiments.[4,5]

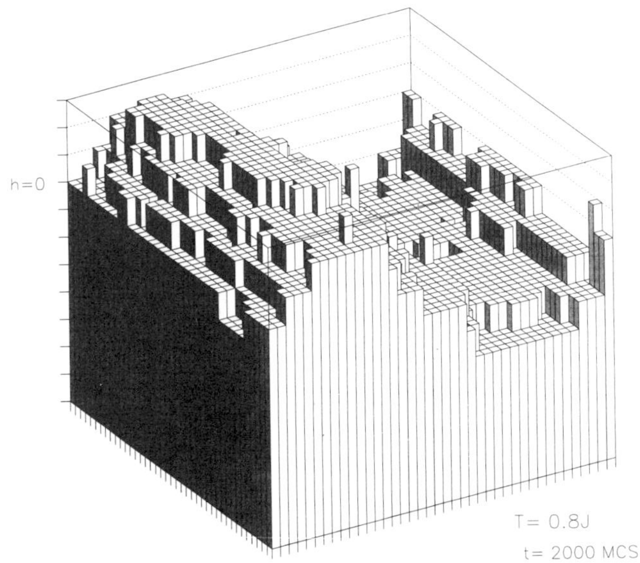

Figure 4. Monte Carlo configuration of a grating below the roughening transition temperature of the standard SOS model, at $t=$ 2000 MCS, using evaporation kinetics. The initial amplitude of the gratings is five lattice spacings.

Below roughening, pronounced lattice effects show up in the simulations, as in the case of wires. The meandering of the top(bottom) steps and the islanding on the top(bottom) terrace leads to slow and fast time scales in the decay of the amplitude. The profile shapes near the top(bottom) broaden at integer values of the amplitude and acquire a nearly sinusoidal form in between. Again, these features are not captured by the continuum theory. For evaporation kinetics, continuum theory[2,3] suggests that the decay of the profile amplitude z_m scales like $z_m(t, L) = Z_m(t/L^g)$, where $g=$

4. First simulations[6] showed that the corresponding effective scaling exponent depends sensitively on the wavelenght L and temperature. It was estimated to be at least 3. Actually, its behavior resembles, for comparable parameters, quite closely that found later in simulations of the top step annihilation process[7] (accordingly, the simulational data may be argued to be consistent with an asymptotic value 4, suggesting that the top step annihilation dominates the dynamics of the relaxation). Indeed, recent simulations[18] on a variant of the SOS model, allowing to compute gratings with much larger wavelenghts, seem to be consistent with $g = 4$ (possibly, logarithmic corrections to the simple power–law may be present[18]).

The discrepancy between the 'standard' continuum theory[2,3], predicting a non–parabolic sharpening of the profiles near the top(bottom), and the simulational findings may be reconciled by introducing a non–vanishing mobility at the top(bottom) of the grating[7,3,18]. Away from the extrema, the continuum theory provides a good description of the profile shapes. The importance of that mobility is also seen by introducing a miscut of the surface, in addition to the periodic corrugation. Suppose a fixed inclination of the surface perpendicular to the direction of modulation.[2,7] The miscut corresponds to additional steps on the top(bottom) terrace, being sources of additional mobility. Accordingly, the broadening at the top(bottom) is enhanced, as predicted by the continuum theory[2] and observed in simulations, see Figure 5.

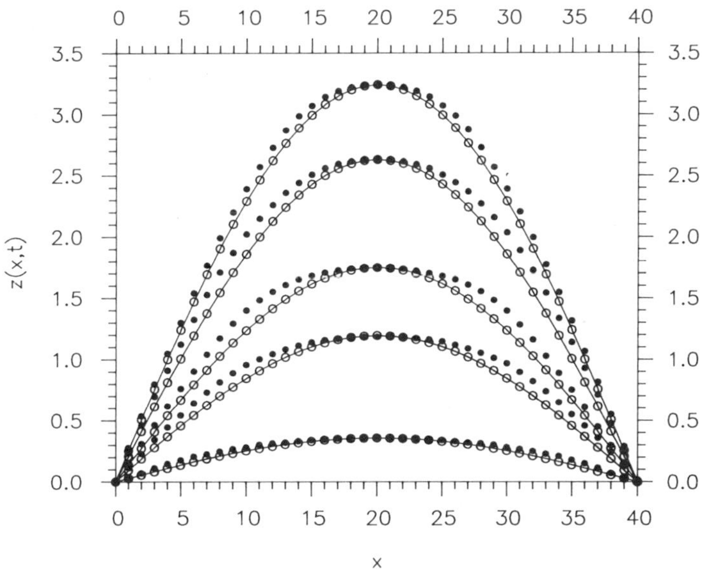

Figure 5. Profiles $z(x,t)$ of a grating below the roughening transition temperature, at increasing time, of the standard SOS model with 80×320 sites and a miscut of a few lattice spacings, using evaporation kinetics in Monte Carlo simulations (full symbols). For comparison, a sinusoid is shown (open symbols). The initial amplitude of the grating is five lattice spacings.

In the case of surface diffusion, conflicting continuum theories exist[2,3,10,14–16]. In the theory of Bonzel et al.[4,15,16], thermodynamic singularities below roughening are regularised. The approach, motivated by experiments showing the broadening of profile shapes at the maxima and minima of gratings, circumvents the non-analyticity problem, addressed by the other theories. However, those theories yield contradictory results.

For instance, Rettori and Villain[2] predict a sharpening of the shapes at the top(bottom) and a scaling exponent $g=5$, while Spohn et al.[3] obtain a facetting at the top(bottom) and a finite decay time for the amplitude. Ozdemir and Zangwill[10] emphasise the relevance of the rates of attachment and detachment of atoms at the steps, leading to different decay laws of the profile amplitude.

Unfortunately, the simulational data[8,17,12,13] are neither conclusive, suggesting scaling exponents between 4 and 6. Obviously, finite–size and finite–time effects are important, giving rise to gradual crossovers[17,8] (similar to what we observed in the evaporation case[6]). In addition, details of the dynamics, for instance, fast or slow terrace diffusion (as discussed above), should be viewed carefully.

Additional insight may be gained by considering the time scales introduced by the meandering of isolated top steps, by the step-step repulsion stemming from the steps below and by the top step annihilation. Indeed, in the evaporation case, the top step annihilation seems to determine the scaling behavior, with the step–step repulsion acting on the same time scale.

The effect of a miscut of the surface in the case of surface diffusion has been recently studied[31], which seems to be important in interpreting experiments.[5]

Above roughening (see Figure 6), the decay of the gratings is well described by the classical continuum theory[1] for sufficiently small ratios amplitude/wavelenght, with $g=2$ for evaporation, and 4 for surface diffusion. Deviations, observed otherwise, can be explained mostly by the anisotropy of the surface tension.[7,15,16]

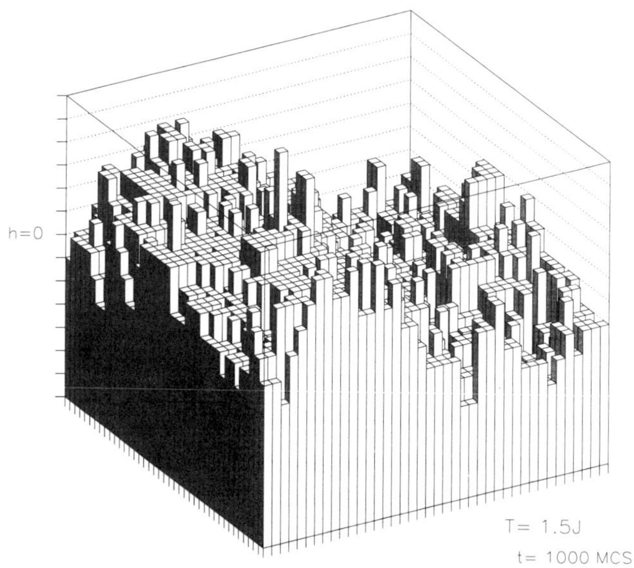

Figure 6. Monte Carlo configuration of a grating above the roughening transition temperature of the standard SOS model, at $t=1000$ MCS, using evaporation kinetics. The initial amplitude of the grating is five lattice spacings.

So far, we considered ensembles of indefinitely long steps. A different situation is encountered for surface perturbations of finite extent along both directions of the surface. A simple example is a **square bump**. In that case the surface atoms are assumed to be initially at height h_0 on a square of $L \times L$ sites, and on level 0 otherwise. During thermal relaxation closed steps will be formed, see Figure 7. Related experiments have been performed.[32]

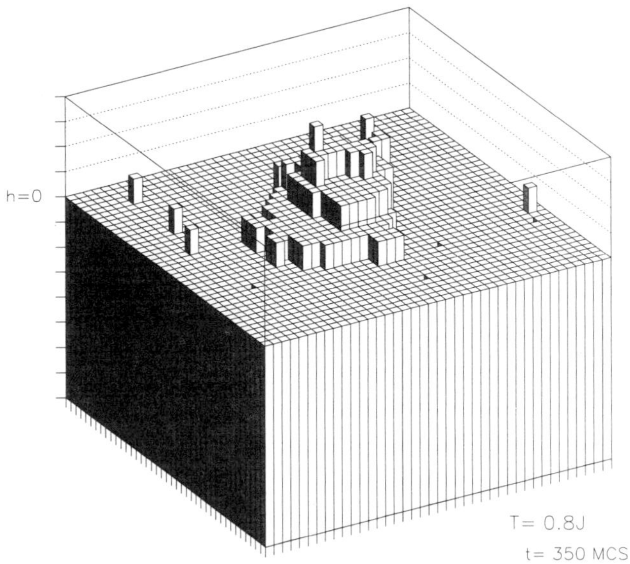

Figure 7. Monte Carlo configuration of a square bump below the roughening transition temperature of the standard SOS model, at t= 350 MCS, using evaporation kinetics. The initial height of the bump is four, the initial extent 12 × 12 lattice spacings.

Below roughening, the relaxation is driven by the lowering of the line tension of the curved steps. For evaporation kinetics, continuum theory[3] and simulations[7] show a shrinking of the bumps in the late stages of the decay. At small amplitudes, the radially symmetric profile scales with $z(r,t) \sim Z(\sqrt{(ct+r^2)})$, where r is the distance from the center, and c is a constant. The continuum theory fails to describe the layerwise relaxation monitored in the simulations.[7]

Above roughening, the simulations[7] confirm the classical continuum theory.[27] The width of the bumps spreads with time as t^b, where b= 1/2 for evaporation–condensation, and b= 1/4 for surface diffusion. In the latter case, the profile shows an oscillatory behavior away from the foot of the bump, as for pairs of steps and wires.[7,27]

SUMMARY

Results of Monte Carlo simulations on the healing of profiles imprinted on a crystal surface have been briefly reviewed, together with experimental findings and predictions of analytic theories.

In particular, the equilibration of isolated and pairs of steps as well as ensembles of steps as realized in gratings, wires and bumps has been discussed, for different types of kinetics such as step diffusion, evaporation–condensation and surface diffusion.

Special attention has been paid to the profiles shapes, the asymptotic decay laws of the amplitude, and related scaling behavior. The roughening transition temperature of the relevant crystal surface plays a crucial role for these properties, for a given type of transport mechanism.

Most cases are well understood by now. Additional work seems to be needed for clarifying the flattening of gratings below roughening in the case of surface diffusion.– It is also desirable to elucidate the effect of (long–range) step–step interactions, induced, for example, by elastic forces, on step fluctuations.

ACKNOWLEDGEMENTS

It is a pleasure to thank especially Phil Duxbury, one of the chairmen of this conference, for a productive and enjoyable cooperation. I should also like to thank several colleagues, most of whom are mentioned in the references, for pleasant joint work as well as very useful information and discussions.

REFERENCES

1. W. W. Mullins, J. Appl. Phys. **30**, 77 (1959)
2. A. Rettori and J. Villain, J. Phys. (Paris) **49**, 257 (1988); F. Lançon and J. Villain, in:*Kinetics of Ordering and Growth at Surfaces*, M. G. Lagally (ed.), New York: Plenum 1990
3. H. Spohn, J. Phys. I (France) **3**, 69 (1993); J. Hager and H. Spohn, Surf. Sci. **324**, 365 (1995)
4. H. P. Bonzel, U. Breuer, B. Voigtländer, and E. Zeldov, Surf. Sci. **272**, 10 (1992)
5. C. C. Umbach, M. E. Keeffe, and J. M. Blakely, J. Vac. Sci. Technol. A**9**, 1014 (1991); M. E. Keeffe, C. C. Umbach, and J. M. Blakely, J. Phys. Chem. Solids **55**, 965 (1994)
6. W. Selke and J. Oitmaa, Surf. Sci. **198**, L346 (1988); W. Selke and T. Bieker, Surf. Sci. **281**, 163 (1993)
7. W. Selke and P. M. Duxbury, Acta Physica Slovaca **44**, 215 (1994); Phys. Rev. **52**, 17468 (1995); Z. Phys. B**94**, 311 (1994); W. Selke, Proceedings of the 8th Joint EPS-APS International Conference on Physics Computing, eds. P. Borcherds, M. Bubak and A. Maksymowicz, CYFRONET Krakow, p. 443 (1996)
8. Z. Jiang and C. Ebner, Phys. Rev. B**40**, 316 (1989); Phys. Rev. B**53**, 11146 (1996)
9. J. Krug, H. T. Dobbs, and S. Majaniemi, Z. Phys. B**97**, 281 (1995)
10. M. Ozdemir and A. Zangwill, Phys. Rev. B**42**, 5013 (1990)
11. M. A. Dubson, M. Kalke and J. Hwang, Phys. Rev. **B47**, 10044 (1993); M.A. Dubson and G. Jeffers, Phys. Rev. **B49**, 8347 (1994)
12. M. V. Ramana Murty and B. H. Cooper, Phys. Rev. B (1996) in print
13. J. D. Erlebacher and M. J. Aziz, Preprint (1996)
14. J. P. Straley and E. B. Kolomeisky, Preprint (1996)
15. H. P Bonzel, E. Preuss, and B. Steffen, Appl Phys. A**35**, 1 (1984)
16. H. P. Bonzel and E. Preuss, Sur. Sci. **336**, 209 (1995)
17. P. C. Searson, R. Li, and K. Sieradzki, Phys. Rev. Lett. **74**, 1395 (1995)
18. L.-H. Tang, private communication; these proceedings
19. N. C. Bartelt, J. L. Goldberg, T. L. Einstein, and E. D. Williams, Surf. Sci. **273**,

252 (1992)

20. A. Pimpinelli, J. Villain, D.E. Wolf, J.J. Metois, J.C. Heyraud, I. Elkinani, and G. Uimin, Surf. Sci. **295**, 143 (1993)
21. N.C. Bartelt, T. L. Einstein, and E. D. Williams, Surf. Sci. **312**, 411 (1994)
22. P. J. Upton and D. B. Abraham, Phys. Rev. **B39**, 9650 (1989); R. Lipowsky, J. Phys. **A18** L585 (1985); S. F. Edwards and D. R. Wilkinson, Proc. R. Soc. Lond. **A381**, 17 (1982)
23. L. Kuipers, M. S. Hoogeman, and J. W. M. Frenken, Phys. Rev. Lett. **71**, 3517 (1993)
24. M. Giesen-Seibert, R. Jentjens, M. Poensgen, and H. Ibach, Phys. Rev. Lett. **71**, 3521 (1993); M. Giesen, G. Schulze Icking-Konert, D. Stapel, and H. Ibach, Surf. Sci. **366**, 229 (1996)
25. J. Li, R. Berndt, and W.-D. Schneider, Phys. Rev. Lett. **76**, 1888 (1996)
26. B. Blagojevic and P.M. Duxbury, Preprint (1996)
27. W. W. Mullins, in *Metal Surfaces: Structure, Energetics and Kinetics*, R. Vanselow and R. Howe (eds.), New York: Springer 1963, p. 17
28. S. G. J. Mochrie, S.Song, M. Yoon, D. L. Abernathy, and G. B. Stephenson, Physica B **221**, 105 (1996)
29. E. D. Williams, Surf. Sci. **299/300**, 502 (1994); for step doubling, see, e.g., S. V. Khare, T. L. Einstein, and N. C. Bartelt, Surf. Sci. **339**, 353 (1995)
30. A. Sánchez, A. R. Bishop, D. Cai, and N. Gronbech-Jensen, Phys. Rev. B **52**, 5433 (1995)
31. H. P. Bonzel and W. W. Mullins, Surf. Sci. **350**, 285 (1996); S. Surnev, P. Coenen, H. P. Bonzel, and W. W. Mullins, Prog. Surf. Sci. (1996) in print
32. D. R. Peale and B. H. Cooper, J. Vac. Sci. Tech. **A10**, 2210 (1992); L. Kuipers, Thesis, University of Amsterdam (1994)

COARSENING OF MBE STRUCTURES IN 2+1 DIMENSIONS

Ellák Somfai and Leonard M. Sander

Department of Physics
University of Michigan
Ann Arbor, MI 48109-1120, USA.

INTRODUCTION

Molecular beam epitaxy is a widely used technique for growing structures on crystal surfaces. One of the goals is to be able to control the growth process to such extent that one can make the nanostructures complex enough for a particular purpose. An ambitious example is a quantum computer.[1]

In this paper we investigate the effect of the growth parameters (near equilibrium or far from equilibrium) on the growth process. In many materials the diffusion of an adatom on a singular surface (terrace) is faster than across a step edge: the downward diffusion is suppressed by the Ehrlich–Schwoebel barrier.[2] This barrier gives rise to an instability in the growth process against mound formation, as proposed by Villain,[3] and investigated by Johnson et al.[4] These 3-dimensional mounds coarsen: their lateral size L increases in time according to $L \sim t^n$. In this paper we focus on the question how n depends on the growth conditions. We also show that the continuum equation proposed by the Michigan group[4,5] is a reasonable description of the process.

MONTE–CARLO SIMULATION

One of our tools is an activated hopping Monte-Carlo simulation. We implemented a simple model for growth processes: the solid-on-solid model on cubic lattice. Although one expects that a model reflecting the microscopic details of the crystal structure (e.g. fcc(001)) would better describe the real situation,[6] even this very simple model captures correctly the phenomenon of coarsening. Also, the cubic lattice model appears to simulate very well the more complicated semiconductor growth, where one cube stands for e.g. a 2x4 block of GaAs. In our model the atoms land on the surface at rate F. All atoms on the surface undergo activated diffusion with rate $\nu \cdot \exp(-E/kT)$, where $\nu = kT/h$ is the attempt frequency (k is the Boltzmann constant, h is the Planck constant), and the energy barrier, E, depends on the local neighborhood of the diffusing atom. We make a further simplification: E depends linearly on the number of nearest neighbors, n, of the atom at the initial position (the number of bonds to break), and also on the change in next-nearest neighbors, Δnn, if it is negative:

$$\Delta nn = \begin{cases} nn^f - nn^i, & \text{if } nn^f - nn^i < 0 \\ 0, & \text{otherwise.} \end{cases}$$

where nn^f and nn^i is the number of next-nearest neighbors in the final and initial state. The first factor takes care of island formation and attachment of adatoms to step-edges, while the second models the Ehrlich–Schwoebel barrier to some extent. The entire form of the barrier is

$$E = E_0 + n \cdot E_n - \Delta nn \cdot E_{nn} \tag{1}$$

where E_0 is the barrier of a flat terrace. Details of the model can be found in Johnson et al.[4]

Fig. 1 shows surfaces obtained by this simulation. The surface is symmetric under the transformation $h \to -h$ in case of high flux and high Ehrlich–Schwoebel barrier, while for other regions of the parameter space this symmetry is broken: the

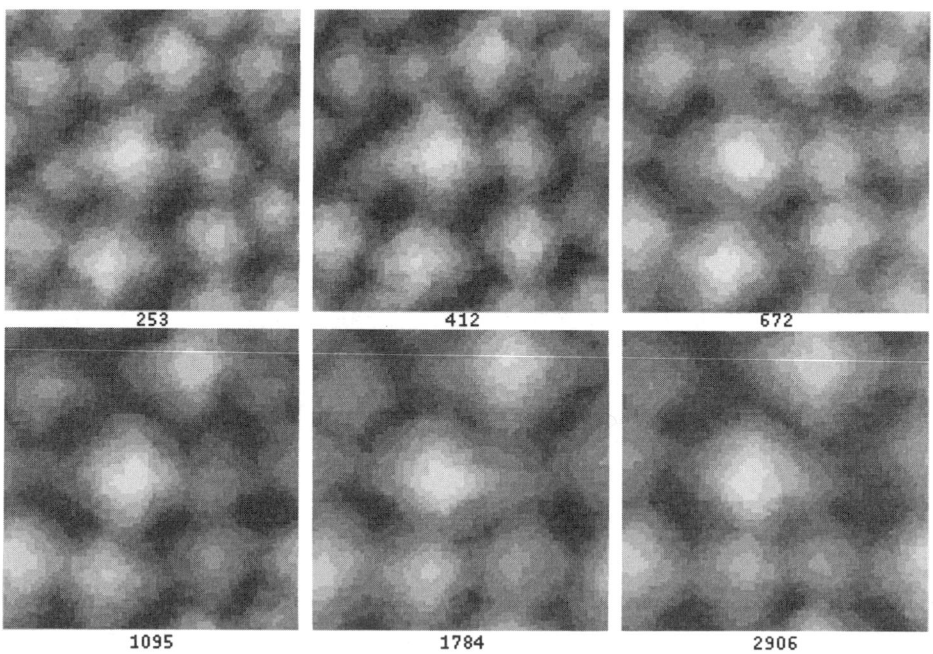

Figure 1. Snapshots of the time evolution of the surface obtained by the simulation, the flux of the incoming atoms is $F = 1\frac{\text{ML}}{\text{sec}}$, and the strength of the Ehrlich–Schwoebel barrier is $S = 0.653$. In this regime of the parameter space the surface grows symmetrically with respect to the transformation $h \to -h$. The number of deposited monolayers is shown under the images.

mounds have flat tops, and the valleys between them are narrow and deep. We can trace the process of coarsening. Initially the surface consists of many small mounds, they can be characterized by the maxima (tops) and the saddle points between mounds. We now focus on the coarsening, i.e. how two of these mounds merge. The mounds grow in time, and they compete with each other for the base area. Eventually one of them becomes large by a fluctuation. Then the saddle point between the mounds approaches the smaller maximum and annihilates it (Fig. 2): this is the most important step of the coarsening process. What is left is one large mound, which rearranges itself to be more-or-less symmetric. There can be fluctuations in the other direction: the top of a mound can split and merge again as observed by Šmilauer and Vvedenski.[7]

The coarsening process (growth of dominant wavelength) takes place during the approach to equilibrium as well. We carried out simulations verifying this. We started the process with 2 dimensional sine wave initial conditions, and with no incident flux, so that the surface relaxed towards a plane. After a short transient (during which the

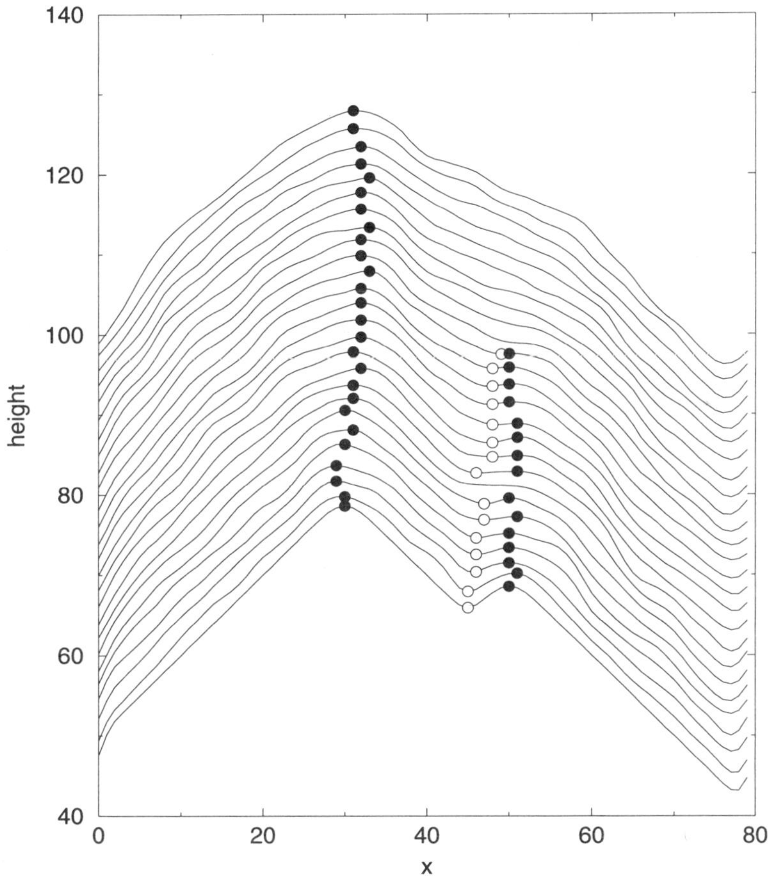

Figure 2. Time evolution of the top (•) and the saddle point (∘) on a cross-sectional view of two evolving mounds. The saddle point annihilates the top of the small mound, while the maximum of the large mound only fluctuates. Both the height and the lateral position are measured in lattice units.

surface rearranged its profile according to the strength of the Ehrlich–Schwoebel barrier) the root-mean-square width decreased as $w \sim \exp(-st)$, where the wavelength-dependence of the exponent, s, is

$$s = 4D_4 \left(\frac{2\pi}{L}\right)^4 , \tag{2}$$

showing that the equilibrium coarsening is dominated by diffusion processes.[8] That is, we have $\partial h/\partial t = -D_4 \nabla^4 h$, the Mullins equation. We have verified that this equation holds for our Monte-Carlo model, and we have measured D_4.

CONTINUUM EQUATION

A different way to approach the coarsening process is to concentrate on the long wavelength properties and analyze them in the framework of a continuum equation. The height $h(\vec{x}, t)$ of the surface is measured from its mean height. Changes in height arise from the divergence of the surface mass current:

$$\frac{\partial h}{\partial t} = -\nabla \vec{j} \, . \tag{3}$$

The surface current consists of a non-equilibrium part driven by the incident flux and the Ehrlich–Schwoebel barrier, and the equilibrium part driven by capillary forces:

$$\vec{j} = \frac{FS\sigma^2 \nabla h}{1 + (\sigma \nabla h)^2} + D_4 \cdot \nabla \nabla^2 h + D_6 \cdot \nabla \nabla^2 \nabla^2 h + \ldots \qquad (4).$$

The parameters are the flux, F, of the incoming atoms; the strength, S, of the Ehrlich–Schwoebel barrier ($S = R - T$ with R being the probability of reflection and T the probability of transmission across a step edge). The length, σ, is the mean distance between nucleation centers on a terrace, and D_4 is the strength of the equilibrium capillary forces. The last term in the current is the next term in the expansion series consistent with the symmetry. D_4 corresponds to healing of edges, while D_6 may correspond to the healing of corners. This last term is negligible near equilibrium compared to the D_4 term, this is why the classic Mullins theory takes the leading D_4 term only.

To give a physical interpretation of the non-equilibrium part of the current,[4] let us assume that the Ehrlich–Schwoebel barrier is infinitely strong (no atoms can jump down a step), and the up steps are perfect sinks. In this case for small slopes ∇h, only the atoms that land in strips of width σ will reach the up steps, the rest will attach to islands on the terrace, and does not contribute to the net mass current. In this case the total current is the fraction of the incoming flux which contributes to the current, multiplied by the mean migration length of the adatoms: $\vec{j} = F\sigma^2 \nabla h$. For the large slope limit, every incoming atom contributes to the current, but the migration length is only $1/|\nabla h|$, yielding $j = F/|\nabla h|$. The generalization for finite Ehrlich–Schwoebel barriers ($S < 1$) is simply multiplying the current with S. A convenient interpolation between these expressions for the non-equilibrium current is given in Eq. (4).

The present form (4) of the surface current is odd in h, so a surface growing from this current will be symmetric under the transformation $h \rightarrow -h$. Another feature of this current is that in the absence of curvature it is finite for nonzero slopes ∇h.

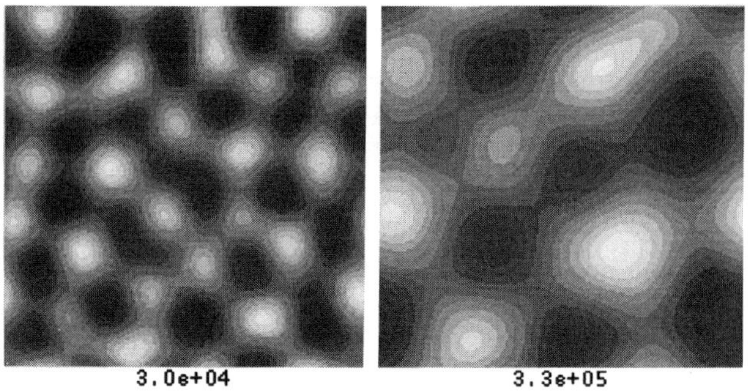

Figure 3. Typical images of the surfaces obtained by the integration of the continuum equation. The rescaled time is shown below the images.

Other terms in the current, which we did not include, could make the current zero[9] for a given slope m_0. This stabilizes the slope of the growing structures around m_0, and explains the phenomena of selected slope. Although the selected (or "magic") slope has been observed in many experiments, it is not necessarily present in every case and is not believed to be important in the coarsening process.

By integrating Equations (3) and (4), neglecting the D_6 term, with random initial conditions, mounds similar to those of the simulation can be obtained (Fig. 3). These mounds also coarsen in time. However, there has not been direct test of this equation as a description of multilayer growth. In particular, Eq. (4) was derived by fitting to Monte-Carlo data in the submonolayer regime. In this paper we show that certain aspects of multilayer growth by the Monte-Carlo model are well represented by Eq. (3) and (4).

Results of Ref. 5 show that integrating these equations generates coarsening: the time dependence of the lateral size, r_c, of the mounds scales with a power of time, $r_c \sim t^n$, with exponent $n = 1/4$. This exponent is associated with the leading D_4 term. Similarly, Stroscio et al[10] found $n = 1/6$ numerically when only the D_6 term was present. A detailed analytical proof is given by Golubovic.[11]

GENERALIZED FREE ENERGY

In order to quantitatively compare the simulation with the continuum equation, we introduce a generalized free energy of the surface, and write the equation of motion in variational form.

By appropriate rescaling of the variables $\vec{x}$, t and h to $\vec{X}$, T and H, the parameters in the equation of motion can be scaled out, leading to a dimensionless equation. Considering the present form (4) of the current, without the D_6 term, the equation of motion of the surface can be written in form of a functional derivative:

$$\frac{\partial H}{\partial T} = -\frac{\delta \mathcal{F}[H]}{\delta H} \tag{5}$$

where

$$\mathcal{F}[H(X,T)] = \frac{1}{2} \int \left(-\log(1 + (\nabla_X H)^2) + (\nabla_X^2 H)^2 \right) d^2X \tag{6}$$

which we call free energy. It is a functional of the rescaled height H. $\mathcal{F}$ consists of a non-equilibrium and an equilibrium term, like the current. This free energy does not coincide with the conventional free energy of the surface, but has the similar property that in our non-equilibrium growth conditions $\mathcal{F}$ is the quantity driving the system, and is a monotonically decreasing function of time. It is important to point out, that contrary to the previous claims,[5] in case of a current that produces asymmetric surfaces, no free energy can be found. (The lowest order term in the free energy density, which breaks the $H \to -H$ symmetry but conserves the $\vec{X} \to -\vec{X}$ symmetry and the translation invariance in H and X would be 3rd order in H and 6th order in X, thus can be excluded from our consideration.)

At this point we are able to compare the coarsening process of the Monte-Carlo simulation with that of the continuum equation, (5). We rescale the surfaces obtained in the simulation to the dimensionless variables $H(\vec{X}, T)$, and compare the time evolution of the free energy associated with the rescaled surface of the simulation (with different parameter values) with the free energy of the continuum equation. As we expect, the free energy (Fig. 4) decreases in time. But it turns out that the

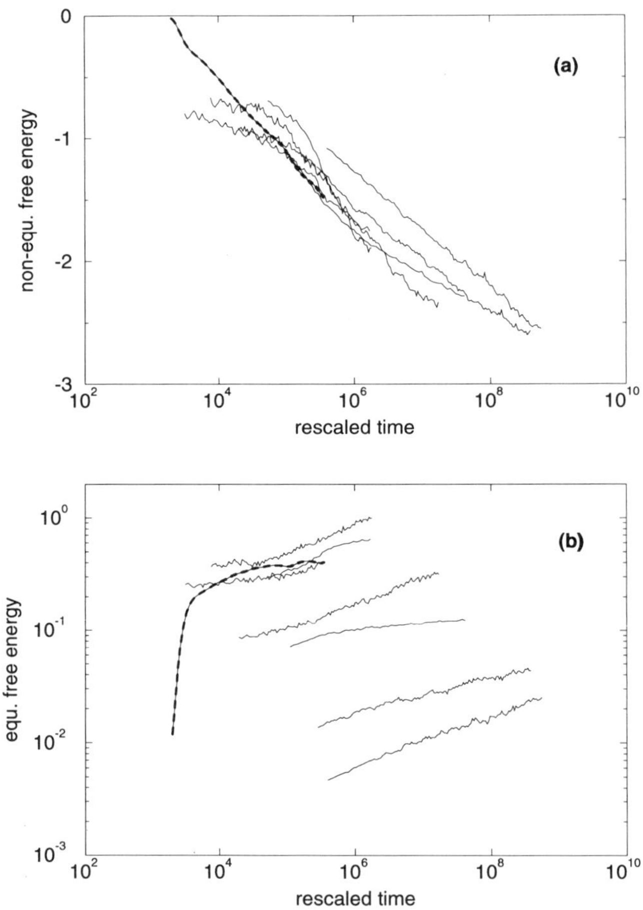

Figure 4. The (a) nonequilibrium and the (b) equilibrium part of the free energy. (The free energy itself is the sum of these two.) The dashed line corresponds to the continuum equation, the solid lines are the rescaled curves of the simulation for different parameter values (F, S).

equilibrium part itself increases, so it cannot drive the process. This is different from the case of spinodal decomposition, which also shows power law coarsening and which can be mapped on to this problem in 1+1 dimensions. In that case minimization of the surface energy drives the process. Here the surface energy increases in time.

It can be seen from the comparison that the non-equilibrium part (Fig. 4a), which in most cases dominates the free energy, is consistent with that of the continuum equation. But on the other hand, although the equilibrium part (Fig. 4b) more-or-less coincides with the result of the continuum equation for some parameter values of the simulation, for an another domain of the parameter space it does not. This could mean (and later we will argue that it does) that the D_6 term of Eq. (4) *is* important in those cases. We will give an explanation for this later in this paper.

COARSENING

An another way of describing the coarsening process is to study the time dependence of the characteristic feature separation r_c (the lateral size of the mounds, defined as the first zero crossing of the correlation function $\langle h(0)h(\vec{r})\rangle$). In most cases r_c scales as a power of time,

$$r_c \sim t^n \tag{7}$$

where n is the coarsening exponent. As we mentioned before, for diffusion dominated growth $n = 1/4$, and this is the case for the continuum equation, in the absence of the D_6 term.[5,9]

But in the coarsening process of the Monte-Carlo simulations (Fig. 5) the exponent is $n = 1/6$ for a domain of the parameter space (F, S), while for other regions it is $n = 1/4$. The region where the value of the coarsening exponent does not equal to the

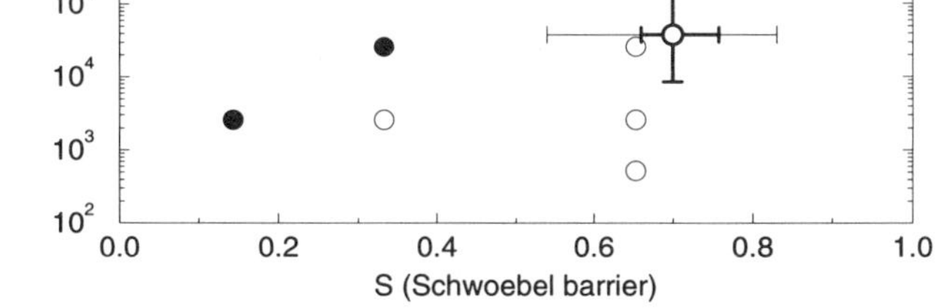

Figure 5. The value of the coarsening exponent n of the Monte-Carlo surfaces and two experimental surfaces as a function of the growth parameters. The points (○) where $n = 1/6$ coincide with those simulations where the equilibrium part of the free energy *did not* match that of the continuum equation. The error bars show the parameter range/uncertanity of an Fe/Fe(100) experiment of Ref. 10 (○: measured $n = 0.16 \pm 0.04$) and Ref. 12 (•: measured $n = 0.23 \pm 0.02$). The estimate of the Ehrlich–Schwoebel barrier is taken from Ref. 13 (thin line) and Ref. 14 (thick line).

value of the diffusion dominated case is the high flux, high Ehrlich–Schwoebel barrier corner of the parameter space. In this regime the growth conditions are strongly out of equilibrium: the high Ehrlich–Schwoebel barrier restricts the free diffusion of the adatoms, and the high flux also keeps the system far from equilibrium, the adatoms do not have enough time to find a preferred equilibrium position. This coincides with the region where the equilibrium free energy did not match the equilibrium free energy of the continuum equation. In this far from equilibrium regime the D_6 term of the current could be significant, D_6 can depend on the flux and the Ehrlich–Schwoebel barrier. This term can dominate the coarsening, and give $n = 1/6$ if the D_4 term of Eq. (4) is missing, as first suggested by Stroscio et al.[10] The presence of the D_6 term, which we neglected in our continuum equation, is the explanation for the fact, that the equilibrium part of the free energy of the far from equilibrium simulations deviates from that of the continuum equation.

Experimental results support this parameter-space dependence of the coarsening exponent as well. In case of Fe/Fe(100) homoepitaxial growth (where there are estimates for the value of the Ehrlich–Schwoebel barrier), at room temperature $n = 1/6$ has been measured[10] ($n = 0.16 \pm 0.02$), while at elevated temperature[12] the exponent is 1/4 ($n = 0.23 \pm 0.02$). These results are in excellent agreement with our predictions (Fig. 5).

In the near-equilibrium regime, $n = 1/4$ and both parts of the free energy (thus the free energy itself also) coincide with that of the continuum equation. In these cases Equations (3), (4) and (functional) give a good description of the evolution of the surface, the correct roughening behavior, and the correct functional form of the free energy. It should be noted that in order to fit the free energy of the continuum equation to that of the simulations, we used only one free fitting parameter (the time offset of the continuum equation, since its random initial conditions are arbitrary), all other parameters were either input parameters of the Monte-Carlo model or measured directly in the simulation, as in the case of D_4, above.

SUMMARY

We have investigated the coarsening process in homoepitaxial systems using Monte-Carlo simulations and continuum equations. From the geometric aspects of coarsening, the scenario for merging two mounds is the following: after initial competition between the mounds, the saddle point between them annihilates with the maximum of the smaller mound, then the one big mound rearranges itself to be symmetric.

In case of relaxation to equilibrium, the process is diffusion-dominated and the presence of the D_4 term is verified. For non-equilibrium conditions we have two cases: For weakly out of equilibrium (low flux, low Ehrlich–Schwoebel barrier) the D_4 term is still present and dominates the long-time coarsening, characterized by $n = 1/4$. However, for strongly out of equilibrium cases (high flux, high Ehrlich–Schwoebel barrier) the D_4 term seems to be dominated by the D_6 term, causing coarsening with exponent $n = 1/6$.

Acknowledgements

We would like to thank Brad Orr, Chris Orme and Ji Li for useful discussions, and also Mark Johnson and Chris Orme for letting us to use their computer programs.

This work was supported by NSF grant DMR 9420335.

REFERENCES

1. S. Lloyd, A potentially realizable quantum computer, *Science* **261**, 1569 (1993); J.I. Cirac, P. Zoller, Quantum computations with cold trapped ions, *Phys. Rev. Lett.* **74**, 4091 (1995).
2. G. Ehrlich and F.G. Hudda, Atomic view of surface self-diffusion: tungsten on tungsten, *J. Chem. Phys.* **44**, 1039 (1966); R.L. Schwoebel, Step motion on crystal surfaces II. *J. Appl. Phys.* **40**, 614 (1969).
3. J. Villain, Continuum models of crystal growth from atomic beams with and without desorption, *J. Phys.* (Paris) I **1**, 19 (1991).
4. M.D. Johnson, C. Orme, A.W. Hunt, D. Graff, J.L. Sudijono, L.M. Sander and B.G. Orr, Stable and unstable growth in molecular beam epitaxy, *Phys. Rev. Lett.* **72**, 116 (1994).
5. A.W. Hunt, C. Orme, D.R.M. Williams, B.G. Orr and L.M. Sander, Instabilities in MBE growth, *Europhys. Lett.* **27**, 611 (1994).
6. J.G. Amar and F. Family, Molecular beam epitaxy simulations with correct crystal geometry, preprint (1995).
7. P. Šmilauer and D.D. Vvedenski, Coarsening and slope evolution during unstable epitaxial growth, *Phys. Rev. B* **52**, 14263 (1995).
8. W.W. Mullins, Theory of thermal grooving, *J. Appl. Phys.* **28**, 333 (1957).
9. M. Siegert and M. Plischke, Slope selection and coarsening in molecular beam epitaxy, *Phys. Rev. Lett.* **73**, 1517 (1994); M. Siegert, in: *Scale Invariance, Interfaces and Non-Equilibrium Dynamics*, A. J. McKane, ed., Plenum, New York (1994).
10. J.A. Stroscio, D.T. Pierce, M. Stiles, A. Zangwill and L.M. Sander, Coarsening of unstable surface features during Fe(001) homoepitaxy, *Phys. Rev. Lett.* **75**, 4246 (1995).
11. L. Golubovic, Interfacial coarsening in MBE growth models without slope selection, preprint (1996).
12. K. Thürmer, R. Koch, M. Weber and K.H. Rieder, Dynamic evolution of pyramid structures during growth of epitaxial Fe(001) films, *Phys. Rev. Lett.* **75**, 1767 (1995).
13. J.G. Amar and F. Family, Step barrier for interlayer-diffusion in Fe/Fe(100) epitaxial growth, *Phys. Rev. B* **52**, 13801 (1995).
14. M.C. Bartelt and J.W. Evans, Transition to multilayer kinetic roughening for metal (100) homoepitaxy, *Phys. Rev. Lett.* **75**, 4250 (1995).

FLATTENNING OF GROOVES:
FROM STEP DYNAMICS TO CONTINUUM THEORY

Lei-Han Tang

Condensed Matter Theory
The Blackett Laboratory
Imperial College
London SW7 2BZ, UK

INTRODUCTION

Crystals are often associated with their unique polyhedral shapes. Their smooth, shining facets provide a constant source of fascination to ancient and modern beings alike. Advances in microscopy have not tainted the myth a bit: interesting surface structures are found down to the atomic level, be it surface reconstruction, surface ripples and dimples, or snow-flakish fractals...

Most surface structures we see are nonequilibrium patterns. This is because the relaxational dynamics of a crystal surface is usually quite slow unless one is very close to the melting point. In addition, a qualitative change in the relaxation behavior is expected at the roughening temperature T_R. The actual value of T_R depends not only on the substance one is dealing with, but also on the crystallographic orientation of the surface. For a given material, T_R is highest for low Miller index surfaces, and becomes vanishingly small for high Miller index surfaces (known as vicinals).

The classical theory of near-equilibrium surface relaxation was reviewed by Mullins (1963). In this theory, evolution of the surface profile $h(\mathbf{x})$ is driven by variations of an excess chemical potential

$$\mu(\mathbf{x}) \sim -\nabla^2 h. \tag{1}$$

The actual rate of relaxation depends on the mode of mass transport operative. The three basic modes are:

i) Exchange of atoms with the vapor which serves as a reservoir of constant chemical potential (due to the large mean free path of atoms), commonly known as *evaporation-condensation dynamics.* In this case, the local surface velocity is directly proportional to the excess chemical potential,

$$\partial h/\partial t = -m_s\mu, \tag{2}$$

where m_s is known as the surface mobility.

Dynamics of Crystal Surfaces and Interfaces
Edited by Duxbury and Pence, Plenum Press, New York, 1997

ii) Exchange of atoms with other parts of the surface through diffusion in the solid, known as *bulk diffusion dynamics*.

iii) Exchange of atoms with other parts of the surface through surface transport, known as *surface diffusion dynamics*. Due to mass conservation, the equation of motion for h takes the form,

$$\partial h/\partial t = -\nabla \cdot \mathbf{j}_s = \nabla \cdot \mathbf{D} \cdot \nabla \mu, \tag{3}$$

where $\mathbf{D}$ is the diffusion matrix which becomes a scalar for an isotropic surface.

Case i) is important close to the melting temperature, while case iii) dominates at lower temperatures. Under Eq. (1), Eqs. (2) and (3) are both linear in h (the same is true in the bulk diffusion case). Consequently, an initially sinusoidal modulation remains sinusoidal during relaxation. The amplitude $A(t)$ of modulation decays exponentially with time t, with a time constant τ which grows as a power-law function of the wavelength L,

$$\tau(L) \sim L^z. \tag{4}$$

The "dynamic exponent" depends on the transport mechanism (Mullins 1963),

$$z = \begin{cases} 2, & \text{evaporation-condensation,} \\ 3, & \text{bulk diffusion,} \\ 4, & \text{surface diffusion.} \end{cases} \tag{5}$$

There is considerable doubt whether Eq. (1), which assumes a smooth and continuous height function $h(\mathbf{x})$, applies below the roughening temperature T_R, where the solid surface decomposes into a set of flat terraces separated by well-defined steps. One might make a case that, under low resolution, the surface still looks continuous, so that a continuum description should be possible. This point of view has indeed been taken by several authors, but different approaches have led to different results, which is unsettling (see, e.g., Selke and Duxbury, 1995, and references therein). Ideally, one would like to start from the dynamics on the atomic level and using methods of statistical mechanics to derive the continuum equation through a suitable coarse-graining procedure. Well below the roughening temperature, most of the action happens or is controlled by surface steps, so an alternative starting point could be the dynamics of individual steps and interactions between them, which one knows a lot about (e.g., Pimpinelli et al., 1993). This latter approach was taken by Villain and his collaborators in treating the relaxation problem (Rettori and Villain, 1988, Lançon and Villain, 1990). Unfortunately, some of the most important predictions of their theory have not been supported by numerical simulations (Selke and Duxbury, 1995), which are free of complications in a real laboratory experiment (see, e.g., Bonzel et al., 1992, and references therein). The work reported below is an attempt to modify their theory to reach a better agreement with simulation results on the most-discussed case of one-dimensional grooves on a high-symmetry surface.

DRIVING FORCE TOWARDS EQUILIBRIUM

The second law of thermodynamics states that an isolated system in equilibrium has maximum entropy. This is the basis for a variational principle often used in determining the equilibrium state of a system. When the system contains several elements which are allowed to exchange mass with each other, the variational principle yields the condition that all elements must have equal chemical potential once equilibrium is established.

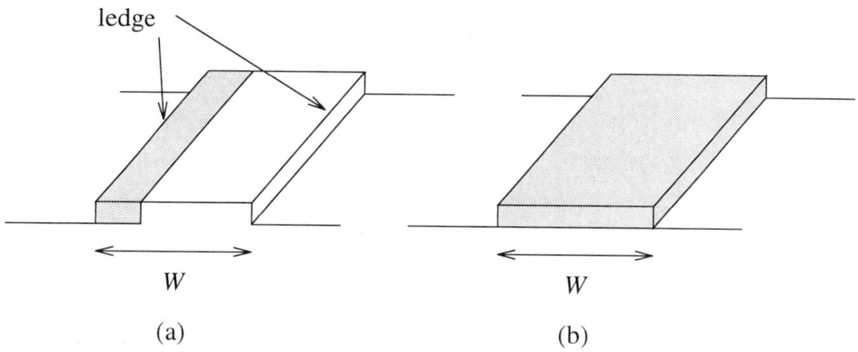

Figure 1: Monolayer strip on a flat surface. The shaded area in each case indicates the material removed in a virtual displacement of the left ledge.

The above property of the chemical potential offers a convenient device for discussing the relaxational dynamics of near-equilibrium systems. By *near-equilibrium* one usually has in mind a system where each part has reached equilibrium locally, but long wavelength modes have not been completely relaxed. This allows one to define a local chemical potential $\mu(\mathbf{x})$ whose spatial variation serves as the driving force towards global equilibrium.

For liquid-vapor interfaces, the correlation length in the bulk is of the order of atomic distance unless one is close to the critical point. Hence the concept of local equilibrium is well justified in most practical circumstances. For solid surfaces above the roughening temperature, the concept also makes sense. Since the surface is rough, adding (or removing) an atom to a particular part of the surface does not disturb the local equilibrium state very much, and this sampling procedure can be used to determine the local chemical potential. This is the essence of the Gibbs-Thomson relation (1).

It is equally evident that a solid surface below T_R requires a different treatment. Since the surface is mostly made up of flat terraces, the work done by adding (or removing) a particle to a particular part of the surface becomes a very sensitive function of the local surface configuration. To expand this point further, let us consider the extreme case of a high-symmetry surface at $T = 0$. The question is what chemical potential should we assign to the local configuration depicted in Fig. 1, a monolayer strip of width W on an otherwise flat surface. Figure 1(a) illustrates a move which eliminates part of the strip, but keeps the two ledges present. The change in the free energy of the system comes only from a change in the interaction between the two ledges. From this contribution alone one would conclude that the excess chemical potential μ, which is defined as the change of free energy per atom removed, is nearly zero. [1] On the other hand, if the variation corresponds to removing the whole strip as depicted in Fig. 1(b), a free energy equal to the free energy of the two ledges plus their interaction energy has been gained. Assuming the contribution from the interaction is much smaller than the ledge energy, we obtain,

$$\mu_L \simeq 2\gamma_L\Omega_s/W, \qquad (6)$$

where γ_L is the free energy of the ledge per unit length, and Ω_s is the surface area per atom. Thus, depending on the kind of local change we make to the surface, we obtain different answers.

[1] In a realistic situation, one should consider the elastic interaction between the two ledges. This, however, does not alter the qualitative conclusion of our argument.

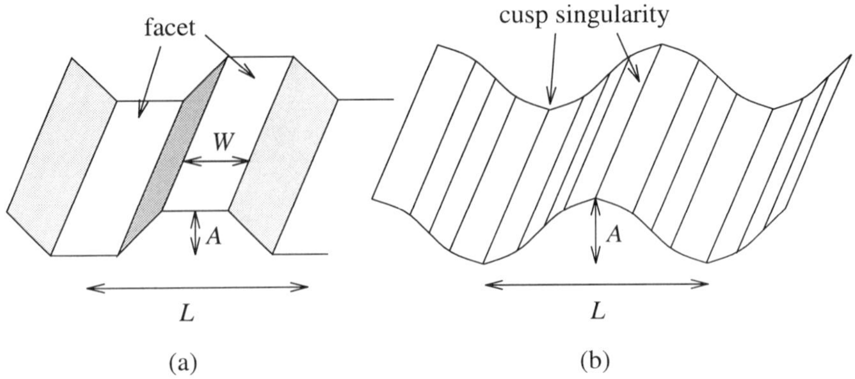

Figure 2: One-dimensional grooves during relaxation. (a) facetted top and bottom; (b) cusp singularity at top and bottom.

The ambiguity in the definition of the local chemical potential illustrated above is actually the origin of the discrepancy between previous theories (Bonzel and Preuss, 1995, Rettori and Villain, 1988, Spohn, 1993) proposed to describe the relaxation of one-dimensional grooved patterns on a high symmetry surface [e.g. Cu (111)] below its roughening temperature T_R (see Fig. 2). One class of theories (Bonzel and Preuss, 1995, Spohn, 1993) predict that, after some initial transient period, facets of the high symmetry surface appear at the top and bottom parts of the grooves, as shown in Fig. 2(b). Although different authors have used different formulations of the problem, the facetting aspect can be basically understood by taking Eq. (6) as the excess chemical potential of the top terrace. [2] (The bottom terrace has an excess chemical potential of opposite sign.) The tendency for facet formation is clear: when the width W of the top terrace is small, the chemical potential for the top-most layer is very high. This leads to a rapid disappearance of the top layers until W has reached a value comparable to the width of the groove, L. Beyond this point, the chemical potentials at the top and bottom of the groove do not vary significant. As a result, the subsequent decrease of the groove height $2A(t)$ with t is approximately linear, as explicit solution of the equations shows. The relaxation time, however, depends on the transport mechanism that is operative. For evaporation-condensation dynamics, assuming a finite desorption and recondensation rate on the terrace, a linear relation $\tau \sim A_0\mu_L^{-1} \sim A_0 L$ is obtained, where A_0 is the initial amplitude of the groove. Mass conservation in the surface diffusion case changes the behavior to $\tau \sim A_0 L^3$.

An alternative theory, first proposed by Rettori and Villain (1988), takes the point of view which corresponds to the variation depicted in Fig. 1(a), in which case the excess chemical potential of the top terrace is much lower than Eq. (6) suggests. In fact, in their treatment, the shrinkage of the top terrace is driven by the *pressure* from the step trains on either side, which has its origin in the repulsive interaction (of elastic or entropic origin) of like steps. This yields an effective chemical potential for the top terrace,

$$\mu_{RV} \simeq G_3/W^3, \tag{7}$$

where G_3 is a constant proportional to the strength of the step-step interaction. The ledge free energy γ_L does not play any role in the process. Because μ_{RV} decreases much faster with increasing W than in the previous case, the driving force towards equilibrium

[2]Due to the up-down symmetry usually assumed during relaxation of small amplitude modulations, we shall focus mostly on the upper half of the groove.

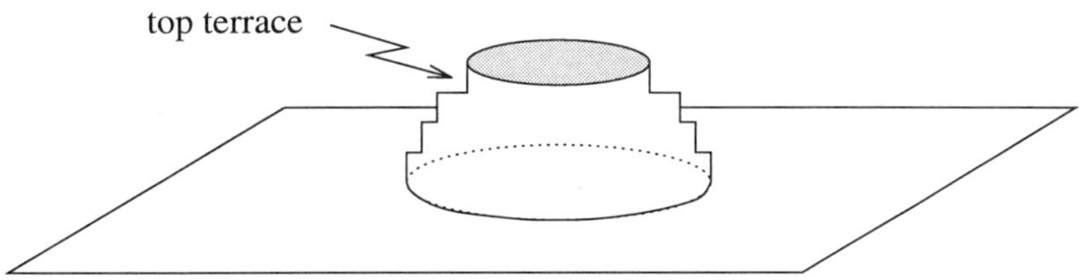

Figure 3: Two-dimensional bump, or wedding cake.

is much smaller. This leads to a much slower relaxation dynamics. As the top terrace broadens (which has to happen after the initial transient which eliminates high gradients of the chemical potential), removal of each additional layer takes progressively longer time, giving rise to a power-law decay of the amplitude,

$$A(t) \simeq A_0(t/t_0)^{-\alpha}. \tag{8}$$

In fact, in this picture, the relaxation time τ is dominated by the disappearance of the last few layers, where W becomes of order L. For surface diffusion dynamics, Rettori and Villain (see also Ozdemir and Zangwill, 1990) showed that

$$\tau \sim L^2 \delta\mu_{RV}^{-1} \sim L^5. \qquad \text{(surface diffusion, Rettori and Villain)} \tag{9}$$

For evaporation-condensation dynamics, assuming a surface mobility proportional to step density, Lançon and Villain (1990) arrived at

$$\tau \sim L \delta\mu_{RV}^{-1} \sim L^4. \qquad \text{(evaporation-condensation, Lançon and Villain)} \tag{10}$$

The initial amplitude does not play a significant role in determining τ, if one is patient enough to wait for the surface to become completely flat!

From an atomistic point of view, the second approach appears to be more reasonable, as surface evolution is a continuous process. Although Eq. (6) is the correct expression to use in the discussion of the equilibrium shape of a crystal which contains facets, it does not really help the case for the first approach because one is in a nonequilibrium situation where time scales are of ultimate importance. On the other hand, it is quite puzzling that simulation results have not lent support to the scaling proposed by Rettori and Villain, and it is not clear whether there is scaling (let alone universal scaling) at all (Bonzel and Preuss, 1995). So what has gone wrong? I will present my version of the story below. But before I do that, I would like to mention another case where different approaches do agree with each other.

The geometry we consider now is illustrated in Fig. 3, known as two-dimensional bump according to Selke and Duxbury (1995). The argument given below is borrowed from Lançon and Villain (1990). The top terrace of the dome-shaped object is a circle which can be continuously shrunk to zero. Let R be the radius of the circle at a given time. The free energy decrease associated with a decrease of R by δR is roughly equal to the change in the length of the circle times γ_L, the line tension,

$$\delta F \simeq \gamma_L 2\pi \delta R. \tag{11}$$

The number of particles involved in this change is

$$\delta N = 2\pi R \delta R / \Omega_s. \tag{12}$$

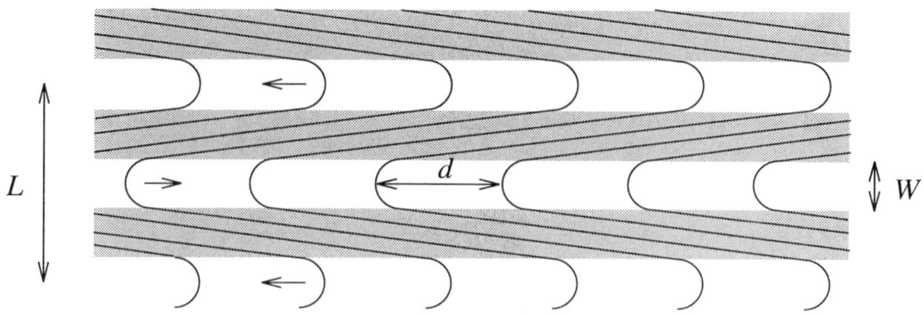

Figure 4: Top view of grooves on a miscut surface. The top and bottom parts of the groove contain a sequence of finger-like ascending steps which move in opposite directions so as to straighten up the undulation. Sloping parts of the grooves are shaded.

Dividing (11) by (12), we obtain,

$$\mu = \gamma_L \Omega_s / R. \tag{13}$$

[Taking away the whole top layer all at once would yield a value twice as big as (13), but is still qualitatively correct.] The dependence of μ on R is rather similar to Eq. (6). Hence we expect the top part of the dome to be facetted after an initial transient, and the time dependence of the bump height to be approximately linear.

An intermediate case between the groove and bump on singular surfaces is that of grooves on a miscut surface (Lançon and Villain, 1990), as illustrated in Fig. 4. Unlike the case shown in Fig. 2, steps are now closed in the direction perpendicular to the grooves, forming finger-like patterns. The chemical potential of the top terrace now becomes nonuniform. At the tip of a finger, (13) is reasonable, while far away from the tip, (7) is more appropriate. When the distance d between the finger tips is much larger than the width W of the top terrace, one may define an "average" chemical potential for the top terrace which takes the form,

$$\mu_{\text{avg}} \simeq \mu_L(W/d) + \mu_{RV}(1 - W/d) \simeq 2\gamma_L \Omega_s d^{-1} + G_3 W^{-3}. \tag{14}$$

Assuming the two coefficients are of the same order of magnitude (in atomic units), the first term dominates when $d < W^3$.

LIFETIME OF THE TOP TERRACE

In an attempt to explain the discrepancy between predictions of the Rettori and Villain theory and simulation results for the one-dimensional groove problem, Selke and Duxbury proposed step meandering as an important process which shortens the lifetime of the top terrace, thereby yielding a faster decay of the groove amplitude and a more rounded shape of the surface profile (see Selke and Duxbury, 1995, and references therein). Step meandering is a well-known phenomenon in statistical mechanical treatment of solid surfaces, and have also been observed recently in experiments (see Williams, 1994 for a recent review). During the shrinkage of the top terrace, the bounding steps fluctuate away from their average position (Fig. 5). Pinching points can thus form before the average position of the bounding steps collide with each other, and break up the top terrace into a sequence of elongated islands. The tips of the islands then shrink relatively rapidly due to its high chemical potential, in much the same way as the finger tips on a miscut surface move.

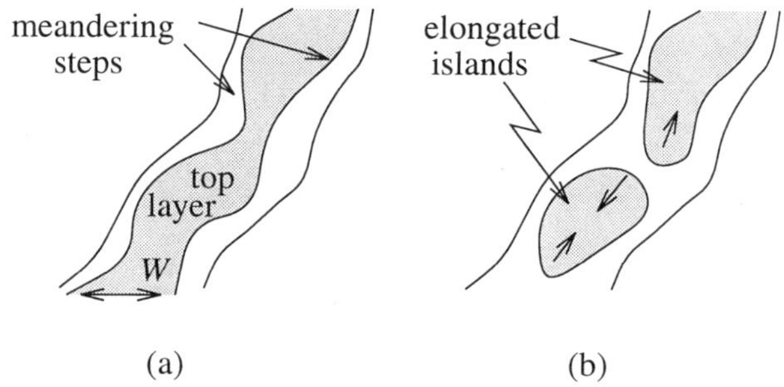

Figure 5: Dissociation of top terrace: (a) formation of pinching points through step-meandering; (b) shrinkage of elongated islands.

Obviously, in order for the pinching points to dominate the dissociation of top terrace, they must reach sufficient density. Equation (14) suggests that the critical density $\rho_c = 1/d_c$ is given by the condition

$$2\gamma_L \Omega_s \rho_c = G_3 W^{-3}. \tag{15}$$

To see if this picture yields qualitatively different conclusions from that of Rettori and Villain, we need to estimate the time it takes to reach the critical density, starting from two well-separated steps of minimum distance W.

Evaporation-Condensation Dynamics

The monolayer strip with straight edges as depicted in Fig. 1 can be taken as an idealised initial configuration of the top terrace. We are interested in calculating the typical time for the top terrace to disappear (or most part of it, to be precise). This is actually a problem which I have not been able to find an exact answer. Qualitatively, however, an estimate for the lifetime of the strip can be made as follows.

Assuming the width of the terrace W is much larger than the lattice constant of the crystal, we can use a continuum description of the step motion. The problem can be simplified by using the relative separation between the two lines, $x(y,t)$, where y runs along the strip. The equation of motion for x is given by,

$$\dot{x} = \nu \partial_y^2 x + \eta(y,t), \tag{16}$$

where $\eta(y,t)$ is a gaussian white noise with

$$\langle \eta(y,t) \rangle = 0, \qquad \langle \eta(y,t)\eta(y',t') \rangle = 2D\delta(y-y')\delta(t-t'). \tag{17}$$

The fluctuation-dissipation theorem demands,

$$D/\nu = 2k_B T/\Gamma, \tag{18}$$

where Γ is the line stiffness constant which equals to γ_L if the surface is isotropic. [3] The value of D can be approximately related to the time τ_0 it takes for a kink to move by one lattice constant a,

$$D \simeq a^3/\tau_0. \tag{19}$$

[3]The extra factor 2 comes from the fact that we are dealing with two fluctuating lines.

Equation (16) stops to be valid when x reaches zero, i.e., the two lines collide. Beyond this point, the line breaks up into a number of segments, each segment ending at a pair of pinching points. The velocity of the pinching points should be of the order of

$$v_{\text{tip}} \simeq (a/\tau_0)(\mu_L/k_BT) \simeq 2\nu/W, \tag{20}$$

a result which would have been guessed from (16) if surface anisotropy is negligible.

The density of pinching points at a given time t can be estimated as follows. Consider the transverse displacement $u_l(t) = x(l,t) - x(0,t)$ of two points on the line at $y = 0$ and $y = l$. Obviously, $\langle u_l(t)\rangle = 0$. The standard deviation of $u_l(t)$ is given by,

$$\sigma_l(t) = \langle u_l^2(t)\rangle = (D/\nu)\ l\ \Phi(\nu t/l^2), \tag{21}$$

where the "scaling function" $\Phi(x) \simeq (8x/\pi)^{1/2}$ for $x \ll 1$ and $\Phi(x) \simeq 1$ for $x \gg 1$. Solution of the Langevin equation (16) shows that σ_l reaches 80% of the saturated value $\sigma_l(\infty) = Dl/\nu$ at $t_l = l^2/\nu$ (see, e.g., Nattermann and Tang, 1992). At this time, fluctuations of wavelength much bigger than l have not reached equilibrium, so that the line can be considered as independently fluctuating segments, each of length l. Since (16) is linear, the distribution of $u_l(t)$ is gaussian for gaussian noise,

$$P_l(u) = (2\pi\sigma_l)^{-1/2}\exp[-u^2/(2\sigma_l)]. \tag{22}$$

The probability P_c that pinch-off has occurred in a given segment of length l is approximately the same as the probability that u_l has reached a value greater or equal to the width of the strip W. Hence we obtain the density of pinching points at time $t = t_l$,

$$\rho(t) \simeq l^{-1}\int_W^\infty du P_l(u) = \frac{1}{2}(\nu t)^{-1/2}\Big[1 - \operatorname{erf}\Big(\frac{W}{t^{1/4}}\frac{\nu^{1/4}}{(2D)^{1/2}}\Big)\Big], \tag{23}$$

which, for small t (i.e., $\sigma_l < W^2$), reduces to

$$\rho(t) \simeq \frac{1}{\sqrt{2\pi}}\Big(\frac{D}{\nu}\Big)^{1/2}\frac{1}{(\nu t)^{1/4}W}\exp\Big(-\frac{\nu^{1/2}W^2}{2Dt^{1/2}}\Big). \tag{24}$$

Due to the exponential factor, the density of pinching points increases very rapidly at small t, and switches over to a much slower increase as σ_l approaches W^2. The lifetime τ of the strip is given by the time t_1 it takes to reach a sufficient density $\rho(t_1)$ of pinching points (which is the same as the density of elongated islands), plus the time $t_2 = \rho^{-1}(t_1)/(2v_{\text{tip}})$ for the elongated islands to shrink to zero, where t_1 is chosen as to minimise the sum $t_1 + t_2$. Hence we have,

$$\tau = \min_t\Big[t + \frac{W}{4\nu}\rho^{-1}(t)\Big] = \frac{\nu W^4}{4D^2}\alpha\Big(\frac{\nu W}{D}\Big), \qquad \text{(evaporation-condensation)} \tag{25}$$

where, to the order of accuracy considered, $\alpha(w)$ is given by the solution of the equation

$$w = \frac{\pi^{1/2}}{2\alpha^{5/4}}\exp(\alpha^{-1/2}). \tag{26}$$

For large w, Eq. (26) yields

$$\alpha(w) \simeq \ln^{-2}(w/w_0), \tag{27}$$

where $w_0(w)$ is a slow function of x.

A few remarks on Eq. (25) are in order.

i) The "typical" crossing time of two free lines at a given y is given by a similar formula as Eq. (25), but with $\alpha = 1$. This result is in fact well-known and has been rederived recently by a number of groups (see, e.g., Pimpinelli et al, 1993, Williams, 1994, Selke and Duxbury, 1995). Such a lifetime for the top terrace (or the strip in our case) coincides with (10) obtained from the entropic step repulsion argument. In fact, it is natural to expect the two approaches to give similar results (apart from numerical prefactors) since in both cases the underlying driving force for the collision of the two top ledges is thermal fluctuations.

ii) The new ingredient of our estimate is the extra factor $\alpha(w)$. This comes about because we need only one pinching point per segment of length $l_{\text{island}} \simeq \tau v_{\text{tip}} \sim W^3$ to initiate the dissociation process. In comparison, the length of "correlated" segments are of the order of $l(\tau) \simeq (\nu\tau)^{1/2} \sim W^2$. Since l_{island} is much bigger than $l(\tau)$ for large W, the nucleation of pinching points invokes rare-event statistics, i.e., the tail of the gaussian distribution for u_l.

iii) In the original groove problem, repulsion from neighboring steps is likely to reduce the lifetime of the top terrace further. However, since (25) is determined by rare fluctuations of the two steps toward each other, it is plausible that repulsion from other steps does not play a significant role, though a more quantitative discussion on this point would be helpful. In my opinion, the main uncertainty in applying (25) to the groove problem comes from a not-so-sharply defined W. An operational definition of W from an instantaneous surface profile, averaged along the groove, will be given later. There is certainly room for a slightly different numerical factor in (25) due to this uncertainty.

iv) In typical experiments and simulations, the parameter $w = \nu W/D \simeq W\gamma_L/(2k_BT)$, which is approximately the ratio between W and the typical linear size of thermal islands $\xi = k_BT/\gamma_L$, can vary between 1 to 10^2–10^3. For this range of w, $\alpha(w)$ is actually a quite strong function of w. A log-log plot of τ against W can yield a range of exponents centered around 3 (see below). This may well be one of the origins of nonuniversality in numerical tests of scaling (Bonzel and Preuss, 1995).

Surface Diffusion Dynamics

The above analysis can in principle be applied to the surface diffusion case. There are, however, a few complications. Consider again the simpler geometry of one monolayer strip of width W, as illustrated in Fig. 1. Unlike the evaporation-condensation dynamics, the total number of atoms in the system is now conserved. The final equilibrium configuration should thus be a disc of monolayer thickness. At short times, however, one can expect the strip to first break into a series of elongated islands through step meandering. The density of pinching points becomes substantial when fluctuations of the distance between two lines at a given y become of order W. If terrace diffusion is the dominant mechanism for surface transport, the time it takes for this to happen is proportional to W^6.[4] On the other hand, if ledge diffusion is the dominant mechanism, a time constant proportional to W^8 is expected.

The presence of other steps in the groove problem can significantly alter the above results, in contrast to the evaporation-condensation case. If the step bounding the second layer is completely reflective, as in the case of infinite Schwoeble barrier, terrace diffusion is possible only for hops smaller than a distance of order W. A crossover

[4]This case belongs to bulk diffusion and hence $z = 3$, see Eq. (5). Note that L in Eq. (4) is replaced by W^2 here.

to the more slow ledge diffusion case is expected beyond a distance of order W along the step, and hence a time constant which scales as W^7. On the other hand, if the Schwoeble barrier is absent, then a crossover to a faster dynamics (9) is expected.

Thus, depending on the mode of transport which is operative on the length and time scales of interest, any value for the dynamic exponent z between 5 and 8 can be expected for the surface diffusion case. Smaller values of z are also conceivable if the rare-event dominated top terrace dissociation or a miscut enters the game. A detailed analysis, however, is beyond the scope of present article.

GROOVE PROFILE

Rettori and Villain (1988), and Lançon and Villain (1990) have written down equations of motion for the one-dimensional groove profile in both discrete and continuous forms. In the discrete form, the variables are the average positions $x_n(t)$ of step n in the step train leading from the top to bottom of the groove (or vice versa). In the continuous form, the surface profile is specified by a height function $h(x,t)$. The equation for $h(x,t)$ can be obtained from the equations for $x_n(t)$ by taking a suitable continuum limit.

In writing down these equations, it was assumed that the interaction energy between neighboring steps is given by the equilibrium form, u^{-2}, where u is the distance between two neighboring steps. This is a reasonable assumption even if the interaction is of purely thermodynamic origin. Local equilibrium is established when the typical size of step meandering is of the order of step-step separation, and collisions have been made with neighboring steps. The system is, however, somewhat off-equilibrium close to the top and bottom of the groove.

For evaporation-condensation dynamics, there is a simple way to convert the equation of motion (16) for a single step to the equation of motion for a vicinal surface at a slope $s \ll 1$, assuming steps have a contact repulsion. The idea is that, if we scale x by the terrace width a/s, y by the distance l over which the saturated value of σ_l reaches the terrace width squared, $(a/s)^2$, and t by the corresponding saturation time t_l, we should arrive at an isotropic, dimensionless equation, i.e.,

$$\frac{\partial h}{\partial(t/t_l)} = \frac{\partial^2 h}{\partial(sx/a)^2} + \frac{\partial^2 h}{\partial(y/l)^2} + \text{noise.} \tag{28}$$

Using the relations given previously, we obtain,

$$\frac{\partial h}{\partial t} = \tilde{\nu} s^2 \frac{\partial^2 h}{\partial x^2} + \nu \frac{\partial^2 h}{\partial y^2} + \text{noise,} \tag{29}$$

where $\tilde{\nu} = D^2/(\nu a^2) \simeq \nu(k_B T/a\gamma_L)^2$. [5]

Applying Eq. (29) to the groove problem, we obtain the equation of motion for the surface profile averaged over y,

$$\partial \bar{h}/\partial t = \tilde{\nu}|\partial_x \bar{h}|^2 \partial_x^2 \bar{h}. \qquad \text{(Lançon and Villain, 1990)} \tag{30}$$

This equation, however, is not very accurate for the top and bottom terrace. Taking $\delta x = W$ and $\Delta \bar{h} = a$, the life time of the top terrace is estimated to be $\tau \simeq W^4/(\tilde{\nu} a^2) = \nu W^4/D^2$ which agrees with (10) but not with the improved estimate (25).

[5]The expression for $\tilde{\nu}$ would be different if the step-step interaction is dominated by elasticity or other effects.

To fix the deficiency of (30), we may introduce a modification which is effective only close to the top and bottom part of the groove. One recipe has been suggested by Selke and Duxbury (1995). Here we introduce a different modification by adding an offset to the $(\partial_x h)^2$ term,

$$\partial\bar{h}/\partial t = \tilde{\nu}[s_0^2(t) + |\partial_x \bar{h}|^2]\partial_x^2 \bar{h}. \tag{31}$$

The value of $s_0(t)$ is determined by the condition that, at time t, a lifetime for the top terrace corresponding to (25) is obtained. This condition yields $\tau^{-1} = \tilde{\nu} s_0^2 (W/2)^{-2}$ or

$$s_0 = \alpha^{-1/2}\frac{a}{W} \simeq \frac{a}{W}\ln\frac{\nu W}{w_0 D}. \tag{32}$$

Here $W(t)$ is the width of the top terrace at time t, which can be extracted from the profile $\bar{h}(x,t)$ via the equation

$$\bar{h}(x_m \pm W/2, t) = \bar{h}(x_m, t) - a, \tag{33}$$

where x_m is the point where $\bar{h}(x,t)$ achieves its maximum. Thus Eqs. (32) and (33) offer a self-consistent way of determining $s_0(t)$. [6]

The s_0^2 term has the obvious effect of speeding up the relaxation process, and produces a more rounded profile at the top and bottom of a groove. The modification is significant only so far as $s_0 \geq |\partial_x h|$. It is clear from Eq. (32) that the effect extends only up to a few layers from the top and bottom of the groove. Nevertheless, this is sufficient to account for the discrepancy between the Lançon and Villain theory and simulation results. In relatively terms, the difference between (30) and (31) decreases as one increases the width of the groove, L. It can be shown that, without the logarithmic correction, the scaling invariance of Eq. (30), i.e.,

$$x \to Lx, \quad \bar{h} \to A\bar{h}, \quad t \to (L^4/A^2)t, \tag{34}$$

is preserved. This is however spoiled by the logarithmic correction.

The surface diffusion case is even more complicated due to a whole set of time scales involved. There is no difficulty in principle to analyse this case along similar lines as illustrated above, though one can easily be tangled up in all sorts of possibilities unless a particular set of parameters are given.

NUMERICAL COMPARISON

To test the validity of Eqs. (25) and (31), we have performed simulations of groove relaxation under evaporation-condensation dynamics and numerical integration of (31). Below we describe briefly the model simulated and present data for the averaged surface profile and the lifetime of the top terrace during relaxation. Details of this study will be reported elsewhere.

The simulation is performed on the (111) surface of a face-centered-cubic crystal. Figure 6 shows the top three layers of an ideal surface. We say an atom is a *surface atom* if there is no atom sitting right above it along the (111) direction. The number of surface atoms (or better to say, surface sites) is thus conserved. [7] We demand in

[6] The s_0^2 term in Eq. (31) can be motivated by taking (29) with $s = \partial_x h$ as the dynamical equation with noise. In general s_0 should also acquire a dependence on x, which we choose to ignore as the term is significant around the top and bottom of the groove anyway. It is not clear, however, whether the logarithmic correction to the lifetime of the top terrace would come out correctly in such an approach.

[7] According to this definition, atoms in the third layer shown in Fig. 6 are still considered to be surface atoms, although they are fully coordinated.

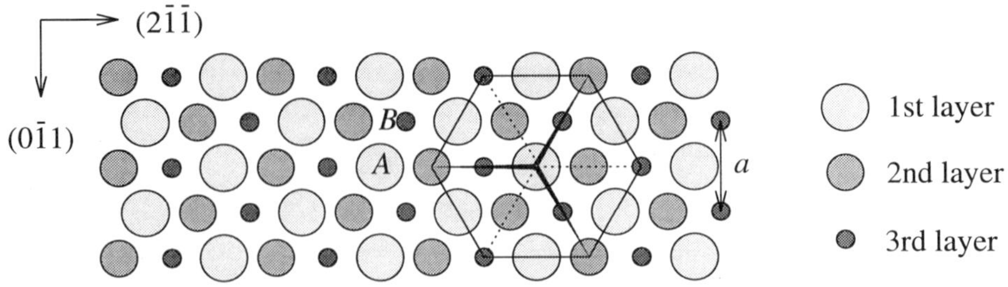

Figure 6: The perfect (111) surface of an fcc crystal. All atoms shown are considered as surface atoms. The surface sites form a triangular lattice of lattice constant $a/\sqrt{3}$, where a is the nearest neighbor distance of the fcc crystal. Atom at site A is allowed to evaporate, while site B is ready to host a condensation event. Also shown is the cubic cell of a sublattice which defines the parallel updating algorithm.

addition that the crystal beneath the surface has no vacancies. Hence each surface atom rests at the center of a triangle formed by three atoms in the layer below, and has a nearest neighbor coordination number of at least three. The nearest neighbor coordination number of an atom in the bulk is 12, three from atoms in the layer below, three from atoms in the layer above, and six from atoms in the same layer. A surface atom which can be removed should not have any nearest neighbors in the layer above, so its coordination number is at most nine.

We assign an excess energy to the surface as

$$E_s = (\epsilon/2)N_s, \tag{35}$$

where N_s is the total number of *missing* coordinations of surface atoms, and ϵ is the energy of a nearest neighbor bond. When a surface atom is removed under the conditions specified above, the change in the surface excess energy is given by

$$\Delta E_s^{(-)} = E_{s,\text{new}} - E_{s,\text{old}} = (m-3)\epsilon \tag{36}$$

where m $(0 \le m \le 6)$ is the number of nearest neighbors the atom has in the same layer. Thus if more than half of the nearest neighbors in the same layer are present, evaporation increases surface energy. Otherwise it decreases the surface energy. A positive ϵ thus favors a flat surface.

Our simulations were performed using the activated dynamics. The probability for removing an atom at a given time step follows the Arrhenius law,

$$p_{\text{evap}} = p_0 b^m \tag{37}$$

where $b = \exp(-\epsilon/2k_BT)$ and $m(0 \le m \le 6)$ is the number of nearest neighbors in the same layer for the atom in question. Similarly, the probability for adding an atom (at a site where such move is allowed) at a given time step is given by,

$$p_{\text{cond}} = p_0 b^{(6-m)}. \tag{38}$$

Obviously, the detailed balance condition

$$p_{\text{evap}}/p_{\text{cond}} = b^{2m-6} = \exp(-\Delta E_s^{(-)}/k_BT) \tag{39}$$

is satisfied.

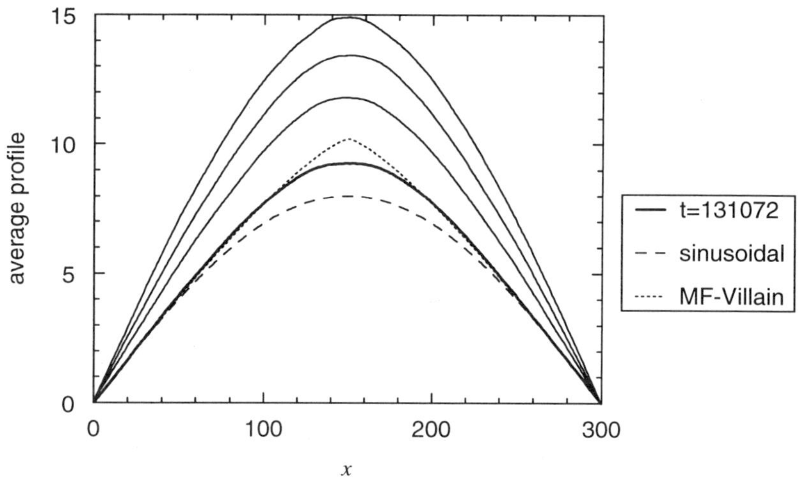

Figure 7: Averaged groove profiles (for half a period) during relaxation at $T = 0.68T_R$. The length of the system along the groove is $M = 19200a/\sqrt{3}$. Also shown are the sine function (dashed line) and prediction of the Lançon-Villain theory (dotted line).

An fcc lattice can be decomposed into four interpenetrating simple cubic lattices. This decomposition is used in our parallel updating scheme, where surface sites on each simple cubic lattice serves as a subset (or sublattice) of atoms for simultaneous updating at a given time step. [8] We have taken advantage of a mapping which enables one to represent the surface configuration by an Ising configuration on a triangular lattice, as used previously for growth simulations (Forrest and Tang, 1990). Multi-spin coding techniques can then be used to save computer memory as well as to speed up the simulation.

By comparing the numerically calculated height-height correlation function to the expected universal behavior at the roughening transition, we have estimated the roughening temperature of the model to be given approximately by

$$T_R \simeq (2\ln 1.6)^{-1}\epsilon/k_B = 1.06\epsilon/k_B. \tag{40}$$

The simulation results reported below are for $b = 1/2$ or $T \simeq 0.68T_R$. Well-defined steps and terraces are indeed seen during the relaxation of the groove, indicating that the surface is well below its roughening temperature.

The initial surface is prepared to be an almost perfect sinusoidal groove with straight ledges running along the $(2\bar{1}\bar{1})$ direction, defined as the y-direction. The modulation is thus along the $(0\bar{1}1)$ direction, defined as the x-direction. Height is measured in units of atomic layers, and x in units of $a/2$ (distance between neighboring atomic planes), where a is the nearest-neighbor distance of the fcc crystal. The system size is typically chosen to fit in one full wavelength. The largest groove simulated has a wavelength $L = 300a$, and the length of the groove $M = 19200a/\sqrt{3}$. [9]

Figure 7 shows a sequence of surface profiles during relaxation for the largest groove simulated. The initial top to bottom distance of the groove is 30 atomic layers. Each solid line represents the surface height averaged along the groove at a given time.

[8] Note that atoms on a given simple cubic lattice are not nearest neighbor atoms of the fcc lattice, hence their updating can be done simultaneously for the energy function chosen.

[9] To test the validity of the step-meandering picture, it is necessary to choose M to be bigger than W^3, where W is the width of the top terrace.

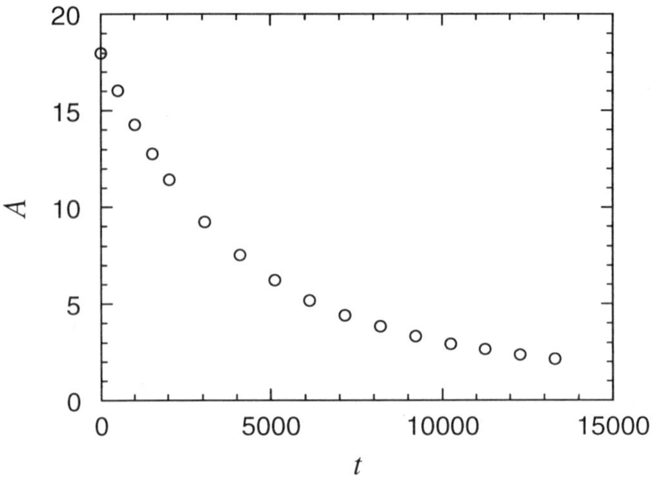

Figure 8: Amplitude of the groove profile against t for $L = 90a$. The lifetime of the top terrace τ can be determined from the local slope of the curve through interpolation.

(Due to symmetry only half of the period is shown.) The curve indicated by the thick solid line is the profile after $t = 131072$ updates per surface site. Also shown are the sine function and the predicted profile from the Lançon-Villain theory, chosen in such a way that the slopes at the steepest part of the profile agree with one another. It is seen that the Lançon-Villain curve agrees quite well with the measured profile apart from the top one or two layers, where the latter has more rounded shape. The rounding tendency at the top of the groove is more pronounced (in relative terms) for smaller systems, in agreement with earlier observations made by Selke and Duxbury (1995).

Figure 8 shows the amplitude A (half the distance between top and bottom terrace, in units of layer spacing) of the surface profile as a function of t for a groove of wavelength $L = 90a$. The decay is slower than a simple exponential as in Mullins' theory. From the function $A(t)$ one can estimate the lifetime of the top terrace $\tau(t)$ using the formula,

$$\tau(t)^{-1} = dA/dt. \tag{41}$$

The width of the top terrace $W(t)$ can be obtained from the surface profile using Eq. (33). This procedure defines a function $\tau(W)$ which is plotted in Figure 9 for many different groove wavelengths and initial amplitudes. All data fall nicely onto a single curve, indicating that the lifetime of the top terrace is mostly controlled by the width of the top terrace. The rest of the surface has very little effect on τ. For the range of W values considered, which is slightly less than one decade, the data can be fitted to a power law $\tau \sim W^3$ (dashed line in Fig. 9). It can also be fitted to Eq. (25), i.e., $\tau \sim W^4/\ln^2(W/W_0)$, where we have taken $W_0 = 6$ (solid line in Fig. 9).

We have also compared the surface profiles from the simulation with results from a numerical integration of Eq. (31). Figure 10 shows a sequence of profiles obtained by taking $s_0 = W^{-1}\ln(W/6)$. The dashed line in Fig. 10 is a reproduction of the thick line in Fig. 7. The agreement between the two profiles is nearly perfect! The procedure also fixes the time unit $\tilde{\nu}^{-1}$ in the continuum equation. Profiles at earlier times also agree reasonably well. There is however one major difference: the simulation took a week to complete on a Sun SPARC-20 workstation, while the numerical integration finished in the twinkling of an eye.

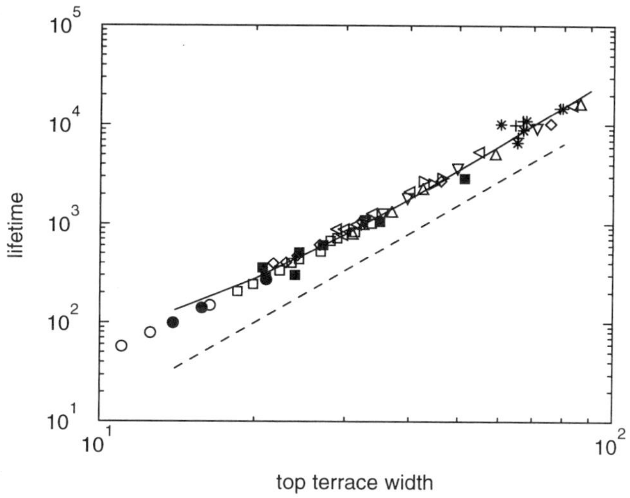

Figure 9: Lifetime of the top terrace versus terrace width for grooves of different width and initial amplitude. The dashed line indicates a power law with an exponent 3. The solid line is a fit to Eq. (25).

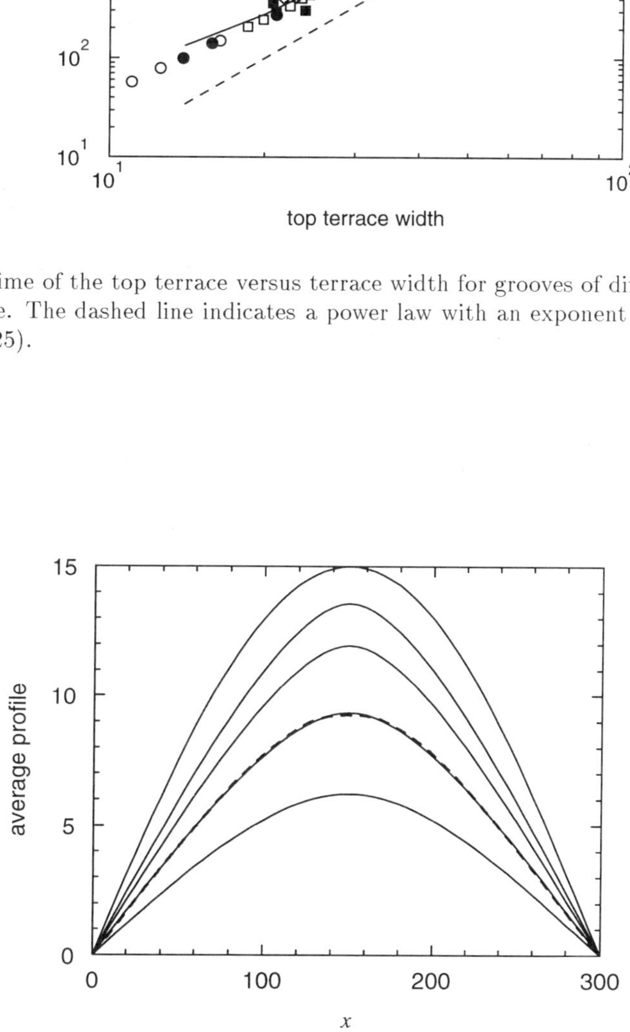

Figure 10: Groove profiles obtained from numerical integration of Eq. (31). The dashed line is the data from the simulation. Times are proportionally scaled to make the profiles to correspond to those shown in Fig. 7, except the topmost one at $t = 0$ and the bottom-most one at twice the time of the one above.

CONCLUDING REMARKS

We have shown in this article that it is possible to derive a continuum equation for the relaxation of one-dimensional grooves on a high-symmetry surface below T_R for the relatively simple case of evaporation-condensation dynamics, starting from the dynamics of meandering steps. The predictions of the continuum equation are in very good agreement with simulations of an idealised surface model. In this sense, a seeming contradiction between previous simulation results and the continuum theory of Lançon and Villain has been resolved. In addition, our analysis suggests that, for not so large wavelengths, the effective dynamic exponent z characterising the dependence of the groove lifetime on the groove width L can assume a broad range of values, especially for surface diffusion dynamics. Presence of a slight miscut in most experimental setups presents additional complications to the analysis of the relaxational behavior. In my view, groove relaxation is one problem where the concept of universal scaling is not to be taken beyond its face value. Nevertheless, the method proposed in this paper for determining the lifetime of the top terrace as a function of terrace width can yield valuable information on surface mass transport for systems of experimental interest.

The continuum equation (31) is somewhat unusual in that it is not completely local. This is because we are interested in grooves of finite wavelength L and a finite number of layers $2A$. In the limit $L, A \to \infty$ but keeping A/L finite, one does recover the Lançon-Villain equation. The equation is however singular at the top and bottom parts of the groove, and some procedure is needed to define more precisely how to treat these singular points. Thus we are back to square one! Lançon and Villain avoided addressing this problem by letting the problem addressing itself. As we have seen in our analysis, the most natural way out turns out not to be the most accurate way out due to the logarithmic correction arising from rare-event statistics.

Acknowledgements: It is my great pleasure to thank editors of this volume for organising a very enjoyable conference at the beautiful lake city of Michigan. Thanks are also due to the Institut für Theoretische Physik, Universität zu Köln, Germany, where part of the work was carried out.

REFERENCES

Bonzel, H.P., Breuer, U., Voigtländer, B., and Zeldov, E., 1992, *Surf. Sci.* 272:10.

Bonzel, H.P., and Preuss, E., 1995, *Surf. Sci.* 336:209.

Forrest, B.M., and Tang, L.-H., 1990, *Phys. Rev. Lett.* 64:1405.

Lançon, F., and Villain, J., 1990, in: *Kinetics of Ordering and Growth at Surfaces*, M. G. Lagally, ed., Plenum, New York.

Mullins, W.W., 1963, in: *Metal Surfaces: Structure, Energetics, and Kinetics*, N.A. Gjostein and W.D. Robertson, ed., Am. Soc. Metals, Metals Park, OH.

Nattermann, T., and Tang, L.-H., 1992, *Phys. Rev. A* 45:7156.

Ozdemir, M., and Zangwill, A., 1990, *Phys. Rev. B* 42:5013.

Pimpinelli, A., Villain, J., Wolf, D.E., Metois, J.J., Heyraud, J.C., Elkinani, I., Uimin, G., 1993, *Surf. Sci.* 295:143.

Rettori, A., and Villain, J., 1988, *J. Physique* 49:257.

Selke, W., and Duxbury, P.M., 1995, *Phys. Rev. B* 52:17468.

Spohn, H., 1993, *J. Phys. I (France)*, 3:69.

Williams, E.D., 1994, *Surf. Sci.* 299/300:502.

NUCLEATION AND GROWTH IN ELASTODYNAMICS

Lev Truskinovsky

Department of Aerospace Engineering and Mechanics
University of Minnesota
Minneapolis, MN 55455

INTRODUCTION

Structures made of transforming materials exhibit a striking capacity to hysteretically recover significant deformation with a controllable amount of energy absorbed in the process. The unusual properties of these materials are due to the fact that large deformations and inelastic behavior are accomplished by coordinated migration of mobile phase or domain boundaries. Intensive research in recent years has led to well-defined static continuum theories for some of the transforming materials (see Pitteri and Zanzotto (1997) for a recent review). Within the context of these theories, the main unresolved issues include history and rate sensitivity in the constitutive structure.

In this overview we focus on the elastodynamical aspects of the transformation and intentionally exclude phase changes controlled by diffusion of heat or constituent. To emphasize ideas we use a one dimensional model which reduces to a nonlinear wave equation. Following Ericksen (1975) and James (1980), we interpret the behavior of transforming material as associated with the nonconvexity of elastic energy and demonstrate that a simplest initial value problem for the wave equation with a non-monotone stress-strain relation exhibits massive failure of uniqueness associated with the phenomena of nucleation and growth.

The multiplicity of solutions at the continuum level can be viewed as arising from a constitutive deficiency in the theory, reflecting the need to specify additional pieces of constitutive information through some kind of phenomenological modeling (see, for instance, Truskinovsky, 1987; Abeyaratne and Knowles, 1991). Here we take a different point of view and interpret the nonuniqueness as an indicator of essential interaction between macro and micro scales.

We recall that our wave equation represents a long wave approximation to the behavior of a structured media (atomic lattice, periodically layered composite, bar of finite thickness), and does not contain information about the processes at small scales which are effectively homogenized out. When the model at the microlevel is nonlinear, one expects essential interaction between different scales which in turn complicates any universal homogenization procedure. In this case, the macro model is often formulated on the basis of some phenomenological constitutive hypotheses; nonlinear elasticity with nonconvex energy is a theory of this type.

Dynamics of Crystal Surfaces and Interfaces
Edited by Duxbury and Pence, Plenum Press, New York, 1997

The well known phenomenon of a finite time blow up in nonlinear elastodynamics is a sign that the phenomenological equations are at least incomplete. In some cases (Dafermos, 1979), the detailed microlevel behavior turns out to be irrelevant and the closure can be achieved by prescribing a single inequality. This means, for instance, that the fine structure of a shock discontinuity does not affect the dynamics and that the localized perturbations in initial data die out instantly. The situation is more complex in the case of material with a generic nonconvex energy where in order to obtain a unique solution at the continuum (or macro) level, one must *"de-homogenize"* the model and introduce additional physical hypotheses about the behavior at the sub-continuum (micro) scale. It is important to remember that the physical picture at micro and macro levels can be quite different. For example, analysis of physically motivated discrete models show, that phase boundary motion at the micro scale requires overcoming a barrier which, as we show, is formally absent in a continuum picture (Slepyan and Trojankina, 1984); the same is true for the nucleation which is barrier-free in the classical elastodynamical setting. The regularization can also be achieved in numerical calculations because of the dissipation and dispersion which formal discretization brings into the model.

For the purpose of illustration, in this paper we use a viscosity-capillarity model (Truskinovsky, 1982; Slemrod, 1983) as an artificial "micromodel", and investigate how the information about the behavior of solutions at the microscale can be used to narrow the nonuniqueness at the macroscale. The viscosity-capillarity model contains a parameter $\sqrt{\varepsilon}$ with a scale of length, and the nonlinear wave equation is viewed as a limit of this "micromodel" obtained when this parameter tends to zero. As we show, the localized perturbations of the form $\phi(x/\sqrt{\varepsilon})$ can influence the choice of attractor; for this type of perturbation, support (but not amplitude) vanishes as the small parameter goes to zero. Another manifestation of this effect is the essential dependence of the limiting solution on the contributions of the type $\varphi((x-Dt)/\sqrt{\varepsilon})$, describing the structure of the jump discontinuity.

Since in this problem not only the limit but also the character of convergence matters we conclude that consistent homogenization of the micromodel should lead to a description in a broader functional space than is currently accepted. One interesting observation is that the concave part of the energy is relevant only in the region with zero measure where the singular, *measure valued* contribution to the solution is nontrivial (different from point mass). We remark that the situation is similar in fracture mechanics where a problem of closure at the continuum level can be addressed through the analysis of a discrete lattice (e.g. Truskinovsky, 1996).

METASTABILITY

Following Ericksen (1975), consider an elastic bar which occupies a segment $[0,1]$ in the reference state. Let $u(x)$ be a displacement field so that $x+u(x)$ is the deformed position of a material particle with the reference coordinate x. The stored elastic energy of the bar has a density $f(w)$ where $w=u_x$ is the longitudinal strain. We assume that $f(w)$ is not convex, in particular, $f''(w)>0$ for $w<\alpha$ (phase 1) and $w>\beta$ (phase 2) and $f''(w)<0$ for $\alpha<w<\beta$ (spinodal region) (see Fig.1a). The corresponding stress-strain relation $\sigma=f'(w)$ is nonmonotone (see Fig.1b), and one can formally define equilibrium stress σ_M and equilibrium strains a and b in such a way that: $\sigma(a)=\sigma(b)=\sigma_M$ and $\int_a^b \sigma dw=\sigma_M(b-a)$ (Maxwell construction). Two regions $a<w<\alpha$ (in phase 1) and $\beta<w<b$ (in phase 2) are called *metastable*.

The concept of metastability in this elementary setting has exactly the same meaning as in the closely related case of van der Waals's fluid. Consider the simplest equilibrium treatment of a bar loaded by a constant stress σ_0. The static problem reduces to the minimization of the functional

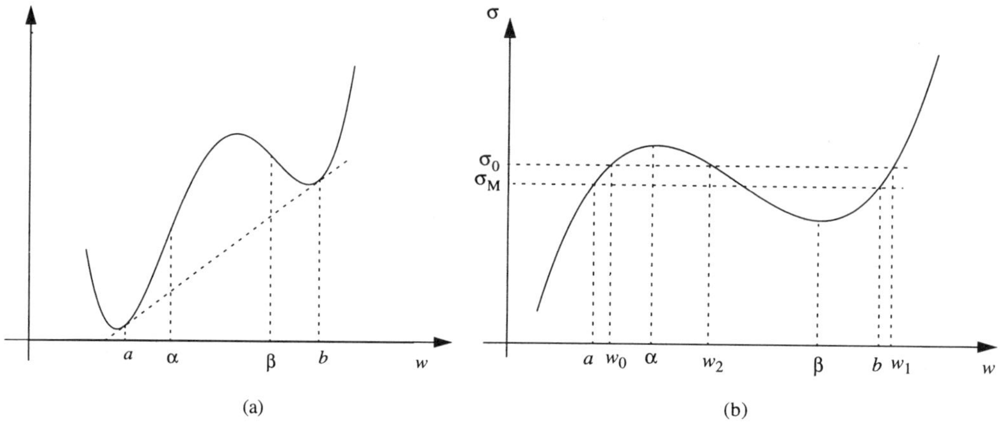

Figure 1. The free energy (a) and the corresponding stress-strain relation (b) for the elastic material supporting two phases.

$$G\{u\} = \int_0^1 g(w;\sigma_0)dx \qquad (2.1)$$

where $g(w;\sigma_0) = f(w) - \sigma_0 w$ is the potential (Gibbs) energy. If we choose the value of stress from a metastable region, say $\sigma_0 \in (\sigma_M, \sigma(\alpha))$ as in Fig.1b, then the function $g(w;\sigma_0)$ has three extrema: two minima (at w_0 and w_1) and a maximum (at w_2) (see Fig.2).

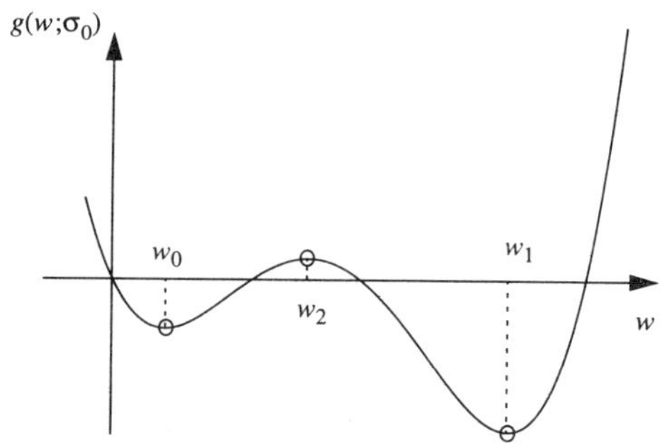

Figure 2. Potential (Gibbs) energy of the bar loaded by the constant stress $\sigma_0 \in (\sigma_M, \sigma(\alpha))$.

One can show (Ericksen, 1975) that the homogenous state $w(x) \equiv w_1$ corresponding to a deeper minimum of $g(w;\sigma_0)$ is the global minimizer of the functional (2.1); another homogenous configuration $w(x) \equiv w_0$ (the metastable state), is only a weak local minimizer which is not even isolated in the strong sense (in $W^{1,p}$). In other words, it is unstable in the class of piecewise smooth competitors: the “dangerous” perturbation is a "Weierstrass needle" a nucleus of state w_1 with an infinitesimal support. In the fully 3D case the situation is essentially similar, the only difference being that the metastable region begins with the failure of quasiconvexity rather than convexity, which is due to the nontrivial constraint of strain compatibility (see, for instance, Ball and James, 1996) .

The absolute instability of the "metastable" states in the framework of classical elasticity manifests itself in dynamics as well. The associated elastodynamical problem reduces to a solution of the nonlinear wave equation $u_{tt} = \sigma'(u_x)u_{xx}$. It is convenient to rewrite it as a mixed type first order system

$$v_t = \sigma(w)_x \, , \; w_t = v_x, \tag{2.2}$$

where we introduced particle velocity $v = u_t$. The elastodynamic problem with initial conditions $w(x,0) = w_0$, $v(x,0) = 0$, corresponding to the metastable state, has a trivial solution $w(x,t) \equiv w_0$, $v(x,t) \equiv 0$. To show that this solution is not unique, choose an arbitrary point $x = x_0$ inside the segment $[0,1]$ and prescribe the same initial data everywhere except for this point. Then, we obtain a (degenerate) Riemann problem with piecewise constant initial data and, at least locally, one expects to find a self-similar solution of the type $w = w(\varsigma)$, $v = v(\varsigma)$, where $\varsigma = (x - x_0)/t$. The elastic field in this case must be a combination of constant states separated by jump discontinuities and/or centered Riemann waves. Classical conservation laws must be satisfied on the discontinuities which leads to the following Rankine Hugoniot jump conditions

$$D[v] + [\sigma] = 0, \; D[w] + [v] = 0. \tag{2.3}$$

Here as usual $[A] = A_+ - A_-$ and D is the discontinuity speed. The entropy inequality yields

$$D([f] - \{\sigma\}[w]) \geq 0 \tag{2.4}$$

where $\{A\} = ½(A_+ + A_-)$ the average value. As was first shown by James (1980), the nontrivial solution satisfying (2.2 , 2.3) has the following form (see Fig.3)

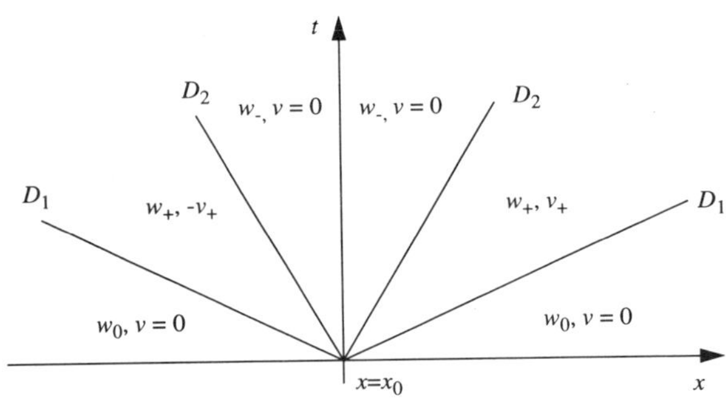

Figure 3. Nontrivial solution of a degenerate Riemann problem with initial data in the metastable area.

$$
w(x,t),v(x,t)=\begin{cases} w_-,0 \quad , & |x-x_0|<D_2t \\ w_+,\pm v_+ & D_2t<|x-x_0|<D_1t \\ w_0,0 \quad , & D_1t<|x-x_0| \end{cases} \qquad (2.5)
$$

Solution (2.5) (see Fig. 4) describes the nucleation of a phase 2 ($w=w_-$) which is accompanied by a generation of shock wave precursors in phase 1 ($w=w_0$) and is satisfactory only until the first shock wave reaches the boundary of the segment $[0,1]$. The entropy inequality (2.4) is automatically satisfied for the precursors (moving with the speed D_1) and is satisfied for the phase boundaries (moving with the speed D_2) if the area **A1** in Fig. 4 is smaller than the area **A2** (phase boundary is dissipation free if **A1= A2**).

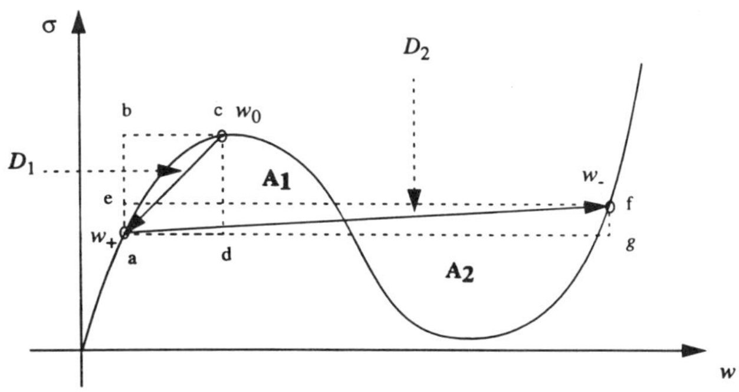

Figure 4. Nontrivial solution of a degenerate Riemann problem with initial data in the metastable area.

The only restriction imposed by the Rankine-Hugoniot conditions (2.3) is that the areas of the rectangles *abcd* and *aefg* in Fig. 4 are equal. This shows that, the information contained in (2.3, 2.4) is not sufficient to find all the unknowns, since the balance equations impose only four restrictions on the five constants w_+, w_-, v_+, D_1, D_2. Moreover, one obtains a two-parameter family of solutions since the nucleation point x_0 is also arbitrary. The two sources of nonuniqueness in this problem, and the necessity to make additional assumptions, have long been recognized by physicists who traditionally distinguished between the theories of *growth* and *nucleation*.

GROWTH

The only way to determine all five constants w_+, w_-, v_+, D_1, D_2 in the above problem is to supply an additional jump condition. This jump condition cannot be universal since, if applied at both discontinuities, it leads to an overdetermined system. We must therefore differentiate between the waves moving with the speeds D_1 and D_2. Notice that only the first one (with the speed D_1) satisfies the Lax condition (Lax, 1957) which means in this context that the wave is subsonic with respect to the state behind and supersonic with respect to the state ahead. We call this discontinuity a "shock". The second discontinuity (moving with the speed D_2) is subsonic with respect to the states both in front and behind; we refer to it as "kink" (Truskinovsky, 1993a). An elementary analysis of the configuration of the characteristic directions shows that it is the subsonic discontinuity (kink) which requires an additional jump condition (see Kulikovsky (1976) for general background on "non-evolutionary" or "undercompressive" shocks).

There is a long history of phenomenological modeling of kinks - phase or domain boundaries - in physics (normal growth hypothesis in crystal growth, Hertz - Knudsen theory of condensation, etc.); in the framework of elasticity theory the phenomenological "kinetic relations" were recently reviewed by Gurtin (1993) and Lin and Pence (1996). Here instead of postulating the missing jump condition we shall focus on its *derivation* from a model for a fine structure of the interface. As shown in Truskinovsky (1993a) this method naturally distinguishes between shocks and kinks.

In order to describe the internal structure of a moving discontinuity, the classical balance equations are supplemented with an additional physical hypothesis regarding the material behavior in the transition region. The principal difference between shocks and kinks in terms of the relevant physical mechanisms can be understood as follows. Consider a generic discontinuity propagating with constant velocity $D > 0$ which transforms the state w_+ into the state w_-. Suppose that equations (2.2) and Rankine-Hugoniot conditions (2.3) are satisfied, in particular $D^2 = (\sigma_+ - \sigma_-)/(w_+ - w_-)$. Then the total energy release rate associated with the moving discontinuity can be written in the form Gdm, where $dm = Ddt$. One can show that $G(w_+,w_-) = f(w_+) - f(w_-) - \frac{1}{2}(\sigma_+ + \sigma_-)(w_+ - w_-)$, where G is a driving (or configurational) force (Knowles, 1979; Truskinovsky, 1987). To calculate the hypothetical "microscopic" variation of the rate of dissipation inside the discontinuity, suppose that the balance of mass and linear momentum is satisfied for every intermediate state between w_+ and w_-, which means that $D(w - w_+) + (v - v_+) = 0$ and $D(v - v_+) + (\sigma - \sigma_+) = 0$. Then the rate of dissipation $R = D\left(f(w_+) - f(w) + \frac{1}{2}\left(v_+^{\,2} - v^2\right)\right) + (\sigma_+ v_+ - \sigma v)$ can be calculated explicitly as a function of w. Introduce $\psi_{w_+}(w) = -R/D$, which may be viewed as a dynamic analog of the potential energy $g(w)$ from (2.1). The straightforward calculation gives

$$\psi_{w_+}(w) = f(w) - f(w_+) - (w - w_+)\left(\sigma_+ + \frac{1}{2}\frac{\sigma_+ - \sigma_-}{w_+ - w_-}(w - w_+)\right), \qquad (3.1)$$

in particular, $\psi_{w_+}(w_-) = -G(w_+, w_-)$.

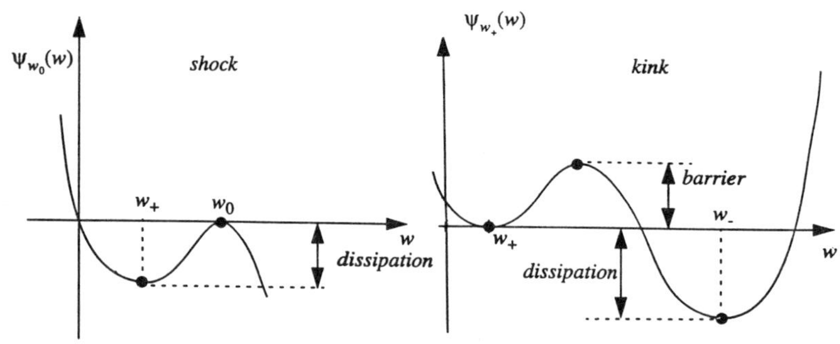

Figure 5. Dynamic driving force in the case of shocks and kinks.

Let us fix the state ahead of the discontinuity w_* and compare the behavior of the function $\psi_{w_*}(w)$ for shocks and kinks involved in solution (2.5). Two important conclusions can be drawn from the analysis of the graphs of $\psi_{w_0}(w)$ and $\psi_{w_+}(w)$ sketched in Fig.5. First, if $G(w_+, w_-) > 0$ (the opposite inequality is prohibited by the second law of thermodynamics) then one has to consider dissipation in the interphase region: for shocks introduction of the dissipation is usually sufficient to describe the structure of the discontinuity. Second, in case of kinks it is necessary to introduce an additional mechanism for crossing the *barrier*. The "no barrier" condition is implicit in the Oleinik (1959) criterion for shocks and the exclusively dissipative regularizations lead to the situation when kinks cannot move (Pego, 1987). On the other hand, in the absence of dissipation, the propagation of a kink can be viewed as an autocatalytic process which does not require extra energy after it is initiated. A process of this type is feasible in principle in dispersive media if group velocity for at least some wave lengths is greater than the speed of the kink. It is also clear that, contrary to the case of shocks, the velocity of the kink must be special to make this "tunneling" process possible.

One of the most interesting micromodels for the nonlinear wave equation is a discrete chain of atoms connected by nonlinear springs; such sytems automatically exhibit "macroscopic" dissipation due to energy transfer between long and short waves. In discrete models, shocks are usually de-localized and classical discontinous waves represent weak but not strong limits of highly oscillatory solutions (Lax et al., 1992)). Autocatalytic barrier crossing in bi-stable chains is also possible (Slepyan and Trojankina, 1984), however, here for simplicity we limit our consideration to mesoscopic continuum models exhibiting dispersion and dissipation.

There are several ways that dispersion (and the corresponding length scale) can be brought at the phenomenological level into the conservative part of the model. The two most well known examples of such theories are: gradient (or van der Waals) models with energy

$$f(u_x) + \varepsilon/2\, u_{xx}{}^2$$

(e.g. Carr et al., 1984), and strongly nonlocal models with energy

$$f(u_x) + \int K(x-y) u_y(y) dy$$

(e.g. Fosdick and Mason, 1996). The nonlocal model can sometimes be reformulated as a local theory with an additional order parameter; an Ericksen-Timoshenko bar with energy

$$f(u_x) + \delta/2 (u_x - \phi)^2 + \varepsilon/2\, \phi_x{}^2$$

is one example (e.g. Rogers and Truskinovsky, 1996). More general phase field models with an energy of the type $f(u_x, \phi, \phi_x)$ have also been considered (e.g. Roshin and Truskinovsky, 1989; Fried and Gurtin, 1994)).

The simplest example of a theory which incorporates both dispersion and dissipation is the so called viscosity-capillarity model (Truskinovsky, 1982, Slemrod, 1983). It combines van der Waals correction to the energy with Kelvin viscoelasticity, which in the present context amounts to the following additional constitutive assumption

$$\eta w_t = \sigma - \sigma(w, w_x).$$

Here $\sigma(w, w_x)$ is the equilibrium stress and η is the effective viscosity coefficient prescribing a rate of interphase kinetics. One can also consider other phenomenological dissipation models like Maxwell viscoelasticity (e.g. Mihailescu-Suliciu and Suliciu, 1992), or internal order parameter relaxation (e.g. Truskinovsky, 1988). As is well known, the van der Waals model cannot be considered as a reasonable long wave description for the simple atomic lattice because of the "wrong" sign to the gradient term (e.g. Kunin, 1982). However, since the group velocity is always larger than phase velocity, the dispersion is of the "right" type for the description of subsonic kinks.

With the introduction of the two new small parameters ε and η, the regularized wave equation takes the form

$$u_{tt} = \sigma'(u_x)u_{xx} + \eta u_{xxt} - \varepsilon u_{xxxx}. \tag{3.2}$$

Consider traveling wave solutions $u(x,t) = u(\xi)$, where $\xi = (x - Dt)/\sqrt{\varepsilon}$. The corresponding boundary value problem in the infinite domain takes the form

$$\left(\sigma(w) - D^2 w - \mathbf{W}Dw' - w''\right)' = 0; \qquad w'(\pm\infty) = 0,\ w(\pm\infty) = w_{\pm}; \tag{3.3}$$

where $\mathbf{W} = \eta/\sqrt{\varepsilon}$. For the given state in front of the discontinuity, the set (spectrum) of admissible velocities D consists of two parts: *continuous*, corresponding to shocks (saddle-node (focus) trajectories) and *discrete*, corresponding to kinks (saddle-saddle trajectories). To be specific take $\sigma(w) = w(w-1)(w-\tfrac{1}{2})$. Then an additional condition selecting kinks can be written explicitly (Truskinovsky, 1987, 1994)

$$3\left(1 - \frac{12}{\mathbf{W}^2}\right)(w_+ + w_- - 1)^2 + (w_- - w_+)^2 = 1 \tag{3.4}$$

We remark that the continuum spectrum does not contain all "supersonic" Lax discontinuities (Shearer and Yang, 1993; Truskinovsky, 1993b). The generic picture of the admissibility domain for both shocks and kinks is presented in Fig.6.

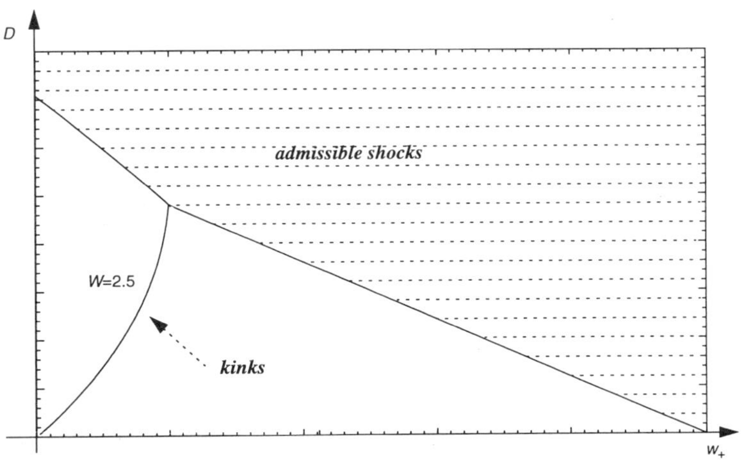

Figure 6. Set of jump discontinuities (kinks and shocks) compatible with the isothermal viscosity-capillarity model; **W**=2.5.

Some interesting aspects of the interface kinetics appear only when temperature and latent heat are included into the model. If the process of heat conductivity is governed by a classical Fourier law, the entropy balance equation takes the form $Ts_t = \kappa T_{xx} + \eta w_t^2$ where $s = -\partial f/\partial T$. Suppose for simplicity that equilibrium stress is cubic in strain and linear in temperature and assume that specific heat at fixed strain is constant. Then in nondimensional variables the system of equations takes the form (see Ngan and Truskinovsky, 1996a)

$$
\begin{aligned}
u_{tt} &= (\sigma(u_x,T))_x + \mathbf{W}_1 u_{xxt} - u_{xxxx} \\
Ts_t &= \mathbf{W}_2 T_{xx} + \mathbf{W}_1 u_{xt}^{\,2} \\
\sigma = 1 + \mathbf{W}_3 T + u_x(u_x - 1)(u_x - \tfrac{1}{2}), &\quad s = -\mathbf{W}_3 u_x + \ln T
\end{aligned}
\tag{3.5}
$$

where we introduced the following dimensionless numbers: $\mathbf{W}_1 = \eta / \sqrt{\varepsilon}$ - the ratio of viscosity to nonlocality, $\mathbf{W}_2 = \kappa / \sqrt{\varepsilon}$ - the ratio of heat conductivity to nonlocality and $\mathbf{W}_3$ - the measure the latent heat. A step-type traveling wave in this model describes an adiabatic shock or kink; the behavior of the corresponding heteroclinic trajectory in 3D phase space is similar to that for the 2D phase space of the isothermal system. Fig. 7 illustrates a numerical example of how the appropriate driving force

$$
G = [f] - \{\sigma\}[w] + [T]\{s\}
$$

(see Abeyaratne and Knowles, 1995) is then related to the speed of a kink D (i.e. a kinetic relation).

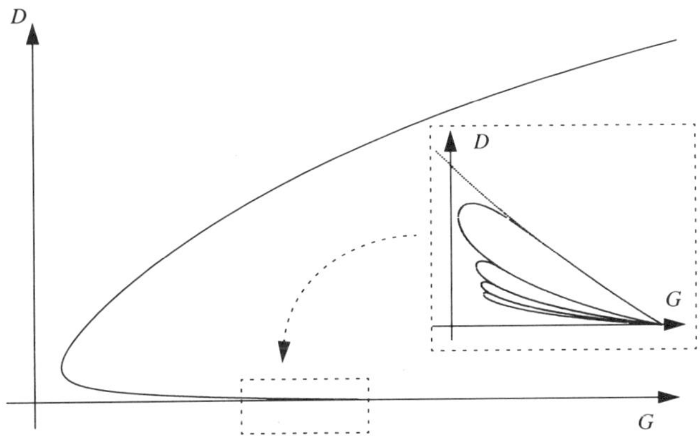

Figure 7. Multivalued kinetic relations for adiabatic kinks; $\mathbf{W}_1$=1,$\mathbf{W}_2$=0.025,$\mathbf{W}_3$=0.03.

Two important effects distinguish adiabatic kinks from isothermal ones. First, the kinetic curve does not originate from the point $G = 0$ due to the negative feedback provided by the latent heat (cf. Patashinskii and Chertkov, 1990 and Turteltaub, 1997). The second effect is the multivaluedness of the kinetic relation at small speeds, which has also been found in lattice models of fracture (Marder, 1995; Slepyan, 1996). Although most of the slow regimes are probably unstable, the general nonmonotonicity of the curve $G(D)$, can give rise to an interesting stick-slip behavior (Rosakis and Knowles, 1997). We also remark that the viscosity-capillarity model in the nonisothermal setting does not provide a kinetic relation of the form $G = G(D; \mathbf{W}_1, \mathbf{W}_2, \mathbf{W}_3)$ because of an implicit dependence on one additional parameter prescribed at infinity; this observation casts some doubts concerning the existence of the simple universal constitutive relations in a force-flux form for the configurational forces.

We conclude that the growth of a new phase is controlled by the rate of dissipation at a moving kink. This dissipation is taking place at the microlevel and must be prescribed in order for the macro-description to be complete. The incompleteness of the continuum model manifests itself through the sensitivity of the solution to the singular (measure-valued) contributions describing fine structure of the subsonic jump discontinuities (kinks).

NUCLEATION

We now turn to consider the other source of nonuniqueness in the breakdown of a metastable state, namely the ambiguity of the nucleation point(s) and the necessity for each point to select between the two solutions - trivial and nontrivial. The degenerate Riemann problem considered here is, of course, only the most elementary example where such a problem arises. For instance, similar nonuniqueness may be responsible for the instability of the moving phase boundary (Truskinovsky, 1993b). The phenomenological nucleation criterion suggested by Abeyaratne and Knowles (1991) selects a resolution based on the size of the static energy barrier shown in Fig.2. Here, again, we consider fine structure arguments for *deriving* a nucleation criterion.

In order to understand better what happens when a nucleation point, say $x = x_0$, is selected, let us focus on the small time behavior of the nontrivial self-similar solution. Consider a solution (2.5) at time $t = \Delta t$. It is convenient to parametrize the functions $w(x, \Delta t)$ and $v(x, \Delta t)$ by x and present them as a curve in the (w, v) plane. It is not hard to see that one then obtains a loop, beginning and ending in a point $(w_0, 0)$ (see Fig. 8b); the details of the loop depend, of course, on the fine internal structures of shocks and kinks (see Fig. 8a).

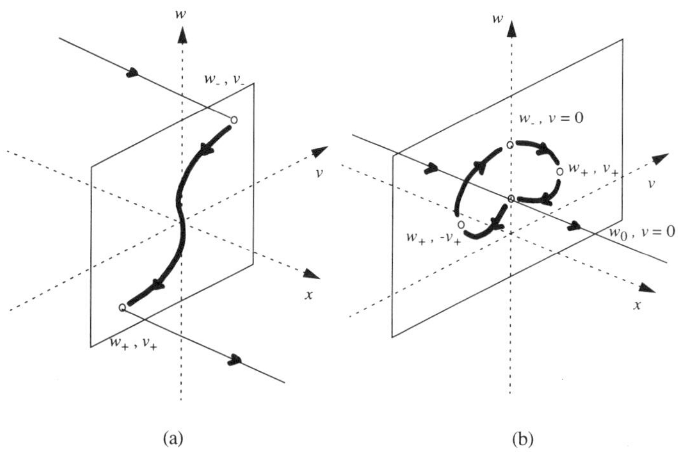

Figure 8. Schematic representation of the singular components of the solution and of the initial data: (a) - fine structure of the kink; (b) - fine structure of the nucleus.

In view of the self-similar character of the solution, the loop does not change as $\Delta t \to 0$ even though the strain and velocity fields converge to the constant initial data everywhere outside the point $x = x_0$. This means, that by selecting the point x_0 we have supplemented constant initial data with a singular part represented by a parametric measure (in the state space) located at $x = x_0$. We conclude that, contrary to the behavior of, say, genuinely nonlinear systems ($\sigma''(w) \neq 0$) (see Di Perna, 1985), the choice of a short time dynamic attractor in this problem (trivial solution vs. self similar dynamic regime (2.5)) is affected by a singular contribution to the initial data and may depend on the structure of the loop.

Since the energy of the nucleus is identically zero, the integral impact of this localized contribution to the initial data can be measured by the corresponding energy density which is finite. For our self-similar solution (2.5) one can equivalently calculate the rate of dissipation R (Dafermos, 1973)

$$R = D_1([f]-\{\sigma\}[w])_{shock} + D_2([f]-\{\sigma\}[w])_{kink}.$$

If the kinetic relation is known then the energy release rate R , which does not depend on t, can be calculated as a function of w_0; the fact that $R \neq 0$ at $t = 0$ means that the initial data with the superimposed loop are in fact instantly "dissipative". This observation, however, does not give insight into the associated barrier separating the uniform initial state and the state with the superimposed loop. As we have seen earlier such a barrier does not exist in the "homogenized " description. It can be calculated, however, in the framework of a regularized model which describes the initial stage of the transformation when internal length scales can not be neglected.

Suppose again that the isothermal viscosity-capillarity model (3.2) describes the "de-homogenized" structured material. Consider the initial value problem corresponding to a metastable state with the fixed strain w_0 and zero velocity but now add a finite perturbation with a small support. Numerical experiments based on the high-order accurate difference scheme developed by Cockburn and Gao (1996) show that sufficiently large perturbations evolve into a regime which closely resembles the self-similar dynamic solution (2.5), while small perturbations gradually decay (see Fig.9). This confirms the existence of the two dynamic attractors and makes it natural to relate the nucleation criterion to the size of the trivial regime's domain of attraction (basin).

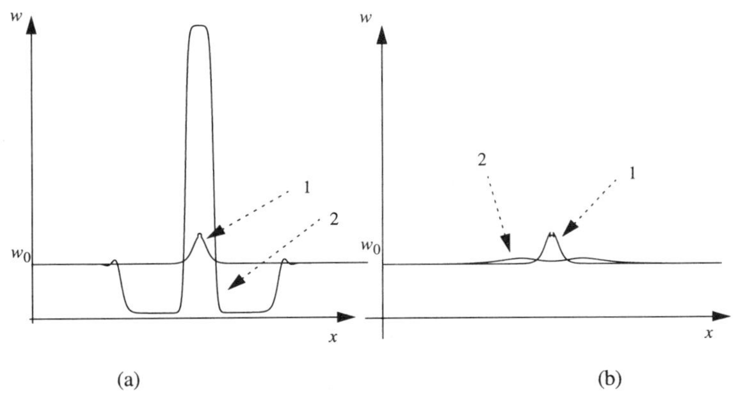

Figure 9. Two regimes of evolution for the initial data corresponding to slightly perturbed critical nucleus: (a) - perturbation leading to the growth of a new phase; (b)- perturbation which eventually decays; 1 - initial data; 2 - solution after finite time.

One representative point on the stability boundary is the so called *critical nucleus*, a saddle point (with a one dimensional unstable manifold) of the static energy functional

$$\int g(w) + \varepsilon/_2 w_x{}^2 dx,$$

where

$$g(w) = f(w) - f(w_0) - \sigma_0(w - w_0).$$

The critical nucleus, which can be found explicitly, is described by a homoclinic trajectory of the Euler-Lagrange equation $\varepsilon w_{xx} = g'(w)$ (see, for instance Bates and Fife, 1993). The fact that this perturbation plays a role of a threshold is clear from Fig.9 which demonstrates extreme sensitivity of the problem to slight variations around the critical nucleus representing particular initial data (see Ngan and Truskinovsky (1996b) for details).

We note that both the energy of the critical nucleus and the size of its support are proportional to $\sqrt{\varepsilon}$. In the limit $\varepsilon \to 0$ the energy of this perturbation goes to zero, however the associated energy density

$$\bar{g}(w_0) = \lim_{\varepsilon \to 0} \varepsilon^{-1/2} \int (g(w) + \varepsilon/_2 w_x^{\ 2}) dx$$

is a function of w_0 (or applied stress) only. The limiting perturbation can therefore be described by a probability measure ν_{x_0}, which in turn suggests that the nucleation criterion should be formulated in terms of the intensity of the exterior, measure-valued "noise". We note that such noise is *invisible* at the continuum level.

CONCLUSIONS

Solids undergoing martensitic phase transformations are currently a subject of intense interest in mechanics. In spite of recent progress in understanding the absolute stability of elastic phases under applied loads, the presence of metastable configurations remains a major puzzle. In this overview we presented the simplest possible discussion of nucleation and growth phenomena in the framework of the dynamical theory of elastic rods. We argue that the resolution of an apparent nonuniqueness at the continuum level requires "de-homogenization" of the main system of equations and the detailed description of the processes at micro scale.

Acknowledgments. This paper is dedicated to R. Fosdick on his sixtieth birthday. I would like to thank T. Pence for most helpful comments, S. C. Ngan for assistance in preparation of the pictures, and B. Cockburn for the permission to use his computer program. This work was supported by the NSF grant DMS-9501433.

REFERENCES

Abeyaratne, R., and Knowles, J.K., 1991, Kinetic relations and the propagation of phase boundaries in solids, *Arch. Rat. Mech. Anal.* 114:119.

Abeyaratne, R., and Knowles, J.K., 1995, Impact-induced phase transitions in thermoelastic solids, Caltech, Preprint.

Ball, J.M., and James, R.D., 1996, Local minimizers and phase transformations, Preprint .

Bates, P.W., and Fife, P.C., 1993, The dynamics of nucleation for the Cahn-Hilliard equation, *SIAM J. Appl. Math.* 53:90.

Carr, J., Gurtin, M.E., and Slemrod, M., 1984, Structured phase transitions on a finite interval, *Arch. Rat. Mech. Anal.* 86:317.

Cockburn, B., and Gao, H., 1996, A model numerical scheme for the propagation of phase transitions in solids, *SIAM J. Sci. Compt.* 17:1092.

Dafermos, C.M., 1973, The entropy rate admissibility criterion for solutions of hyperbolic conservation laws, *J. Diff.Eq.*, 14:202.

Dafermos, C.M.,1979, The second law of thermodynamics and stability, *Arch. Rat. Mech. Anal.* 70:167.

Di Perna, R.J., 1985, Measure valued solutions to conservation laws, *Arch. Rat. Mech. Anal.* 88:223.

Ericksen, J.L., 1975, Equilibrium of bars, *J. Elast.* 5:191.

Fosdick, R.L., and Mason, D.E., 1996, Singular phase energy minimizers for materials with nonlocal spatial dependence, *Quart. Appl. Math.* 54:161.

Fried, E., and Gurtin, M., 1994, Dynamic solid-solid transitions with phase characterized by an order parameter, *Physica D*, 72:287.

Gurtin M., 1993, The dynamics of solid-solid phase transitions 1. Coherent interfaces, *Arch. Rat. Mech. Anal.* 123:305.
James, R.D., 1980, The propagation of phase boundaries in elastic bars, *Arch. Rat. Mech. Anal.* 73:125.
Knowles, J.K., 1979, On the dissipation associated with equilibrium shocks in finite elasticity, *J.Elast.* 9:131.
Kulikovsky, A.G., 1976, On the discontinous solutions in mechanics of continuous media, in: *Theoretical and Applied Mechanics*, W.T. Koiter, ed., North Holland.
Kunin, I.A., 1982, *Elastic Media With Microstructure 1 (One Dimensional Model),* Springer.
Lax, P.D., 1957, Hyperbolic systems of conservation laws II, *Com. Pure Appl. Math.* 10:537.
Lax P.D., Levermore C.D., and Venakides S., 1992,The generation and propagation of oscillations in dispersive IVPs and their limiting behavior, in: *Important developements in soliton theory 1980-1990,* T. Fokas, V.E. Zakharov, ed., Springer-Verlag.
Lin J., and Pence T.J., 1996, Wave attenuation by kinetically active phase boundary scattering during displacive phase transformations, Preprint.
Marder, M., and Gross S., 1995, Origin of crack tip instabilities, *J.Mech.Phys.Solids* 43:1.
Mihailescu-Suliciu, M., and Suliciu, I., 1992, On the method of characteristics in rate type viscoelasticity with non-monotone equilibrium curve, *ZAMM* 72:667.
Ngan, S.C., and Truskinovsky, L., 1996a, Kinetics of adiabatic phase boundaries. To be submitted.
Ngan, S.C., and Truskinovsky, L., 1996b, Dynamics of nucleation in solids. To be submitted.
Oleinik, O.A.,1959, Uniqueness and stability of the generalized solution of the Cauchy problem for a quasilinear equation, *Usp. Mat. Nauk* 14:165.
Patashinskii, A.Z., and Chertkov, M.V., 1990, Motion of the front of a phase transition under strong supercooling conditions, *Sov.Phys.Solid State* 36:295.
Pego, R., 1987, Phase transitions in a one dimensional nonlinear viscoelasticity: admissibility and stability, *Arch. Rat. Mech. Anal.* 97:353.
Pitteri, M., and Zanzotto G., 1997, *Continuum Models for Phase Transitions and Twinning in Crystals*, Chapman and Hall .
Rogers, R., and Truskinovsky, L., 1996, Discretization and hysteresis, *Physica D.* Submitted.
Roshin, A. B., and Truskinovsky, L., 1989, Model of weakly Nonlocal Relaxing Compressible Media, *J. Appl. Math. Mech., (PMM)* 53:904.
Rosakis, P., and Knowles, J.K., 1997, Unstable kinetic relations and the dynamics of solid-solid phase transitions. In preparation.
Shearer, M., and Y. Yang, 1993, The Riemann problem for a p-system of conservation laws of mixed type with cubic nonlinearity, *Proc.Roy.Soc. Edinburgh A* .
Slemrod, M., 1983, Admissibility criteria for propagating phase boundaries in a van der Waals fluid, *Arch. Rat. Mech. Anal.* 81:301.
Slepyan, L. I., 1996, Crack dynamics, in: *Fracture: A Topical Encyclopedia of Current Knowledge*, G. Cherepanov, ed., Krieger.
Slepyan, L. I., and Trojankina, L. V., 1984, Fracture wave in a chain structure, *J. Appl. Mech. Techn. Phys.*, 25:921.
Truskinovsky, L., 1982, Equilibrium phase boundaries, *Sov. Phys. Doklady*, 27:551.
Truskinovsky, L., 1987, Dynamics of nonequilibrium phase boundaries in a heat conducting nonlinear elastic medium, *J.Appl. Math. and Mech. (PMM)* 51:777.
Truskinovsky, L., 1988, Nonequilibrium phase boundaries, *Dokl. Akad. Nauk. SSSR*, 303:1337.
Truskinovsky, L., 1993a, Kinks versus shocks, in: *Shock Induced Transitions and Phase Structures in General Media, R. Fosdick*, E.Dunn and M.Slemrod, ed., IMA 52, Springer-Verlag.
Truskinovsky, L., 1993b, Transition to detonation in the dynamic phase changes, *Arch. Rat. Mech. Anal.* 125:375.
Truskinovsky, L., 1994, About the normal growth approximation in the dynamical theory of phase transitions, Cont. Mech. and Thermodyn. 6:185.
Truskinovsky, L., 1996, Fracture as a phase transformation, in: *Contemporary research in the mechanics and mathematics of materials* , R. C. Batra and M. F. Beatty, eds., CIMNE, Barcelona.
Turteltaub, S. , 1997, Dynamics of phase transitions in thermoelastic solids, Ph.D.Thesis, Caltech.

TWO-DIMENSIONAL MODELS FOR STEP DYNAMICS

John D. Weeks, Da-Jiang Liu, Hyeong-Chai Jeong

Institute for Physical Science and Technology
University of Maryland
College Park, Maryland 20742

1 INTRODUCTION

In this paper we review some of our recent work on the dynamics of step bunching and faceting on vicinal surfaces below the roughening temperature, concentrating on several cases where interesting two dimensional (2D) step patterns form as a result of kinetic processes. We show that they can be understood from a unified point of view, based on an approximate but physically motivated extension to 2D of the kind of 1D step models studied by a number of workers. For some early examples, see refs. [1-5]. We have tried to make the conceptual and physical foundations of our own approach clear, but have made no attempt to provide a comprehensive review of work in this active area. More general discussions from a similar perspective and a guide to the literature can be found in recent reviews by Williams[6] and Williams and Bartelt[7].

We consider conditions where there is no significant island or void nucleation on the terraces, and surface mass transport is associated with the addition and removal of material from the preexisting surface steps. This provides an important simplification since the number of steps is now a conserved quantity and we do not have to deal with problems arising from the annihilation of steps of different sign.

We operate on a *mesoscopic* scale, intermediate between the atomic scale and continuum theory, taking individual surface steps as the fundamental objects of interest. One way to achieve this is to imagine an anisotropic coarse-graining of the surface much like that produced in REM experiments[8]. We average along the nominal step direction (the y-direction) to a scale large compared to the atomic scale but small compared to the step patterns of interest, while maintaining an atomic scale resolution normal to the steps (the steps ascend in the x-direction) and in the z-direction. When viewed from above, smooth long wavelength configurations of steps and terraces can be resolved but not microscopic objects like adatoms and kinks.

This scale is directly relevant to STM and REM experiments, and we believe it offers significant theoretical advantages over approaches that consider a more general coarse-graining using the step density as a variable. We can provide a more intuitive

Dynamics of Crystal Surfaces and Interfaces
Edited by Duxbury and Pence, Plenum Press, New York, 1997

and physically motivated description of changes in the surface morphology using the basic entity involved, the step. Moreover, we can examine the properties of individual steps as they move and bunch together and the step patterns that form.

To proceed, we must describe the effective driving force and the effective interactions between steps on this mesoscopic scale. We focus here on two cases of recent experimental and theoretical interest: current-induced step bunching on Si(111) surfaces[8–10] and reconstruction-induced faceting as seen a number of systems including the O/Ag(110) and Si(111) surfaces[11,12]. In both cases interesting 2D step patterns can arise from the competition between a driving force that promotes step bunching, and the effects of step repulsions, which tend to keep steps uniformly spaced.

2 TREATMENT OF STEP REPULSIONS AND FLUCTUATIONS

2.1 Physical origin

We first discuss the simpler situation that arises in the *absence* of the driving force, where the physics is dominated by the effects of step repulsions. The resulting equations can describe, e.g., the relaxation of initially nonuniform step configurations towards the equilibrium state[13,14]. The origin of the repulsive step interactions can be understood from the following. Although steps of the same sign can in principle bunch together, possibly even producing multiple-height steps, transverse fluctuations of a step in such a bunch are suppressed because of the prohibitively high energy cost associated with step crossings or overhangs. On averaging or coarse-graining, these constrained arrangements have a lower entropy (higher free energy) than that found when steps are further apart. This produces an (entropically-driven) effective *repulsion* between steps in the coarse-grained model, which favors uniform step spacing at equilibrium[6].

Fluctuations of an isolated step are also suppressed by the microscopic energy cost to form kinks. On coarse-graining, this translates into an effective *stiffness* or *line tension* that tends to keep the step straight. Standard microscopic 2D models of step arrays incorporating both of these physical effects include the free-fermion model and the Terrace-Step-Kink (TSK) model[5,15]. Both models have proved very useful, though their microscopic nature makes detailed calculations difficult.

Perhaps the most important conclusion arising from a study of such models is that the projected free energy density* of a *uniform* vicinal surface with slope s is given by the familiar Gruber-Mullins[16] expression:

$$f(s) = f^0 + \frac{\beta}{h}s + gs^3 . \tag{1}$$

Here f^0 is the free energy density of the flat surface, β the creation energy per unit length of an isolated step of height h (taken as the unit of length hereafter) and g the step interaction parameter, whose magnitude in general takes account of both entropic and possible direct elastic interactions. Eq. (1) can equally well be thought of as describing the free energy in a 1D model of straight uniformly spaced steps with effective repulsive interactions, or as the free energy in a 2D model where all steps are in their straight average (equilibrium) positions.

*This is defined as the surface free energy per unit area projected onto the low index facet plane. The use of the projected free energy allows a direct analogy with the thermodynamics of a liquid-vapor system. See, e.g., Williams et al.[6] for a clear discussion.

2.2 Step Hamiltonian for repulsive interactions

Our goal here is to provide a simpler 2D description of the mesoscopic scale physics, consistent with Eq. (1), in a form useful for practical calculations. To that end, in analogy with density functional methods for inhomogeneous fluids[17], we introduce an *intrinsic* (or configurational) *free energy functional* for the stepped surface. This gives the free energy of a macroscopic surface with N_s steps as a functional of the positions $\{x_n(y)\}$ of all the steps.

To obtain a physical picture, imagine first producing a nonuniform step configuration with the aid of an external field. Then the field is turned off. The intrinsic free energy can be crudely thought of as the free energy of such a perturbed system before it relaxes back to the uniform state. Although we will not use it here, the density functional formalism allows for a precise and formally exact treatment of this idea. Arbitrary equilibrium step configurations $\{x_n(y)\}$ can be produced formally by applying an appropriately chosen conjugate external field $\{\phi(x,y)\}$ that couples linearly to the step positions; in the uniform system with straight steps described by Eq. (1), ϕ is zero everywhere. The intrinsic free energy results when the linear contribution to the free energy that depends *explicitly* on the external field is subtracted from the (Helmholtz) free energy of the nonuniform system in the field. Technically, this generates a Legendre transform giving the free energy as a functional of the configurations $\{x_n(y)\}$ rather than the field $\{\phi(x,y)\}$.

In this paper we will not pursue such formal developments any further, and instead use mean field ideas and heuristic arguments to motivate the choice of the appropriate free energy functional. We represent the intrinsic free energy functional in the form of an *effective 2D step Hamiltonian* H and imagine on physical grounds that it has the following form:

$$H(\{x_n\}) = \int dy \sum_{n}^{N_s} \left[\frac{\tilde{\beta}}{2} \left(\frac{\partial x_n(y)}{\partial y} \right)^2 + V(w_n(y)) \right]. \tag{2}$$

The summation is over all N_s steps and the integration is over the coarse-grained y-direction. Thus the magnitude of H depends on the particular values of step positions $x_n(y)$ for all N_s steps and for all y positions. The first term on the r.h.s. describes the energetics of distorting an individual step, controlled by a line tension $\tilde{\beta}$. Everything else is incorporated into an *effective step interaction* V, taken here to be a function only of the *local* nearest neighbor terrace widths $w_n(y) \equiv x_{n+1}(y) - x_n(y)$. This seemingly natural approximation can introduce some notable errors in some cases† and more general expressions for the interaction can (and often should!) be used. However, this simple form will prove adequate for our purposes here.

We can determine the interaction term V in Eq. (2) through the requirement that H reproduce the macroscopic free energy in Eq. (1) in the limit of straight steps with uniform spacing w. Thus if L_y is the length of the system in the y-direction we require

$$L_y N_s V(w) = L_y N_s w f(1/w). \tag{3}$$

The l.h.s. simply evaluates H in this limit and the r.h.s. is the surface area $L_y N_s w$ of the flat reference plane times the projected free energy density $f(s)$ for a uniform system with slope $s = 1/w$. Thus we find

$$V(w) = w f(1/w). \tag{4}$$

†In particular, the energetics of the last step in a bunch next to a reconstructed terrace are incorrectly described. For a 2D model with more general step interaction terms that treats these "corner energy" corrections more accurately see Liu and Weeks[18].

We use this simple expression for V in the rest of this paper.

There are several points worth emphasizing. By using Eq. (4) in Eq. (2), we have made a *local* free energy approximation, relating the free energy of each terrace to that of a uniform system, even when neighboring terrace widths can vary. We have also implicitly assumed that the coarse-graining scale in the y-direction is large enough or that the distortions away from straight steps are small enough that we can use the same functional form to describe repulsive interactions in V in our 2D model as in the 1D model described by Eq. (1). Moreover, we have evaluated this interaction at the same-y position and have used a quadratic approximation for the line tension term. When the steps are relatively straight, these should be reasonable approximations, but we will use them even in more general cases as a physically motivated model.

2.3 Step chemical potential

Near equilibrium, we expect that the dynamics will be controlled by the energetics of small variations of the step positions in Eq. (2). By definition, the change in H to linear order induced by a small variation $\Delta x_n(y)$ in the position of the nth step is

$$\Delta H = \int \left[\frac{\delta H}{\delta x_n(y)}\right] \Delta x_n(y) dy, \tag{5}$$

where $\delta H/\delta x_n(y)$ is the *functional derivative* of H. Although adatoms are not considered explicitly in our coarse-grained model, we physically relate the area change $\Delta x_n(y)dy$ in Eq. (5) generated by the step displacement to the adsorption and emission of atoms at the step edge. Thus if Ω is the area occupied by an atom near the step edge, we define the *step edge chemical potential* $\mu_n(y)$ — the change in free energy per atom for adding atoms to the step at coarse grained position y — as:

$$\begin{aligned} \mu_n(y) &\equiv -\Omega \frac{\delta H}{\delta x_n(y)} \\ &= \Omega[V'(w_n) - V'(w_{n-1}) + \tilde{\beta}\partial^2 x_n/\partial y^2], \end{aligned} \tag{6}$$

where $V'(w)$ is the derivative of $V(w)$ with respect to w. The last line in (6) arises from functionally differentiating Eq. (2), integrating the variation of the $[\partial x_n(y)/\partial y]^2$ term by parts to arrive at the standard form in (6). The term $V'(w_n)$ has dimensions of force per unit length and can be interpreted as an effective *pressure* on the step associated with terrace n. Thus $\mu_n(y)$ depends on the local (linearized) curvature $\partial^2 x_n/\partial y^2$ of the step and on the difference in pressure from terraces behind and in front of the step. Note that constant terms in V and in the pressure V' do not contribute to $\mu_n(y)$.

2.4 Dynamics from step repulsions

We now study the dynamics arising from the step repulsions. While the classic treatment of BCF[19] assumed local equilibrium at step edges, with steps acting as perfect sinks for adatoms, in many cases a different *attachment/detachment rate limited regime* seems more appropriate. Here we assume that the diffusion rate for adatoms on terraces is much larger than the effective adatom exchange rate between step edges and terraces. To model the step motion, we now make a linear kinetics approximation, assuming that the velocity of a step is proportional to the change in free energy[‡]

[‡]Thermal fluctuations could also be taken into account through the addition of noise, leading to Langevin-type equations like those studied by Bartelt, et al.[20], but in the applications we consider here, the systems are far from equilibrium, either from initial conditions or because of an explicit driving force, so the simpler deterministic equations will prove adequate.

produced by its motion. There are then two limiting cases, depending on how the associated mass flow can take place.

2.4.1 Non-local mass exchange The effective mass flow is *non-local* (Case A) when atoms at a step edge can directly exchange with a vapor reservoir (through evaporation-condensation) or with an overall terrace reservoir that forms by fast direct adatom hops between different terraces. In such cases, we assume that step velocity is proportional to the chemical potential difference between the step and the reservoir:

$$\frac{\partial x_n(y)}{\partial t} = \frac{\Gamma_A}{\Omega k_B T}[\mu_n(y) - \mu_{res}], \tag{7}$$

where the proportionality constant is written in terms of Γ_A, the mobility of the step edge, as defined by Bartelt, et al.[20,21]. The chemical potential μ_{res} of the reservoir is set to be zero when there is no net motion of steps. Since atoms from a given step can go to distant regions through the reservoir, we expect that this mass flow will induce no direct correlation between the motion of neighboring steps.

2.4.2 Local mass exchange A second limit arises when the mass movement occurs *locally* through surface diffusion without direct adatom hops between different terraces. Here the effective adatom exchange is between neighboring step positions only. In the limiting case of *local mass exchange* (Case B), the current between step n and step $n+1$ is assumed to be proportional to the difference in step edge chemical potentials $(\mu_n - \mu_{n+1})$; the net velocity of step n is then given by

$$\frac{\partial x_n(y)}{\partial t} = \frac{\Gamma_B}{\Omega k_B T}[\{\mu_n(y) - \mu_{n+1}(y)\} + \{\mu_n(y) - \mu_{n-1}(y)\}], \tag{8}$$

where Γ_B is the step mobility in this conserved model§. This causes a *coupling* of the motion of neighboring steps and, as we will see, can produce interesting step patterns in certain cases.

3 RECONSTRUCTION INDUCED FACETING

In the 1D limit, Eqs. (7) and (8) and related equations have been used to analyze the relaxation of non-equilibrium step profiles[13,14] and in a variety of other applications[3,4]. We will not review this work here, but instead turn directly to two cases where characteristic 2D step patterns and step bunching are found as a result of the competition between the step repulsions and a driving force favoring step bunching. Perhaps the simplest application arises as a result of surface reconstruction.

Surface reconstruction or adsorption can often cause a vicinal surface with a single macroscopic orientation to facet into surfaces with different orientations[6]. Generally the reconstruction occurs on a particular low-index "flat" face, and lowers its free energy relative to that of an unreconstructed surface with the same orientation. However the same reconstruction that produces the lower free energy for the flat face generally *increases* the energy of surface distortions such as steps that disturb the reconstruction. Thus reconstruction is often observed only on terraces wider than some *critical terrace width* w_c. When steps are uniformly distributed initially and if w_c is much greater

§We follow Nozieres[2], assuming Γ_B is independent of the terrace width, and neglecting possible "Schwoebel" asymmetries in our description of step repulsions. Other choices could be made[3], but few qualitative changes are seen. For our purposes here the main point is to describe step repulsions at short distances in a reasonable way so that step crossing is prevented.

than the average terrace spacing w_a, step fluctuations leading to the formation of a sufficiently wide terrace — a "critical nucleus" — are required for the reconstruction to begin. Continued growth of the reconstructed region can make the vicinal surface facet into a "flat" reconstructed surface and a much more sharply inclined unreconstructed surface with closely bunched steps.

Experimental examples include faceting associated with the 7×7 reconstruction on Si(111) surfaces[11] and with the formation of $(n \times 1)$ oxygen chains on O/Ag(110) surfaces[12]. In both cases, reconstruction has been observed only on large ($w > w_c$) terraces, where the critical width w_c depends on temperature, pressure and some other parameters. However, experiments on these and some other systems such as Pt(111) and Au(111), show a noticeable regularity in the size and spacing of the flat facets[11,22,23], though the extent of the ordering varies from system to system. In any case, it seems hard to reconcile these regularities with a picture of random nucleation of the reconstructed regions.

3.1 Two state critical width model

To begin a theoretical discussion, it is clear that the existence of a sufficiently large reconstructed region can provide a driving force favoring step bunching in unreconstructed regions. The situation is still relatively simple since we can think of the vicinal surface as moving towards a new equilibrium state in the presence of reconstruction. Moreover, we expect that the *fully* reconstructed surface can be described by the same basic formalism involving step repulsions discussed earlier in Sec. 2.4.

The dynamics of the step motion leading to faceting would be very complicated if it were strongly coupled to the dynamics of reconstruction. Fortunately, in most cases the growth of a reconstructed region occurs much more rapidly than the characteristic time for step motion[24]. Thus we will not consider the dynamics of reconstruction explicitly here, and instead use a simple *two state model*[14,25,26] where each terrace is either reconstructed or unreconstructed, depending only on its *local width.* We first examine the consequences of the two state model when steps are straight, and then incorporate it into the 2D step models.

3.2 Free energies incorporating reconstruction

The reconstruction on the only large terraces, observed in experiments, can be understood by assuming that the free energy of the fully reconstructed flat surface has a lower value, ($-\epsilon_r$ per unit area) than the unreconstructed one but effectively a higher energy cost, (ϵ_s per unit length) for forming an isolated step[6,27]. Thus letting the previous free energy expression, Eq. (1), represent the unreconstructed surface (denoted by the subscript u) we assume the fully reconstructed surface is described by

$$f_r(s) = (f_u^0 - \epsilon_r) + (\beta_u + \epsilon_s)s + gs^3\,. \tag{9}$$

(In principle the g term would also change but this will have little effect on what follows.) The free energy of the fully reconstructed surface, $f_r(s)$, is lower than that of the unreconstructed surface, $f_u(s)$, when the slope s is less than $s_c \equiv \epsilon_r/\epsilon_s$. The thick curve in Fig **1**(a), given by

$$\begin{aligned} f(s) &= f_u(s)\,\Theta(s - s_c) + f_r(s)\Theta(s_c - s) \\ &= f_u(s) - \;(1 - s/s_c)\,\epsilon_r\;\Theta(s_c - s), \end{aligned} \tag{10}$$

with Θ the unit step function, represents the free energy of a hypothetical macroscopic system in which *all* terraces are reconstructed (unreconstructed) when the *average* slope s is less than (greater than) s_c.

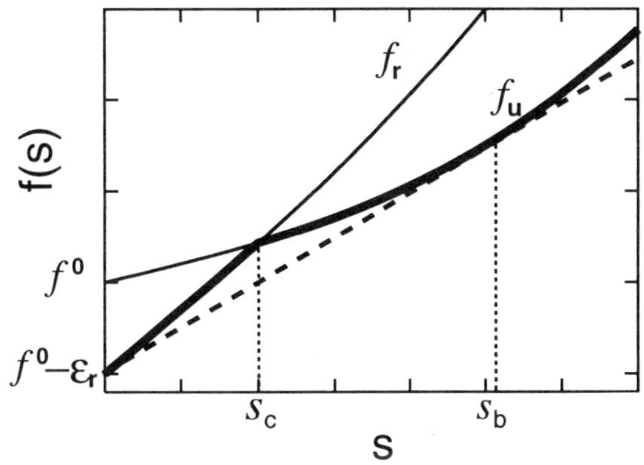

Figure 1: Free energies for unreconstructed surface f_u and reconstructed surface f_r vs slope s. The critical slope, s_c, and the slope of the surface at step bunches, s_b, are given by $s_c = \epsilon_r/\epsilon_s$ and $s_b = (\epsilon_r/2g)^{1/3}$. The thick curve in (a) represents the free energy of a hypothetical system in which all terraces are reconstructed (unreconstructed) when the average slope, s, is less than (greater than) s_c.

However, the situation of interest involves the presence of both reconstructed and unreconstructed regions. In fact, because of thermal fluctuations in a real system, for given average slope s, there exists a *distribution* of terrace widths around the average terrace width $w = 1/s$. Therefore, even in a surface with average slope larger than s_c, we expect to find some terraces with widths larger than $w_c \equiv 1/s_c$ in a large system. These represent regions where reconstruction would nucleate according to our two state model. If the average terrace width w_a is much smaller than w_c, we may have to wait a long time before fluctuations produce a sufficiently wide terrace. However, after such an event, the "critical nucleus" will continue to grow since the reconstructed surface has the lower free energy. Continued growth of the reconstructed terrace can cause neighboring unreconstructed terraces to become narrower and to form step bunches. At sufficiently long times, the surface will in principle facet into a flat reconstructed surface and a high slope (step-bunched) region as predicted by the thermodynamic tie bar construction (dashed line) shown in Fig. **1**.

3.3 2D dynamical equations incorporating reconstruction

We now modify the 2D continuum equations of step motion, Eqs. (7) and (8), in order to study some aspects of the dynamics of faceting. We assume the system is in the nucleation regime where the critical width w_c is much larger than the average step spacing w_a. In the simplest approximation discussed here, we incorporate the physics of the two state critical width model into the definition of the effective interaction term $V(w)$ in Eq. (2), which in turn modifies the step chemical potential terms in Eqs. (7) and (8). Again we set $V(w) = wf(1/w)$ as in Eq. (4) but now we use the f from Eq. (10) that takes account of reconstruction if a terrace is sufficiently wide. Note that this use of the two state model to describe an individual terrace with width w is more accurate than is the use of Eq. (10) to describe the properties of a macroscopic surface with average slope $s = 1/w$.

In this model, reconstruction modifies the effective step interactions and thus the dynamics. Both the presence of a reconstructed terrace behind a given step and repulsive entropic or elastic interactions from the step behind make positive contributions

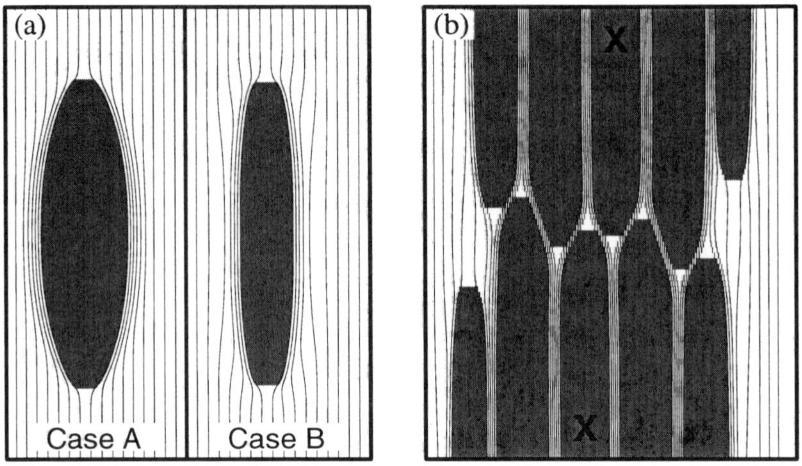

Figure 2: Top view of step configurations near growing nucleus in the early stage (a) and late stage (b). In (a), initial configurations for both non-local mass exchange of Case A and local mass exchange of Case B are shown. In (b), a regular pattern in Case B arising from interaction between two nuclei through an induced nucleation mechanism is shown. The initial positions of the nuclei (created by hand) are on the terraces marked by X but outside the figure.

to the "pressure" moving the given step forward.

We use this simple model with a fixed w_c only to describe the subsequent growth of an initial reconstructed region created by hand¶. Starting from the same initial roughly circular nucleus, we numerically integrated the equations of step motion, Eq. (7) and Eq. (8). Thus we can study the continued evolution of the step configurations under the different modes of mass transport. In both cases after a nucleus is created, it grows much faster in the step edge (y) direction (where step repulsions are relatively small) than it does in the normal (x) direction to the steps. Thus the nucleus quickly forms an elongated cigar-like shape. However, the subsequent temporal and spatial behavior of the faceting process is very different depending on the mechanism of mass transport on the surface.

Fig. **2** (a) shows typical step configurations from both equations in the early stage of the faceting. In Case A, with non-local mass transport, as the reconstructed facet grows it forces neighboring terraces to become *smaller*. There is a smooth relaxation to the average width far from the facet. Step spacings in case B, with local mass transport, show more interesting behavior because of the correlated step motion. While a region of step bunching is again observed close to the facet, on the other side of the bunch there are also some terraces that are *wider* than average. As the facet continues to grow, the number of steps in the bunched region increases but the widths of the wider terraces in front of the step bunch also increase. One of these may become sufficiently wide to serve as new nucleus for reconstruction. This *induced* nucleation process[25] can repeat itself

¶Thermal nucleation cannot be treated properly by the deterministic equations considered here. Even if noise were added, our description of reconstruction using the two state critical width model with fixed w_c is too crude to describe the initial formation of the critical nucleus. However, it does seem adequate to describe the further evolution of the surface once an elongated nucleus has formed. In this context, w_c is the "critical width for continued lateral growth" in y-direction. See Jeong and Weeks[28] for a more satisfactory and general continuum treatment of reconstruction coupled to step motion.

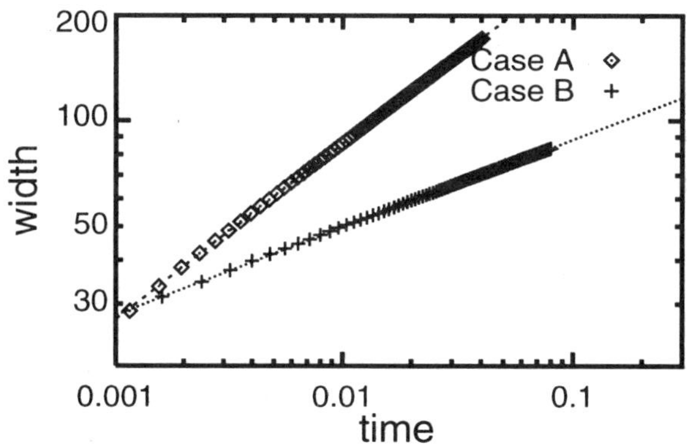

Figure 3: Measured time dependences of the facet widths for Cases A and B are shown in a log-log plot. All data in each case fall into a line indicating that the reconstructed terrace width increases as $w \sim at^{\alpha}$. The α values of the fitting lines are 1/2 for Case A and 1/4 for Cases B. Time and widths are in arbitrary units.

many times, producing a rather regular pattern of faceted and bunched regions that may be relevant for experiment[29,30]. We will discuss this mechanism in more detail later.

3.4 Isolated facet growth

Before doing this, let us first consider the growth rate of an *isolated* facet. We artificially prevent the formation of other (induced) nuclei on all other terraces by using the free energy curve for the unreconstructed surface regardless of the local terrace width. We then measure the time dependence of the facet length and width during growth. Here, we present only the qualitative results of the study. Quantitative results and the comparison with experiments will be presented elsewhere[26,28].

In both Case A and B, the reconstructed region propagates in y-direction with a constant velocity (after an initial transient where it forms the elongated shape). Thus the lateral size of the facet grows *linearly* with time. Linear growth along the step direction has been seen in experiment already[11] and there are rather general theoretical arguments[26] why it is to be expected.

On the other hand, the growth rate of the *normal* width is different in the two cases. Fig. **3** shows the growth of the facet width versus time in a log-log plot. All data in each case fall into a straight line, indicating that the reconstructed terrace width increases as $w \sim at^{\alpha}$. For case A, the facet grows as $t^{1/2}$, while it grows as $t^{1/4}$ for case B. These results are in complete agreement with the predictions of the classic one-dimensional continuum model of Mullins[31].

3.5 Induced nucleation

Let us now relax the constraint forbidding other nuclei from forming. Even if thermal fluctuations were included, this should produce essentially no change in Case A. Since all terraces near the original facet become smaller on average, other thermally nucleated facets are *less* likely to occur nearby. The story is quite different in Case B with local mass transport, where *induced* nuclei can form and inhibit the further growth of the original facet. In Case B, the motion of a step is directly coupled to the motion of neighboring steps. Initially, as the step bounding the reconstructed terrace moves forward to increase the reconstructed terrace's width, the neighboring step must move backward to conserve adatoms locally. Thus both the original reconstructed terrace and the terrace in front of the step that moves backward get wider. When the two steps that move in opposite directions come sufficiently close to each other for step repulsions to become important (with spacing approaching that of the equilibrium step bunch), they "collide" and both begin to move forward together as a bunch because of the driving force from the reconstructed terrace behind. Then the local conservation process repeats itself, causing new steps in front of the advancing step bunch to move backward and making the terraces in front of those steps wider. As the original facet grows, the number of steps in the bunch increases and the widths of the widest terraces in front of the step bunch also increase. Such a sufficiently wide terrace can be a nucleus for the reconstruction of another facet.

A quantitative treatment of this induced nucleation mechanism using a 1D model [the $\tilde{\beta} \to \infty$ limit of Eq. (8)] was carried out by Jeong and Weeks[25]. When the typical distance between steps in a step bunch, $w_b \sim (2g/\epsilon_r)^{1/3}$ is much smaller than the average terrace width w_a, it was shown that only one other terrace, aside from the original facet, is larger than w_a at any given time. In the limit that w_b/w_a goes to zero, the maximum width of the induced wide terrace increases *linearly* with the number of steps, n_b, in the bunch separating it from the original facet. Moreover it remains as the widest terrace for an increasing long time interval, $\Delta t \sim n_b^3$. Once it gets larger than w_c, reconstruction will occur. Further growth of the original facet essentially stops, but the new facet can induce another nucleus on the other side as it continues to grow. Then this new nucleus can induce another one and so on. The velocity of the nucleation front is *linear* in time because it always takes the same amount of time to induce a nucleus. Hence the faceted surfaces arising from this idealized process have a periodic distribution of reconstructed terraces separated by step bunches.

3.6 2D patterns from induced nucleation

An interesting 2D pattern arises from induced nucleation using Eq. (8) when two (thermal) nuclei form that are close in the x-direction but separated by a large distance in the y-direction. Fig. **2** (b) shows a step configuration in Case B arising from two such nuclei (created by hand). As time goes, each nucleus grows as $t^{1/4}$ in the normal direction until it produces its own induced nucleus. In the lateral direction, nuclei grow essentially linearly in t until they "collide" with each other and form a bunch of *crossing steps* between them. After such an encounter, the nuclei stop growing in the y-direction. The number of step in a crossing bunch is determined by how many steps initially separated the two nuclei when they formed. Once this configuration forms, other nuclei induced by the two original facets will produce new crossing steps at essentially the same y-position as the original crossing steps. Hence, an *alignment* of crossing step bunches is formed as shown in Fig. **2** (b). The number of steps in the induced crossing bunches are expected to be the same as that in the original crossing step bunch when the idealized induced nucleation mechanism of the 1D model is accurate. A strong tendency for alignment of crossing steps has been found in some step bunching experiments[32] on vicinal GaAs(001) but we do not know whether this is an equilibrium or purely kinetic phenomena as our model would suggest.

If thermal fluctuations were taken into account, the regular patterns selected by this kinetic mechanism would be expected to be less sharp. In particular, when w_b/w_a is not so small, the effects of mass conservation are spread out over many terraces and several terraces in front of the step bunch become larger than w_a. These would be particularly advantageous sites where thermal nucleation could occur, even before the induced width of the terrace as predicted by the deterministic models would exceed w_c. Thus nucleation sites and times are less precisely determined in this case, and we expect the number of steps in a bunch, n_b, to be smaller than the value predicted by the 1D model or the deterministic 2D model. Nevertheless, as calculations with more realistic models show[28], the qualitative feature of the induced nucleation mechanism as discussed here remain valid. While there are a number of different factors (including in particular elastic interactions[22,23]) that can contribute to the facet spacing in particular systems, induced nucleation represents a very general *kinetic* mechanism that should be considered in analyzing experimental data.

4 CURRENT-INDUCED STEP BUNCHING

We now turn to the second main application. Very interesting step bunching instabilities have been seen on vicinal Si(111) surfaces during evaporation by heating with a direct electric current. The direct current heating acts as a driving force producing a net motion of steps as the surface evaporates. Several experimental groups[8–10,33] have shown that the motion of the steps depends crucially on the direction of the current relative to the step orientation. Current in one direction results in stable step flow with uniform step velocities and spacings, while current in the opposite direction causes the steps to bunch together and form complex two dimensional patterns. An electromigration model, involving the diffusion of adatoms (and possibly advacancies[34]) with an effective charge, has been suggested microscopic model. The microscopic physics responsible for the magnitude and sign of the effective charge must be very complicated: there are three temperature regimes where the stable and unstable current directions change roles[38].

Fortunately, many features of the mesoscopic scale physics and the resulting step patterns do not depend on the details of this microscopic physics. Kandel and Weeks[39] (KW) introduced a very successful mesoscopic model describing the step patterns arising from the consequences of a general driving force leading to step bunching and the effects of step repulsions. However, their treatment of the repulsions was rather crude (they imposed a minimal distance requirement by hand) and they made specific physical assumptions (assuming multistep jumps of adatoms over step bunches) whose general applicability could be questioned. Here we show that results essentially identical to those of KW arise very naturally by adding the appropriate driving force terms to the basic equations (7) and (8) describing the effects of repulsive interactions[||]. The situation is both simpler and more complicated than that discussed in Sec. 3, since we do not have the complications of surface reconstruction to contend with, but now the system of interest is continually driven far from equilibrium by the field. As we will see, this reduces the differences between models that assume local and non local mass flow in the treatment of the repulsive interactions.

We adopt the following simple picture. Initially, we assume that steps are far enough apart that the effects of step repulsions can be ignored. The relevant physics for evaporation involves the detachment of adatoms from step edges, their surface diffusion on the adjacent terraces, and their eventual evaporation. This is quite well described by a generalization of the classical BCF model[19], which considers solutions to the adatom diffusion equation with boundary conditions at the step edges.

4.1 Asymmetric velocity function model

The experiments strongly suggest that there is an *asymmetry* in the step-up and step down directions associated with the direction of the electric field. Such an asym-

[||]For a somewhat different approach that reaches the same conclusion, see Liu et al.[40]

metry can arise both from biased diffusion of adatoms with an effective charge, and by the use of asymmetric kinetic coefficients in the step edge boundary conditions. In a 1D model with straight steps, the result of the generalized BCF model can be expressed quite generally in terms of a set of effective equations of motion for the steps, similar to Eqs. (7) and (8):

$$\frac{\partial x_n}{\partial t} = f_+(w_n) + f_-(w_{n-1}) \; . \tag{11}$$

This relates the motion of step n to *velocity functions* $f_\pm$ of the widths of the terrace in front $[f_+(w_n)]$ and behind $[f_-(w_{n-1})]$ the moving step. A straightforward linear stability analysis of (11) around the uniform step train configuration with terrace width w shows[1] that if

$$f'_-(w) > f'_+(w), \tag{12}$$

the uniform step train is unstable towards step bunching. Here $f'_\pm$ are the derivatives of $f_\pm$.

4.2 Back terrace instability

While explicit expressions for the $f_\pm$ can be derived from the generalized BCF model, many essential features for the electromigration experiments seem to be captured by a simple *linear* model where

$$f^L_\pm(w) = k_\pm w \; . \tag{13}$$

This should be a reasonable approximation when the diffusion length is much larger than typical terrace widths and when the step motion is attachment/detachment limited, as is thought to be the case for Si[41]. This approximation also requires that steps are not so close together that direct entropic or elastic step-step interactions are important. When the asymmetry is such that the step velocity is more sensitive to processes associated with the terrace *in back of* the moving step, i.e., when $k_- > k_+$, Eq. (12) shows that the uniform step train is unstable towards step bunching. For concreteness, we refer to this as a *back-terrace asymmetry.* Note that a strong step edge barrier of the kind envisioned by Schwoebel[42] would yield such an asymmetry on any evaporating surface. However, we emphasize that the effective asymmetry in the velocity functions (13) can originate from a number of different microscopic processes. In particular, in the case of Si(111), the asymmetry is evidently a function of the electric field and the temperature.

4.3 Incorporating repulsive interactions

We assume that such an asymmetry is present in the electromigration experiments when the current is in the unstable direction. The resulting instability will cause some steps to come close to one another. When this happens, the repulsive interactions preventing step crossing must be taken into account and Eq. (13) is inadequate. An attempt to take account of the effects of step repulsions in the context of a generalized BCF model for general 2D step configurations leads to extremely complicated equations. To arrive at a simple 2D model capable of describing the appropriate physics both at large and small step separations, we assume *weak coupling* between the different regimes and simply add the linear driving force terms $f^L_\pm(y)$ in Eq. (13) to the previously derived equations of motion (7) and (8) describing the effects of the repulsions. As with the treatment of repulsions, we evaluate the driving force terms in the 2D model using the same-y approximation.

Note that the $f_\pm$ incorporate both the general driving force leading to overall step motion as well as the source of the instability leading to step bunching. Of course in general one would expect that cross terms describing modifications of the repulsive interactions due to the driving force should arise. However, we believe the simple weak coupling approximation, which prevents step crossing at small separations and which incorporates the fundamental step bunching instability at large separations, will capture the essential physics on large length scales.

We have considered two limiting cases for the kinetics arising from the repulsive interactions in Eqs. (7) and (8). Hence in principle we will find two different model equations for the electromigration experiments when the driving force terms are added. However, as we will see, several basic features of the resulting models are independent of the kinetics when the system is driven far from equilibrium.

4.4 Case A

The simplest model equation capable of describing the electromigration experiments arises from Case A (non local mass flow) by adding the driving force term to Eq. (7). Thus we find our basic result:

$$\frac{\partial x_n(y)}{\partial t} = \frac{\Gamma_A}{\Omega k_B T}\mu_n(y) + k_+ w_n(y) + k_- w_{n-1}(y). \tag{14}$$

Using Eq. (6) this can be written in the 2D velocity function form originally suggested by KW[39]:

$$\frac{\partial x_n(y)}{\partial t} = \frac{\Gamma_A\,\tilde{\beta}}{k_B T}\frac{\partial^2 x_n(y)}{\partial y^2} + f_+^A(w_n(y)) + f_-^A(w_{n-1}(y)), \tag{15}$$

where

$$\begin{aligned} f_+^A(w) &= -\frac{2\Gamma_A g}{k_B T}\left(\frac{1}{w^3}\right) + k_+ w, \\ f_-^A(w) &= +\frac{2\Gamma_A g}{k_B T}\left(\frac{1}{w^3}\right) + k_- w. \end{aligned} \tag{16}$$

Each $f_\pm^A(w)$ in Eq. (16) contains a short ranged direct interaction part preventing step crossing and a long ranged (linear) part describing the effects of diffusion and evaporation. More complicated expressions could be used for each part but we expect much the same qualitative features. For concreteness, we study here the case of evaporation so that the terrace ascends as the step index n increases. The following discussion can also be applied to the growth problem with some straightforward adjustments.

Note that the repulsive step interactions (which generate the short ranged terms proportional to g in $f_\pm^A$) do not change the *total* evaporation rate because on average they cancel each other when summed over all terraces. When the linear approximation is accurate for the $f_\pm$ at larger separations, an even stronger statement can be made. Using Eq. (15) and summing over all terraces, the average velocity of the steps is $(k_+ + k_-)w_a$ where w_a is the average terrace width. The total evaporation rate is then $(k_+ + k_-)$, which is a constant independent of both the average miscut angle and all surface configurations.

With this simplification of the two dimensional step flow problem, we can study the long time behavior of the step train well beyond the initial onset of instability. We start with an array of 40 steps with small perturbations from an initial uniform configuration. We discretize the y coordinate so that each step has 2000 segments. Periodic boundary conditions are used in x and y direction. The time evolution problem of Eqs. (15) using (16) is converted into a set of difference equations. We control the time step so

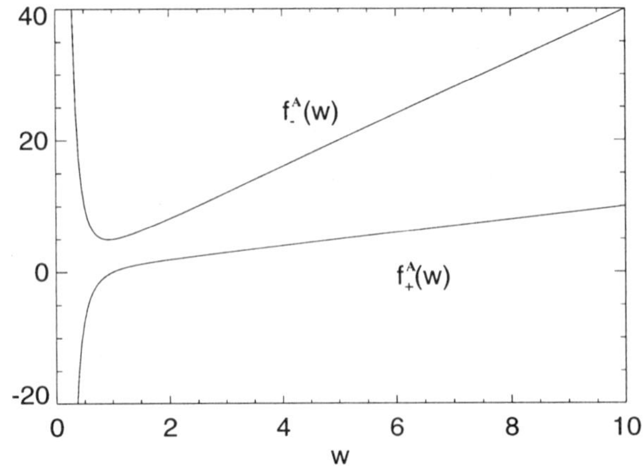

Figure 4: Velocity functions in Case A given by Eq. (16). The parameters used are: $k_+ = 1$, $k_- = 4$, $G = 2\Gamma_a g/k_BT = 1$, $\gamma = 1$ and $w_a = 5$. Parameters are in arbitrary units.

that during any time interval, each step moves only a small amount compared with its neighboring terrace widths. The segment size in the y direction is chosen small enough that the curvature is meaningful but is still large enough to include many features. As in KW[39], we are mainly interested in cases where k_+ and k_- are both positive. As an example we choose $k_+ = 1$, $k_- = 4$, $G \equiv 2\Gamma_a g/k_BT = 1$, $\gamma \equiv \Gamma_A\tilde{\beta}/k_BT = 1$ and $w_a = 5$. The velocity functions $f_\pm$ given by Eq. (16) with these parameter values are shown in Fig. **4**. These parameter values make the effect of direct repulsive interactions very small when steps are separated by the initial spacing w_a.

After the initial pairing instability, some of the steps come much closer to each other and the short ranged interaction terms become important. As the system continues to evolve, step bunches and single *crossing steps*, which leave one bunch and join another, begin to emerge. Thus there is also an instability toward *step debunching* in this model! The crossing steps move at a higher velocity than the step bunches. As successive crossing steps continually escape from the bunch behind and reattach to the bunch in front they usually form nearly equally spaced *crossing arrays* that connect two adjacent step bunches. Fig. **5** shows a snapshot of a system after about 140 monolayers are evaporated. The system continues to coarsen (the bunches gets bigger) until finite size effects become important. The coarsening can happen through the eating away of smaller bunches by debunching or through the merging of neighboring bunches. These patterns have a striking qualitative resemblance to the experimental results. Indeed, quantitative comparisons can be made[43], but this is outside the scope of the present work.

4.5 Debunching instability

The origin of the debunching instability can be understood by considering a group of N_b straight step that are very close to each other and flanked by two very large terraces on each side with widths $w^{(b)}$ and $w^{(f)}$ respectively. The velocity of the step bunch is mainly determined by the width of the large terrace *behind* the step bunch. Defining the average position of the bunch as $X(y) = \sum_{n=1}^{N_b} x_n(y)/N_b$, then from Eqs. (15) and (16) we have

$$\partial X(y)/\partial t \approx \frac{1}{N_b}\left(f_-^L(w^{(b)}) + f_+^L(w^{(f)})\right) , \tag{17}$$

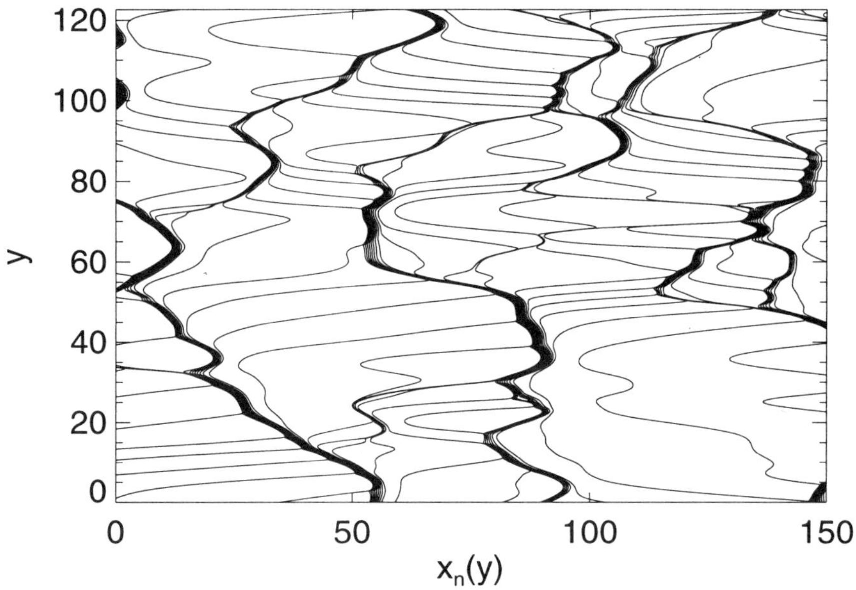

Figure 5: A snapshot of a system of 40 steps after 140 monolayers are evaporated using Case A dynamics. The same parameters as in Fig. 5 are used. Steps flow from left to right. Only a portion of the system is shown.

if we ignore the repulsion from other distant isolated steps and if the width of the step bunch itself is much smaller than the widths of the flanking terraces. The velocity of the first step in the bunch is always larger than $f_+^L(w^{(f)})$ (the other term $f_-^L(w^{(b)})$ is always positive), and hence when $w^{(f)}$ is large enough, the first step's velocity can be *larger* than the average velocity of the bunch, which from Eq. (17) *decreases* as the bunch size increases. Thus the first step can eventually *escape* from the bunch. This leads to the debunching, and the subsequent formation of crossing arrays.

These same features have been observed in the Monte Carlo simulations of the model of KW[39], in which the direct step interactions are treated through the imposition of a minimum distance constraint and allowing multistep jumps. The same long-ranged (linear) velocity functions were used in that model. Both the dynamical behavior and the patterns formed are very similar in the two models. We conclude that the details of the short ranged interactions are not important for the creation of step bunches and crossing arrays in this kind of a model. Indeed, we have verified that modifying the form of the repulsive interaction in Eq. (16) does not change the basic features as long as the interaction is a short ranged repulsion that prevents step overhangs. As would be expected, the main differences are in the details of the step profile in the step bunch. See Lui *et al.*[40,49] for further discussion.

4.6 Case B

The persistence of these basic features is perhaps most dramatically illustrated by considering the more complicated case that arises when the effects of the repulsions are treated with locally conserved dynamics (Case B). Adding the driving force term Eq.

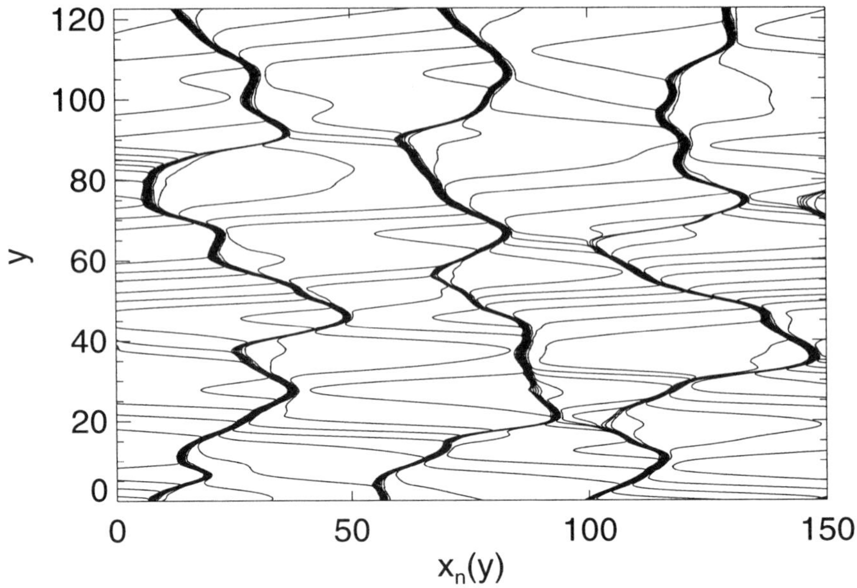

Figure 6: A snapshot of a system of the same size as in Fig. 6 but using Case B dynamics. Parameters used here: $k_+ = 1$, $k_- = 4$, $G = 2\Gamma_b g/k_BT = 1$, $\gamma = \Gamma_b\tilde{\beta}/k_BT = 1$ and $w_a = 5$.

(13) to Eq. (8) yields the new electromigration model

$$\frac{\partial x_n(y)}{\partial t} = \frac{\Gamma_B}{\Omega k_B T}[\{\mu_n(y) - \mu_{n+1}(y)\} + \{\mu_n(y) - \mu_{n-1}(y)\}] + k_+ w_n(y) + k_- w_{n-1}(y). \quad (18)$$

We can write this in a general velocity function form

$$\frac{\partial x_n}{\partial t} = f_+(\mu_n, \mu_{n+1}, w_n) + f_-(\mu_{n-1}, \mu_n, w_{n-1}), \quad (19)$$

but we see that in this case each $f_\pm$ directly couples *four* steps together, rather than two as in the nonconserved case.

Equations similar to (18) could arise from generalizations to 2D of models considered by Uwaha *et al.*[44,45] and Natori[46]. These authors modified the one dimensional BCF model to include direct step interactions, which can affect the adatom chemical potential at the step edges[2,3] (or the equilibrium adatom concentration near the steps). All step motion was still assumed to arise from the modified diffusion fields, so local mass conservation is satisfied. The diffusion fields were determined using the quasistatic approximation. Note that this approximation is less justified when steps are interacting strongly since their velocity may not be very slow compared with the diffusion field[47,48]. They found interesting dynamical behavior in a number of different cases, including some effects of step debunching.

The time evolution of a 2D step array given by Eq. (19) can be again solved for numerically. We find that the basic features are the same as in the nonconserved case. Fig. **6** shows a snap shot of the system using the same anisotropy ratio $k_+/k_- = 4$ as in Fig. **5**. We observe again that when there is enough driving force to move the system far away from equilibrium, the details of how we treat the step interactions are not very important for the formation of the basic crossing array patterns. Quantitative comparison of the two models with each other and with experiment will be given elsewhere[40,49].

In conclusion, the patterns generated by KW[39] seem to arise naturally in 2D systems with unstable step flow resulting from a back-terrace instability. We are able to reproduce most of KW's results by simply modifying the velocity functions at short distances to incorporate the effects of step repulsions, as suggested by the weak coupling picture. These compare favorably with experiments on current induced step bunching in Si(111)[8–10,43]. Other physical limits, e.g., local mass conservation, can be incorporated with some generalization of the forms of the velocity functions.

Acknowledgements: Part of this work was done in collaboration with Daniel Kandel and Ellen Williams, who we also thank for many helpful discussions. We also thank Ted Einstein and Norm Bartelt for useful remarks. This work was supported by the NSF-MRG with continuing support from the NSF-MRSEC under contract DMR-96-32521.

REFERENCES

1. P. Bennema and G. H. Gilmer, in *Crystal Growth: An Introduction*, p. 263, P. Hartman, ed., North Holland, Amsterdam, (1973).
2. P. Nozières, *J. Phys. France* 48:1605 (1987).
3. A. Rettori and J. Villain, *J. Phys. France* 49:257 (1988).
4. M. Ozdemir and A. Zangwill, *Phys. Rev. B* 42:5013 (1990).
5. N. Bartelt, T. Einstein, and E. Williams, *Surf. Sci.* 240:L591 (1990).
6. E. D. Williams, *Surf. Sci.* 299/300:502 (1994).
7. E. D. Williams and N. C. Bartelt, in *Handbook of Surface Science*, W. N. Unertl, ed., North-Holland, Amsterdam, (1996).
8. A. Latyshev, A. Aseev, A. Krasilnikov, and S. Stenin, *Surf. Sci.* 227:24 (1990).
9. A. Latyshev, A. Aseev, A. Krasilnikov, and S. Stenin, *Surf. Sci.* 213:157 (1989).
10. See, for example, Y. Homma, R. J. McClelland, and H. Hibino, *Jpn. J. Appl. Phys.* 29:L2254 (1990); H. Yasunaga, and A. Natori, *Surf. Sci. Rep.* 15:205 (1992) and references therein; M. J. Ramstad, R. J. Birgeneau, K. I. Blum, D. Y. Noh, B. O. Wells, and M. J. Young, *Europhys. Lett.* 24:653 (1993); A. V. Latyshev, A. B. Krasilnikov, and A. L. Aseev, *Surf. Sci.* 311:395 (1994).
11. R. J. Phaneuf, N. C. Bartelt, E. D. Williams, W. Święch, and E. Bauer, *Phys. Rev. Lett.* 67:2986 (1991).
12. J. S. Ozcomert, W. W. Pai, N. C. Bartelt, and J. E. Reutt-Robey, *Phys. Rev. Lett.* 72:258 (1994).
13. E. S. Fu, M. D. Johnson, D.-J. Liu, J. D. Weeks, and E. D. Williams, *Phys. Rev. Lett.* 77:1091 (1996).
14. D.-J. Liu, E. S. Fu, M. D. Johnson, J. D. Weeks, and E. Williams, *J. Vac. Sci. Technol. B* 14:2799 (1996).
15. B. Joós, T. L. Einstein, and N. C. Bartelt, *Phys. Rev. B* 43:8153 (1991).
16. E. E. Gruber and W. W. Mullins, *J. Phys. Chem. Solids* 28:875 (1967).
17. For a general review, see R. Evans, in *Fundamentals of Inhomogeneous Fluids*, D. Henderson ed., Dekker, New York, (1992).
18. D.-J. Liu and J. D. Weeks (unpublished).
19. W. K. Burton, N. Cabrera, and F. C. Frank, *Proc. R. Soc. London, Ser. A* 243:299 (1951).
20. N. C. Bartelt, J. L. Goldberg, T. L. Einstein, and E. D. Williams, *Surf. Sci.* 273:252 (1992).
21. N. C. Bartelt, T. L. Einstein, and E. D. Williams, *Surf. Sci.* 276:308 (1992).
22. M. Yoon, S. G. J. Mochrie, D. M. Zehner, G. M. Watson, and D. Gibbs, *Surf. Sci.* 338:225 (1995).

23. F. Pourmir, S. Rousset, S. Gauthier, M. Sotto, J. Klein, and J. Lecoeur, *Surf. Sci.* 324:L337 (1995).
24. F. Katsuki and K. Kamei, *Applied Surface Science* 94/95:485 (1996).
25. H.-C. Jeong and J. D. Weeks, *Phys. Rev. Lett.* 75:4456 (1995).
26. D.-J. Liu, J. D. Weeks, M. D. Johnson, and E. D. Williams, *Phys. Rev. B* (in press).
27. R. J. Phaneuf, N. C. Bartelt, E. D. Williams, W. Święch, and E. Bauer, *Phys. Rev. Lett.* 71:2284 (1993).
28. H.-C. Jeong and J. D. Weeks (unpublished).
29. H. Hibino, Y. Homma, and T. Ogino, *Phys. Rev. B* 51:7753 (1995).
30. J. R. Heffelfinger, M. W. Bench, and C. B. Carter, *Surf. Sci.* 343:L1161 (1995).
31. W. W. Mullins, *J. Appl. Phys.* 28:333 (1957).
32. K. Hata *et al.*, *Appl. Phys. Lett.* 76:5601 (1994).
33. M. Suzuki, Y. Homma, Y. Kudoh, and R. Kaneko, *Ultramicroscopy* 42-44:940 (1992).
34. C. Misbah, O. Piere-Louis, and A. Pimpinelli, *Phys. Rev. B* 51:17283 (1995).
35. S. Stoyanov, *Jpn. J. Appl. Phys.* 30:1 (1991).
36. S. Stoyanov, *Applied Surface Science* 60/61:55 (1992).
37. S. S. Stoyanov, H. Nakahara, and M. Ichikawa, *Jpn. J. Appl. Phys.* 33:254 (1994).
38. D. Kandel and E. Kaxiras, *Phys. Rev. Lett.* 76:1114 (1996).
39. D. Kandel and J. D. Weeks, *Phys. Rev. Lett.* 74:3632 (1995).
40. D.-J. Liu, J. D. Weeks, and D. Kandel, *Surf. Rev. and Lett.* (in press).
41. Y.-N. Yang and E. Williams, *Phys. Rev. Lett.* 72:1862 (1994).
42. R. L. Schwoebel and E. J. Shipsey, *J. Appl. Phys.* 37:3682 (1966).
43. E. D. Williams, E. Fu, Y.-N. Yang, D. Kandel, and J. D. Weeks, *Surf. Sci.* 336:L746 (1995).
44. M. Uwaha, *J. Crystal Growth* 128:92 (1993).
45. M. Uwaha, Y. Saito, and M. Sato, *J. of Crys. Growth* 146:164 (1995).
46. A. Natori, *Jpn. J. Appl. Phys.* 33:3538 (1994).
47. R. Ghez and S. S. Iyer, *IBM J. Res. Dev.* 32:804 (1988).
48. F. Liu and H. Metiu, *Phys. Rev. E* 49:2601 (1994).
49. D.-J. Liu, J. D. Weeks, and D. Kandel (unpublished).

NETWORKS OF STEPS ON AU AND PT CRYSTALS

Henk van Beijeren

Institute for Theoretical Physics
Universiteit Utrecht
3508-TA Utrecht, The Netherlands

Enrico Carlon

Institute for Theoretical Physics
Katholieke Universiteit Leuven
B-3001 Leuven, Belgium

Abstract

Networks of steps, seen in STM observations of vicinal surfaces on Au and Pt (110), are analyzed. A simple model is introduced for the calculation of the free energy of the networks as function of the slope parameters, valid at low step densities. It predicts that the networks are unstable, or at least metastable, against faceting and gives an equilibrium crystal shape with sharp edges either between the (110) facet and rounded regions or between two rounded regions. Experimental observations of the equilibrium shapes of Au or Pt crystals at sufficiently low temperatures, i.e. below the deconstruction temperature of the (110) facet, could check the validity of these predictions.

INTRODUCTION

(110) surfaces of fcc metals have been intensively studied in the past years by means of several experimental techniques. It was found by scattering experiments, and subsequently confirmed by direct STM observations that some "heavy" metals such as Au, Pt, Ir form a (2×1) reconstructed state. Other metals, e.g. Pb and Al, are not reconstructed.

The (2×1) structure is also known as missing–row reconstruction because one out of two rows of atoms, aligned along the $[1\bar{1}0]$ direction, is missing from the surface layer. We stress also that there are two different realizations of these reconstructed states, in which either the even or the odd rows are missing.

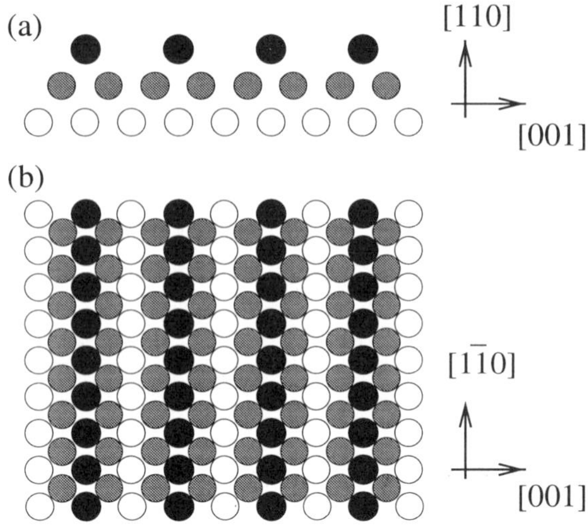

Figure 1: (a) Profile and (b) top view of a missing row reconstructed (110) surface.

Fig. **1** shows a top and a side view of the missing–row reconstructed structure. Along the [001] direction the surface assumes a hill-and-valley profile, where the sides of the hills are actually (111) microfacets. In the (111) orientation, surface atoms are closely packed, therefore such orientations are energetically favored.

In several STM investigations (Gritsch et al., 1991; Gimzewski et al., 1992; Kuipers, 1994) of vicinal orientations of (110) Au and Pt surfaces an unusual pattern of steps, as shown in Fig. **2**, was observed. The figure represents a surface orientation slightly tilted towards the [1$\bar{1}$0] direction with respect to the (110) facet. The missing rows, which are not visible in the figure, run along the vertical direction.

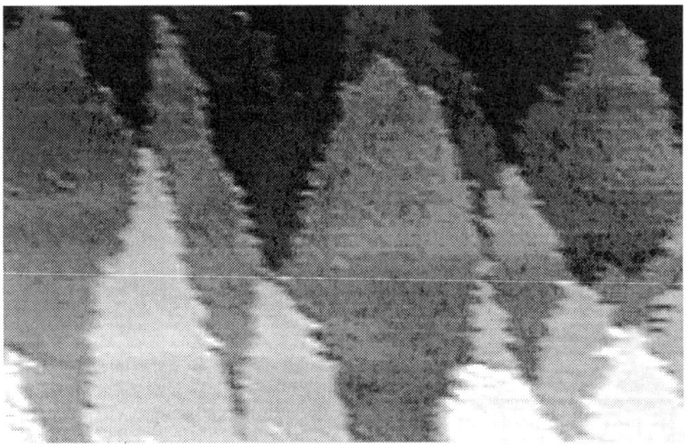

Figure 2: STM images of vicinals of Au(110) (courtesy of M.S. Hoogeman, L. Kuipers and J.W.M. Frenken, AMOLF Amsterdam). The area shown is $190nm \times 120nm$ at $T = 550K$, with a miscut angle of 0.07 degrees.

This pattern of steps is unusual indeed. In normal situations a miscut along the [1$\bar{1}$0] direction is generated by steps that run perpendicular to the missing rows, i.e. along the horizontal direction of Fig. **2**, because such an arrangement minimizes the total length and hence the total free energy of the steps required to produce the miscut. Instead, on missing-row reconstructed surfaces the steps zig-zag and repeatedly touch each other at a collection of contact points; one can also say that they form a *network* of

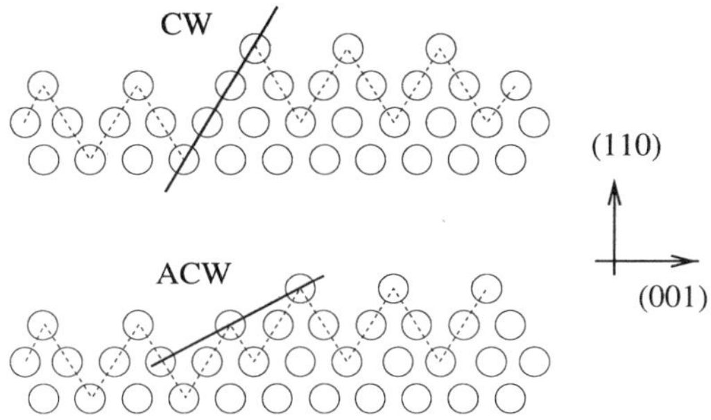

Figure 3: Examples of clockwise (CW) and anticlockwise (ACW) steps.

two arrays of roughly parallel steps crossing each other and forming on average angles ϕ and $-\phi$, with the vertical direction in Fig. **2**. An explanation for the formation of this network has been given already in (Kuipers, 1994). In this article we give a more quantitative description based on a simple, yet, we think, quite realistic model, which describes surface orientations close to the (110) facet. On the basis of this model we discuss the thermodynamical properties of the networks (Carlon and Van Beijeren, 1996). We find that the network is actually unstable or at least metastable: in equilibrium it decays into a combination of stable surface orientations, with the appearance of sharp edges in between. As a consequence the edge of the (110) facet shows cusps, i.e. jumps in the direction of its tangent.

THE MODEL

In a missing–row reconstructed (110) facet one can distinguish two different types of steps parallel to the missing rows, commonly known as clockwise and anticlockwise steps and illustrated in Fig. **3**. In STM observations anticlokwise steps are rarely seen especially at low temperatures, and when present they are mostly pinned by impurities. This indicates that clockwise steps have markably lower free energies per unit of length than the anticlockwise ones; this energy difference can in principle be derived from STM observations, by estimating the relative frequencies at which step segments of given length of either type of step do occur.

On the basis of STM observations on Au(110) surfaces it has been concluded (Kuipers, 1994) that at room temperature and for sufficiently clean samples, anticlockwise steps should be absent from the surface.

Due to the presence of reconstruction not all possible configurations of clockwise steps will actually occur on the surface. As illustrated in Fig. **4**(a) a clockwise step going up followed by another clockwise step going down induces a shift in the reconstruction of the lower terrace. For this reason, a closed clockwise step cannot be formed on a given terrace unless it is accompanied by domain boundaries separating regions of different reconstruction order (Fig. **4**(b)).

This fact may have some important consequences, for instance, on how the (110) facet roughens. In the present work however we will focus on the behavior of the system far from the deconstruction or roughening critical points.

The interplay between step orientation and surface reconstruction is essential for understanding the formation of the network of steps. Indeed, a step perpendicular to the missing rows cannot zig–zag forming clockwise segments parallel to the missing rows, if expensive domain boundaries between opposite reconstruction states are to be

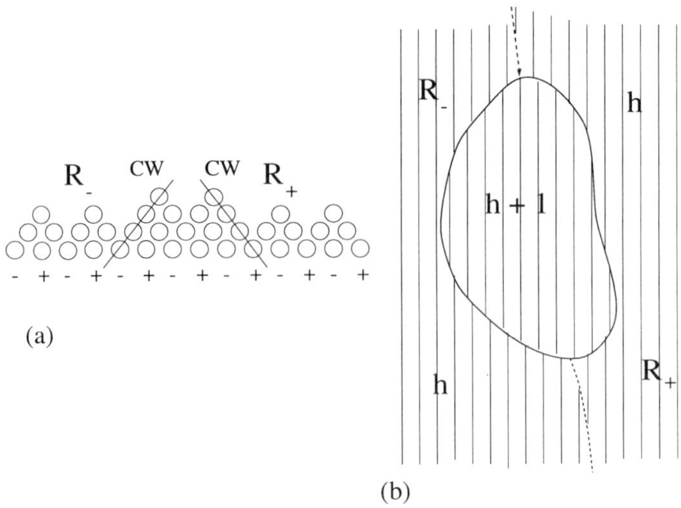

Figure 4: (a) A clockwise step up followed by a clockwise step down generates a shift in the reconstruction order; R_- and R_+ indicate the two possible reconstructed phases. (b) As a consequence of this a closed clockwise step must be accompanied by domain boundaries between the two phases R_- and R_+.

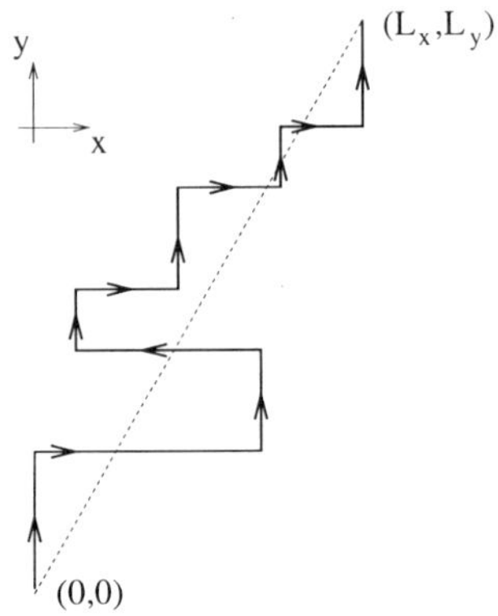

Figure 5: Example of microscopic configuration of a step; if anticlockwise steps are to be avoided only segments along the $+y$ and $\pm x$ directions are allowed.

avoided.

The single step free energy

To clarify and quantify the previous discussion we calculate the free energy of an isolated step as function of the step orientation. For convenience we consider, for the energy of an anticlockwise step segment, the limit $E_{ACW} \to \infty$: only clockwise step segments are allowed and they have energies per unit lengths δ_y and δ_x where y and x are the directions parallel respectively perpendicular to the missing rows. One could eventually introduce more parameters, such as a corner energy, but this is not essential.

Let us consider the step as an oriented walk from a lattice point $(0,0)$ to (L_x, L_y), as shown in Fig. 5. As seen above, if anticlockwise steps are to be avoided, only step segments in the $\pm x$-directions and the $+y$-direction are allowed. A simple calculation (for more details see Appendix I) gives the following free energy of a step per unit of length, tilted over an angle ϕ with respect to the missing rows:

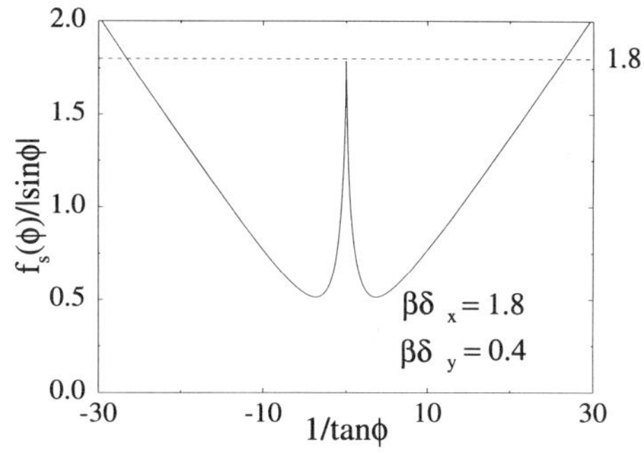

Figure 6: The step free energy per unit of projected length along the x-direction as function of $1/\tan\phi$. Steps with orientations close to $\phi = \pi/2$ are unstable as can be seen from the cusp in the figure.

$$f_s(\phi) = \frac{\ln z(\phi)}{2\beta}\,|\sin\phi| + \left[\delta_y + \frac{\sqrt{2}}{\beta}\ln\left(\frac{2\cosh(2\beta\delta_x) - (z(\phi) + 1/z(\phi))}{2\sinh(2\beta\delta_x)}\right)\right]\cos\phi \qquad (1)$$

where $\beta = 1/(k_B T)$ and:

$$z(\phi) = \frac{\cosh(2\beta\delta_x)\,t(\phi) + \sqrt{1 + \sinh^2(2\beta\delta_x)\,t^2(\phi)}}{1 + t(\phi)} \qquad (2)$$

with $t(\phi) = |\tan\phi|/2\sqrt{2}$. Notice that the limiting value for the free energy of a step running perpendicular to the missing rows is simply $f_s(\pi/2) = \delta_x$: such steps are perfectly straight and their free energy is "frozen", i.e. temperature independent. In the limit $\phi \to 0$, where steps are parallel to the missing rows, the step free energy is:

$$f_s(0) = \delta_y + \frac{\sqrt{2}}{\beta}\ln\left(\tanh(\beta\delta_x)\right) \qquad (3)$$

where the first term on the r.h.s. is the energy of the ground state configuration, while the second term is negative and represents the contribution of thermal excursions along the x-direction.

For investigating the stability of steps of various orientations it is convenient to calculate $f_s(\phi)/|\sin\phi|$, the step free energy per unit of projected length along the x-direction (Van Beijeren and Nolden, 1987). This quantity is shown in Fig. **6**: it has a local maximum with a cusp at $\phi = \pi/2$. For steps slightly inclined with respect to this orientation it decreases, since such steps have a larger entropy. This can be seen expanding $f_s(\phi)/|\sin\phi|$ around $\phi = \pi/2 \pm \epsilon$; such an expansion yields:

$$\left.\frac{f_s(\phi)}{|\sin\phi|}\right|_{\frac{\pi}{2}\pm\epsilon} = f_s(\pi/2) + A|\epsilon| - B|\epsilon\log\epsilon| + \ldots \qquad (4)$$

with A and B some non-negative constants. The second term on the r.h.s. is due to the energy of segments of steps parallel to the y-direction, while the third term is the decrease in free energy due to entropy. The entropic term dominates at sufficiently small $|\epsilon|$, except at $T = 0$, where $B = 0$.

Obviously $f_s(\phi)/|\sin\phi|$ diverges for $\phi \to 0$. It is minimal for an angle ϕ_0 satisfying[1]:

[1]The quantity under the square root in (5) becomes negative at small values of $\beta\delta_x$ and $\beta\delta_y$; this happens at temperatures above the roughening temperature of the (110) surface where the solid-on-solid approximation for the step free energy $f_s(\phi)$ is not positive definite and the simple theory considered here breaks down.

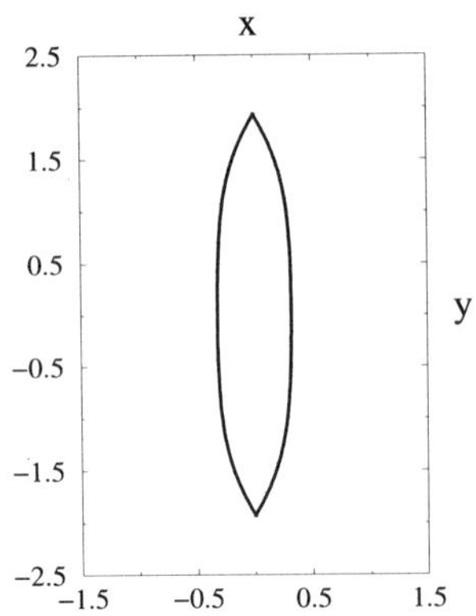

Figure 7: Equilibrium shape of the (110) facet obtained from the Wulff construction, for the same values of $\beta\delta_x$ and $\beta\delta_y$ as shown in Fig. **6**. Notice the cusps, due to instabilities of steps perpendicular to the missing rows.

$$t(\phi_0) \quad = \quad \frac{\sqrt{\left(\cosh(2\beta\delta_x) - e^{-\beta\delta_y/\sqrt{2}}\sinh(2\beta\delta_x)\right)^2 - 1}}{e^{-\beta\delta_y/\sqrt{2}}\sinh(2\beta\delta_x)}. \tag{5}$$

This simple calculation shows that, on a surface tilted along the $[1\bar{1}0]$ direction, steps making a finite angle $\phi_0 \neq \pi/2$ with respect to the missing rows are entropically favored. ϕ_0 is the optimal value of the angle of the network. Its value, given by Eq. (5), could in principle be used to determine the parameters δ_x and δ_y from measured values of ϕ_0 at different temperatures. In fact the temperature dependence of ϕ_0 has been studied in some STM observations of networks on Au(110) surfaces by Hoogeman and Frenken (Hoogeman and Frenken, private communication). They found a characteristic angle that is rather temperature independent and pointed out this may be due to the presence of impurities which act as pinning sites for the network. We think that this point is quite interesting and deserves further experimental investigations.

The equilibrium shape of the (100) facet

The equilibrium shape of the (110) facet can be found by applying a one-dimensional Wulff construction to the step free energy as function of orientation (Van Beijeren and Nolden, 1987). The result of this construction, for a representative choice of step energies and temperature, is shown in Fig. **7**. Steps with orientations close to $\phi = \pi/2$ are unstable and would phase separate into combinations of two steps of orientations ϕ_0 and $-\phi_0$: the shape of the (110) facet resembles that of an almond, with cusps along the $[1\bar{1}0]$ direction.

This type of shape is unusual, because the cusps correpond to first order phase transitions in a one-dimensional system at finite temperatures. In normal situations this could not occur (Van Beijeren and Nolden, 1987): if steps of orientations ϕ_0 and $-\phi_0$ both have free energy, say f_0, steps of intermediate orientations could always be built out of an alternating sequence of segments of orientations ϕ_0 and $-\phi_0$. Then at any non-zero temperature the lowering of free energy through entropy gain outwins the increase through the excess energy at the corners between the successive segments, provided the density of these corners is sufficiently low. This is a variant of the usual

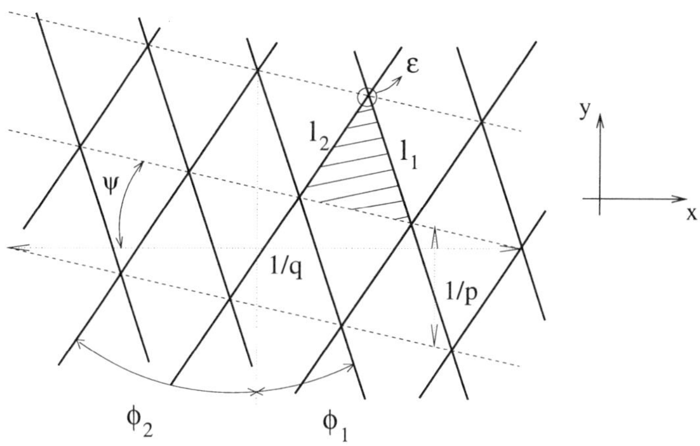

Figure 8: Schematic view of the network of crossing steps.

argument against phase transitions in one-dimensional systems at finite temperatures. In the present case this argument does not work since, due to topological constraints induced by the reconstruction, a configuration of short "zig–zagging" step segments of orientations ϕ_0 and $-\phi_0$ joined together cannot be formed, as pointed out above.

The surface free energy

Next we calculate the surface free energy as function of the slope parameters p and q, which denote the tangents of the tilting angles of the surface with the $[1\bar{1}0]$ and the $[001]$ directions. Both parameters q and p will be considered small, so we restrict ourselves to vicinal surface orientations.

Interactions between steps are introduced in a simple way: a free energy $\tilde{\epsilon}$ is associated to each pair of crossing steps. $\tilde{\epsilon}$ keeps account of both short and long range interactions between the steps and it is defined as the excess free energy associated to the intersection point of two crossing steps forming angles ϕ_1 and ϕ_2 with the y-direction on a (110) facet:

$$\tilde{\epsilon}(\phi_1, \phi_2) \quad = \quad F(\phi_1, \phi_2) - A f_0 - s_1 f_s(\phi_1) - s_2 f_s(\phi_2) \tag{6}$$

Here $F(\phi_1, \phi_2)$ is the free energy of a surface of area A with two crossing steps of lengths s_1 and s_2, f_0 is the free energy of the (110) facet per unit of area (no steps are present), and $f_s(\phi)$ the free energy of a single step, whose properties have been discussed above.

One could try to estimate $\tilde{\epsilon}$ from a microscopic point of view taking into account long range interactions of elastic or entropic type and short range interactions near the crossing point. This estimate is non-trivial and for the general features of the equlibrium crystal shape that we want to address here, only the sign of $\tilde{\epsilon}$ matters. In general, it is expected (Bernasconi and Tosatti, 1993) that elastic interactions between two antiparallel steps (as are the crossing steps of the network, provided ϕ_0 is not too large) yield a negative contribution to $\tilde{\epsilon}$. Entropic interactions (Gruber and Mullins, 1967; Jayaprakash et al., 1984), which become relevant at higher temperatures, are repulsive and give a positive contribution to $\tilde{\epsilon}$. Finally at the crossing point one would expect a positive contribution to $\tilde{\epsilon}$. In the rest of the article we discuss both possibilities for the sign of $\tilde{\epsilon}$, which yield two different scenarios for the equilibrium shape of the crystal.

Let us consider a miscut along the $[1\bar{1}0]$ direction, generated by a pattern of crossing steps forming angles ϕ_1 and ϕ_2 with the y-axis, as shown in Fig. **8**. The dashed inclined lines indicate hypothetical parallel isolated steps that would generate the same macroscopic orientation as the network; $1/q$ and $1/p$ are the average distances between these steps along the x and the y direction.

The free energy per unit of projected area can be written as:

$$\tilde{f}(p,q,\phi_1,\phi_2) \quad = \quad f_0 + \frac{1}{2A}\left[l_1 f_s(\phi_1) + l_2 f_s(\phi_2) + \tilde{\epsilon}(\phi_1,\phi_2)\right] \tag{7}$$

where f_0 is the free energy per unit area of the (110) facet, while l_1 and l_2 are the lengths of the two sides of the dashed triangle, of area A, shown in Fig. **8**.

The actual free energy can be found by minimizing the free energy (7) with respect to the angles ϕ_1 and ϕ_2; to lowest orders in p and q and for $|\psi| < \pi/2 - \phi_0$, where ψ is the angle shown in Fig. **8**, this amounts to minimizing $f_s(\phi)/|\sin\phi|$. Therefore the minimum of (7) is at $\phi_1 \approx -\phi_2 = \phi_0 + O(p,q)$, where ϕ_0 is the angle given by (5). Substituting this back into (7), expressing A, l_1, l_2 as function of p, q, ϕ_1, ϕ_2 and expanding to lowest orders in p and q we find:

$$f(p,q) \quad = \quad f_0 + \frac{f_s(\phi_0)}{\sin\phi_0}p + \frac{\tilde{\epsilon}}{2}\left(\frac{p^2}{\tan\phi_0} - \tan\phi_0\, q^2\right) \tag{8}$$

where $\tilde{\epsilon}$ is the interaction free energy (6) calculated at angles $\phi_1 = -\phi_2 = \phi_0$. As in usual expansions of surface free energies around facets (see e.g. Van Beijeren and Nolden, 1987) the term linear in p represents the contribution of non-interacting steps. The interaction terms are quadratic in the step densities, differently from usual step-step interactions which lead to terms cubic in the step densities (Gruber and Mullins, 1967). The origin of the quadratic term can be understood easily: the number of step crossings per unit area is simply proportional to the product of the densities, $p \pm q\tan\phi_0$, of the two types of steps. In addition there are long range interactions between the parallel steps in the network, but they will only contribute to terms of cubic or higher order in the step density expansion of the free energy (Gruber and Mullins, 1967).

For $|\psi| > \pi/2 - \phi_0$ the expression (7) is minimized by a single array of steps (so $\phi_1 = \phi$ and $l_2 = 0$) and the free energy takes the usual form (Gruber and Mullins, 1967):

$$f(p,q) \quad = \quad f_0 + f_s(\phi)\sqrt{p^2+q^2} + O((p^2+q^2)^{\frac{3}{2}}) \tag{9}$$

Notice that the expression (8) for the free energy of the step network, irrespectively of the sign of $\tilde{\epsilon}$, is a non-convex function of p and q. This result implies that the network is always unstable: some surface orientations disappear from the equilibrium shape of the crystal and are replaced by sharp edges.

EQUILIBRIUM CRYSTAL SHAPES

As is well known (see, for instance, Van Beijeren and Nolden, 1987), the equilibrium crystal shape is the shape that minimizes the total surface free energy at a given fixed volume. From the minimization of the free energy calculated above we can construct the equilibrium shape of the crystal around the (110) facet. This shape depends crucially

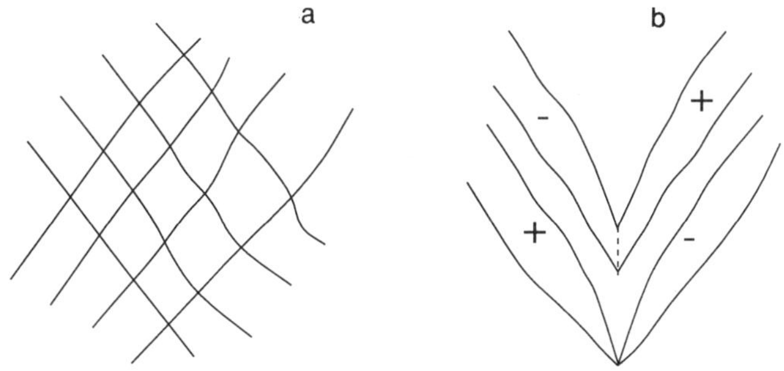

Figure 9: (a) The network of crossing steps, and (b) a combination of two step arrays of different orientation, after the phase separation, predicted by (5) for $\tilde{\epsilon} > 0$, has occurred. The + and − indicate the opposite reconstruction phases induced by the clockwise steps.

on the sign of the interaction free energy $\tilde{\epsilon}$, therefore in the rest of the paper we will distinguish two different cases.

$\tilde{\epsilon} > 0$

When $\tilde{\epsilon}$ is positive the system can decrease the surface free energy reducing the density of crossings. By applying the Maxwell construction to Eqs. (8, 9) one finds that a given surface orientation $(p, 0)$ separates into two orientations of slopes $(p, p\tan\phi_0)$ and $(p, -p\tan\phi_0)$ joining at a ridge, as depicted in Fig. **9**(b). In this combination of two surface orientations all the crossing points have been "eliminated" with a net gain of free energy. To generate the interface between the two surface orientations one has to pay a positive amount of boundary free energy, but this free energy will be proportional to the length of the boundary and therefore negligible in the thermodynamic limit, when compared to terms proportional to the surface area.

The structure of the domain boundary is quite interesting as well. Two clockwise steps running under angles ϕ_0 and $-\phi_0$ with the missing-row direction cannot be simply joined together, because this will cause a mismatch in the reconstruction order. Such joints must be accompanied by a domain boundary.

As we saw already in the previous Section the model predicts that the network is unstable, suggesting that the pattern of steps observed in STM experiments should decay, after a sufficiently long time, into the combination of surface orientations depicted in Fig. **9**(b) In the present case this time may be extremely long. In fact once the network has been formed, the process of disentangling it into stable orientations as the ones shown in Fig. **9**(b), may require the investment of a large amount of free energy to go through very unfavorable states and it may be difficult to observe it experimentally without a careful long annealing of the surface.

Thermodynamically the metastability encountered here is highly unusual. In e.g. a homogeneous gas-liquid system a free energy that is a concave function of density always leads to instability due to spinodal decomposition. In our system this is impeded by topological constraints on the steps, requiring concerted mass transportation over relatively large distances for the decomposition of a network into stable surfaces. Therefore even a non-convex free energy can be metastable. For describing the surface free energy one can use Eq. (8) again. A typical arrangement of steps around the (110) facet in this situation is shown in Fig. **10**(b). In this case the shape profile along the y-axis of the crystal for vicinal orientations is expected to be of the type $z(y) \sim (y - y_0)^2$, due to the term proportional to p^2 in the surface free energy. A free energy expansion

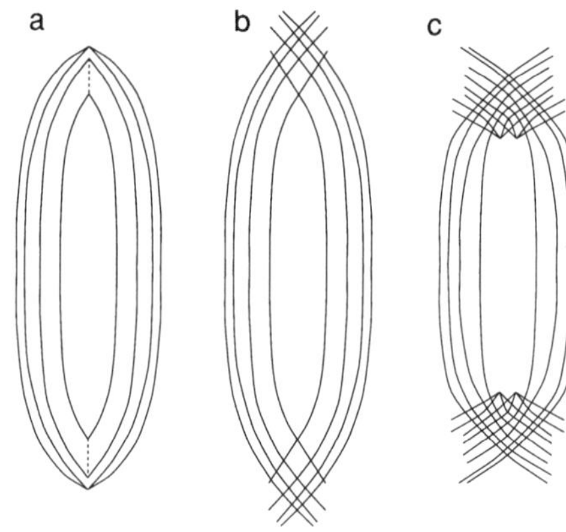

Figure 10: Arrangement of the steps surrounding the (110) facet for $\tilde{\epsilon} > 0$ in the stable phase (a), in the metastable phase (b) and for $\tilde{\epsilon} < 0$ (c).

with a term cubic in the step density would produce a shape profile with an exponent 3/2, the so-called Pokrovsky-Talapov exponent (see, for instance, Van Beijeren and Nolden, 1987).

$\tilde{\epsilon} < 0$

When a negative free energy is associated to a crossing point the network of steps tends to condense to maximize the density of crossings. On the other side, entropic repulsions (Gruber and Mullins, 1967) favor configurations where steps are far apart and tend to stabilize the network.

The competition between repulsions and attractions may give rise to the two following scenarios: 1) if attractions dominate over the whole range of surface orientations the network fully condenses and the stable orientations along the p direction are the (110) and (100) facets, which are directly connected at a sharp edge; 2) if repulsions, possibly of entropic type become dominant, then the surface free energy turns concave at higher step densities and the (110) facet connects to a rounded region, again at an edge.

As pointed out above, the surface free energy $f(p, q)$ contains terms which are of cubic or higher order in the step density. A term cubic in p with a positive coefficient could well stabilize the free energy at not too small values of p. In both scenarios the (110) facet has sharp edges running roughly perpendicular to the missing-row direction.

A simple model with a negative crossing energy (Carlon et al., 1996; Carlon, 1996) the staggered body-centered-solid-on-solid-model (BCSOS model), is discussed in some details in Appendix II. The model reproduces both scenarios 1) and 2), depending upon temperature.

Finally in Fig. **7** the dashed lines show the truncation of the equilibrium shape of the (110) facet by ridges connecting the facet to rounded areas and Fig. **10**(c) shows the expected arrangement of steps around the truncated facet. Notice that in this case there are sharp ridges between rounded regions covered by networks of steps and regions covered by non-crossing step arrays.

CONCLUSION

In the present article we introduced a simple model which describes networks of crossing steps, that have been observed in STM experiments on $Au(110)$ and $Pt(110)$

surfaces. The model shows that these networks are unstable (or at least metastable) with respect to faceting.

From the calculation of the surface free energy for orientations close to the (110) facet the equilibrium shape for this facet has been derived. For repulsive interactions between two crossing steps ($\tilde{\epsilon} > 0$) the facet shape is rather unusual: it is cusped and elongated along the $[1\bar{1}0]$ direction. The cusps are connected to ridges separating rounded regions. If the interaction is attractive ($\tilde{\epsilon} < 0$) the (110) facet is still elongated with sharp edges roughly perpendicular to the missing-row direction. In this case there are much weaker cusps, between these edges and the smooth facet boundaries roughly parallel to the missing–row direction.

The predicted shapes could be observed in experiments on equilibrium shapes of crystals with (110) missing–row reconstructed facets.

Heyraud and Métois (1980) studied shapes of small gold crystals in thermal equilibrium with their vapor; in their samples only the (111) and (100) facets were observed, since the range of temperatures investigated ($T \approx 1000°C$) is above the roughening temperature of the (110) facet. To observe some of the shapes described in this article, temperatures below the roughening and the deconstruction temperatures of the (110) facet should be considered.

We believe that the problems connected to the metastability/instability of the networks of steps deserve further experimental investigation, as well. If $\tilde{\epsilon}$ is positive, as we think should be the case for gold and platinum crystals, one should be able, starting from the metastable network of steps, to observe a nucleation of arrays of parallel steps connected under a ridge. Probably this could be observed in practice, e.g. in STM experiments, only for sufficiently high temperatures and in very pure samples.

Acknowledgements:

We had several stimulating discussions with Joost Frenken, Laurens Kuipers and Misha Hoogeman, who drew our attention to the networks of steps discussed here in the first place. E.C. aknowledges financial support from grant No. CHBGCT940734 from the European Capital and Mobility Programme.

APPENDIX I: CALCULATION OF THE STEP FREE ENERGY

For the calculation of the step free energy we consider first the partition functions G_x and G_y of straight segments of steps along the x and y direction. The segments have energies per unit length equal to δ_x and δ_y. In Fourier space one has:

$$G_x(k_x) = \sum_{n=1}^{+\infty} e^{-2n\beta\delta_x}\left(e^{ink_x} + e^{-ink_x}\right) = \frac{1}{e^{2\beta\delta_x + ik_x} - 1} + \frac{1}{e^{2\beta\delta_x - ik_x} - 1} \quad (10)$$

$$G_y(k_y) = \sum_{n=1}^{+\infty} e^{-n\beta\delta_y/\sqrt{2}} e^{-ink_y} = \frac{1}{e^{\beta\delta_y/\sqrt{2} + ik_y} - 1} \quad (11)$$

Notice that segments of steps along the negative y direction are not allowed. We recall that the factors $\sqrt{2}$ and 2 are the minimal lengths of step segments along the x and y directions. The step is generated by all possible combinations of horizontal and vertical segments:

$$G(\vec{k}) = G_x + G_y + G_xG_y + G_yG_x + \ldots = \frac{G_x + G_y + 2G_xG_y}{1 - G_xG_y} \quad (12)$$

The partition function of a step of lengths L_x and L_y is obtained by Fourier transform:

$$Z(L,\phi) = \int_{-\pi}^{+\pi} \frac{d\vec{k}}{4\pi^2} e^{i(k_y\sqrt{2}L_y + k_x L_x/2)} G(\vec{k}) \tag{13}$$

with $L = \sqrt{L_x^2 + L_y^2}$ and $\phi = \arctan(L_x/L_y)$. The integral over k_y can be rewritten as a contour integral in the variable $z = e^{ik_y}$; the integrand has a simple pole in:

$$z = e^{-\beta\delta_y/\sqrt{2}}(1 + G_x(k_x)) \tag{14}$$

The partition function now becomes:

$$Z(L,\phi) = \int_{-\pi}^{+\pi} \frac{dk_x}{2\pi} \, e^{-L\beta f(k_x,\phi)} \ldots \tag{15}$$

with:

$$\beta f(k_x,\phi) = \frac{ik_x}{2}\sin\phi + \left[\beta\delta_y - \sqrt{2}\log\left(1 + G_x(k_x)\right)\right]\cos\phi \tag{16}$$

The dots in (15) denote terms which are not relevant in the thermodynamic limit $L \to \infty$. Using the saddle point approximation, one can evaluate (15) as:

$$Z(L,\phi) = e^{-L\beta f_s(\phi)} \tag{17}$$

where the step free energy is $f_s(\phi) = f(i\bar{k},\phi)$ and $i\bar{k}$ the stationary point of $f(k_x,\phi)$. Working this out one obtains Eq. 1.

APPENDIX II: THE STAGGERED BCSOS MODEL

Without going into details we present here some relevant results concerning the staggered BCSOS model, which, for certain values of the energy parameters describes networks of crossing steps with a *negative* crossing energy (for more details the reader may consult (Carlon et al., 1996; Carlon and Van Beijeren, 1996 II)).

The model is applicable to surfaces of ionic crystals of bcc type, as for instance CsCl, and describes all surface orientations $(ts1)$ with $|t| + |s| \leq 1$. It does not have the characteristic feature, responsible for the formation of networks in the present paper, that an up step followed by a down step changes a reconstruction order; instead the networks are induced by a large corner free energy, leading to steps with long segments in the principal lattice directions, combined with a negative value of $\tilde{\epsilon}$, which favors crossings of steps. However, the model does illustrate the two different scenario's for step condensation discussed above for $\tilde{\epsilon} < 0$.

At $T = 0$ the crystal has the shape of a truncated piramyd, with a top (100) facet and four side facets of (110) type, as shown in Fig. **11**(a). A step on the (100) facet has an energy per unit of length $e > 0$ and a kink energy $e_K \gg e$: steps consist, especially at low temperatures, of elongated straight segments with few kinks. These segments are oriented parallel to the two axes γ_1 and γ_2 shown in Fig. **11**(b): differently from (110) surfaces of fcc metals these two directions are equivalent. Steps can cross each other, with a gain of energy of $2e$ at each crossing point. In this model, thus, the crossing energy is $\tilde{\epsilon} = -2e < 0$.

The surface free energy has been calculated as function of the slope parameters p, q in mean–field approximation, for all possible values of p and q, also beyond the vicinal orientations. To give an example, we restrict ourselves to considering only $f(p)$, i.e. the

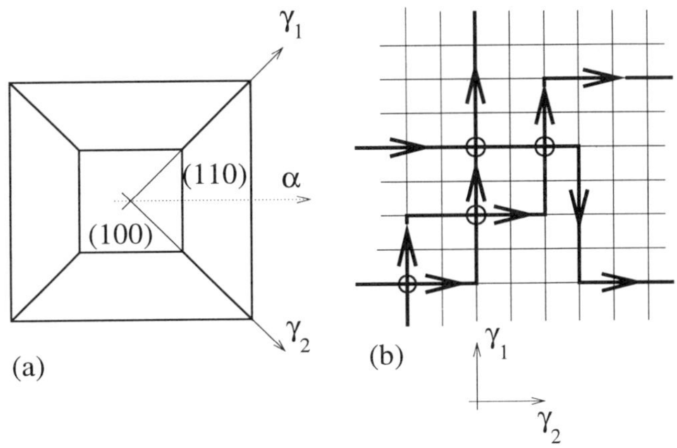

Figure 11: (a) At $T = 0$ the equilibrium shape of the BCSOS model is a truncated pyramid. (b) Steps are composed of elongated segments with few corners since the energy of a corner is much larger then the energy of a straight segment of step $e_K \gg e$. Crossings of steps, indicated by circles in the figure, are energetically favored, i.e. at each crossing there is a gain of energy of $2e$.

surface free energy for surface orientations of the type $(1p0)$, with $0 \leq p \leq 1$; these are all the surface orientations along the axis α of Fig. **11**(a). The mean–field calculation gives a surface free energy that at low temperatures is not stable (Fig. **12**(a)): step attractions, due to the energy gained at crossings, dominate in the whole range of surface orientations and the network condenses until the optimal density of steps is reached, which in the present case corresponds to the (110) facet. In the equilibrium shape the two facets (100) and (110) touch each other at a sharp edge. The free energy instability is of the same form as that predicted in Eq.(8), with a negative value of $\tilde{\epsilon}$.

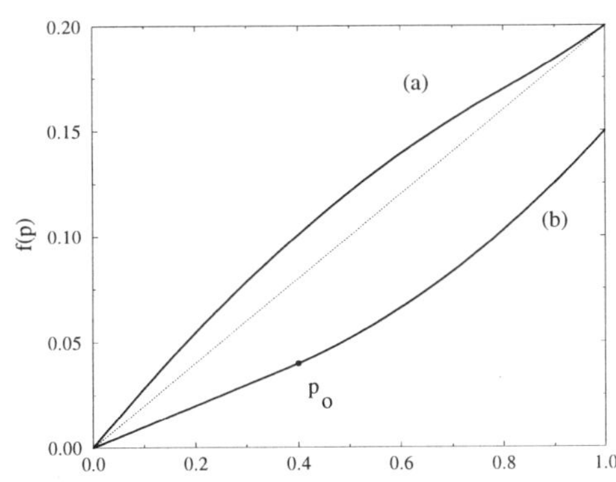

Figure 12: Surface free energy along the axis α of Fig. **11**. (a) At low temperatures the free energy is not stable for the whole range of orientations $0 \leq p \leq 1$; the dotted line denotes the Maxwell construction. (b) At higher temperatures the range of orientations $p_0 \approx 0.4 \leq p \leq 1$ is stable.

We stress that the free energy $f(p)$ of Fig. **12**(a) becomes convex for sufficiently large values of p. This change of sign of the second derivative of $f(p)$ corresponds to a transition from a reconstructed to a non-reconstructed state. In the present case this critical point is of limited relevance since it occurs in an unstable region, and there seem to be no topological restrictions impeding the decay of an unstable orientation into two facets.

The same mean–field analysis at somewhat higher temperatures gives a surface free energy of the type shown in Fig. **12**(b), which is also concave at low p, but convex at higher values of p. At low step densities, attractions still dominate, yielding a concave (i.e. unstable) free energy. Entropic repulsions (Gruber and Mullins, 1967) which become important at high temperatures and high step densities, stabilize the network. The Maxwell construction gives as coexisting surface orientations the (100) facet and a non-faceted orientation of slope p_0. These touch again at a sharp edge.

REFERENCES

Bernasconi, M. and Tosatti, E., 1993, *Surf. Sci. Repts.* 17:363.

Carlon, E., 1996, *Ph.D. Thesis*, Utrecht.

Carlon, E., Van Beijeren, H. and Mazzeo, G., 1996 *Phys. Rev. E* 53:R5549;

Carlon, E. and Van Beijeren, H., 1996, *Phys. Rev. Lett.* 76:4191.

Carlon, E. and Van Beijeren, H., 1996, *preprint.*

Gimzewski, J.K., Berndt, R. and Schittler, R.R., 1992, *Phys. Rev. B* 45:6844.

Gritsch, T., Coulman, D., Behm, R.J. and Ertl, G,, 1991, *Surf. Sci.* 257:297.

Gruber, E.E. and Mullins, W.W., 1967, *J. Phys. Chem. Solids* 28:875.

Heyraud, J.C. and Métois, J.J., 1980, *Acta Metall.* 28:1789.

Jayaprakash, C., Rottman, C. and Saam, W.F., 1984, *Phys. Rev. B* 30:6549.

Kuipers, L., 1994, *Ph.D. Thesis* Amsterdam.

Van Beijeren, H. and Nolden, I., 1987, in: *Structure and Dynamics of Surfaces, Vol. 2*, W. Schommers and P. von Blanckenhagen eds., Springer-Verlag, Berlin, 1987.

COUPLED CHEMICAL AND FACETING SURFACE PHASE TRANSITIONS

Paul Wynblatt

Department of Materials Science and Engineering
Carnegie Mellon University
Pittsburgh, PA 15213

INTRODUCTION

Much work has been performed over the past two decades on phase transitions at surfaces. The majority of this work has focused on structural phase transitions, such as surface reconstruction[1] in pure materials, or reconstructions in the presence of adsorbed layers[2]. More recently, there has also been significant interest on surface disordering transitions in pure metals such as surface roughening-faceting[3] and surface premelting[4]. However, very little effort has been devoted to the study of these phenomena at the surfaces of multi-component materials. At such surfaces, chemical order-disorder transitions[5] may take place, in which a transition can occur from a state where a given chemical species occupies random sites to a state where that species occupies specific sites of a surface superlattice. In addition there can be abrupt changes in surface composition, i.e. in the adsorption or segregation of certain components[6-8]. These latter transitions have been referred to as "surface miscibility gap transitions"[8], or in the context of multi component liquids, as "pre-wetting transitions"[9,10]. Finally, roughening-faceting transitions can also take place in multi-component systems, and their character can be modified from that of the well-studied case of pure materials[11].

In this article, we begin by reviewing the underlying thermodynamic characteristics of the abrupt compositional surface phase transitions in ternary solid solutions, and provide some background regarding the nature of roughening-faceting transitions. We then proceed to describe some experimental results on ternary Pb-Bi-Ni alloys in which these transitions have been observed. In these alloys, these two types of transitions have been found to be closely coupled[11]. The origin of this coupling will also be discussed in terms of the changes in surface energy which are expected to accompany the compositional transitions.

CHARACTERISTICS OF CHEMICAL SURFACE PHASE TRANSITIONS

Abrupt changes in the concentration of segregating solute were first observed in experimental studies of carbon segregation in interstitial metal-carbon alloys by Hamilton and Blakely [6]. More recently, such transitions have also been observed in substitutional Cu-

Ag[7,8] and Pb-Bi-Ni[11,12] alloys The interpretation of the phenomena is somewhat simpler in the case of the substitutional alloys, where both species occupy the same lattice sites.

Statistical thermodynamic descriptions of these transitions in substitutional alloys have been developed for the cases of both binary[13] and ternary alloys[14], using a simple nearest neighbor bond model of the surface segregation phenomenon (including strain energy effects). Results of the model have been evaluated here using model parameters appropriate for a Pb-5at%Bi-0.04at%Ni alloy for which experimental results will be provided below. However, the model can be applied in principle to the computation of equilibrium surface composition of any ternary solution.

Figure 1a gives the equilibrium surface atom fractions of Bi and Ni for the {111} surface of the alloy, as a function of temperature. The results show the presence of a van-der-Waals-like loop, indicating the existence on metastable and unstable surface states. Figure 1b gives the calculated surface excess free energy (subsequently referred to as the surface energy, for the sake of brevity) of this alloy, computed for the equilibrium compositions displayed in Fig. 1a. At high temperatures, Fig. 1a shows only one solution for the surface composition of the alloy, and there is thus only one corresponding value of surface energy. At intermediate temperatures, multiple solutions are obtained in both Figs. 1a and 1b, and their interpretation is a little more clear if one considers the results of Fig. 1b. At temperatures where three values of surface energy are obtained, the lowest value corresponds to an absolute minimum in energy, i.e. the equilibrium state, the next higher value corresponds to a second minimum in energy, representing a metastable state, and the highest energy state corresponds to an energy maximum, or an unstable state of the system. Also, it can be seen that the surface energy branch which corresponds to the equilibrium state at higher temperatures (and which is associated with low surface atom fractions of the two solutes) is intersected by a second branch (associated with higher surface atom fractions of the solutes) which becomes the absolute minimum at lower temperatures. Thus, the point of intersection of these two branches represents an abrupt transition from a low segregation state at high temperatures to a high segregation state at low temperatures. This transition temperature is shown in Fig. 1a, as a vertical line.

Figure 1c compares the results of similar calculations for three different surface orientations, namely: {111}, {100} and {110}. Each of these orientations shows a sharp change in the temperature dependence of the surface energy at its respective transition temperature, where solute surface composition undergoes a discontinuous jump. At higher temperatures, where solute adsorption is low, the temperature dependence of the surface energy is quite weak, whereas at lower temperatures where solute adsorption is high, the temperature dependence of the surface energy is quite strong. This effect is consistent with expectations based on the Gibbs adsorption equation relating surface energy to the magnitude of the adsorption. Furthermore, this large change in the temperature dependence of the surface energy is an important result, and will be used later in the interpretation of experimental results.

CHARACTERISTICS OF ROUGHENING-FACETING TRANSITIONS

Surface roughening-faceting transitions have attracted considerable interest in recent years. The most intensively studied surface from this point of view has been the (110) surface of pure Pb[15,16]. In order to understand roughening-faceting transitions, it is useful to begin by considering the factors which determine whether facets are present on the equilibrium form of a crystal.

In general, the surface energy of crystals is anisotropic. This anisotropy is commonly represented by a so-called γ-plot, which is a polar plot of the surface energy, in which the length of the radius from the origin is proportional to the surface energy of the surface which

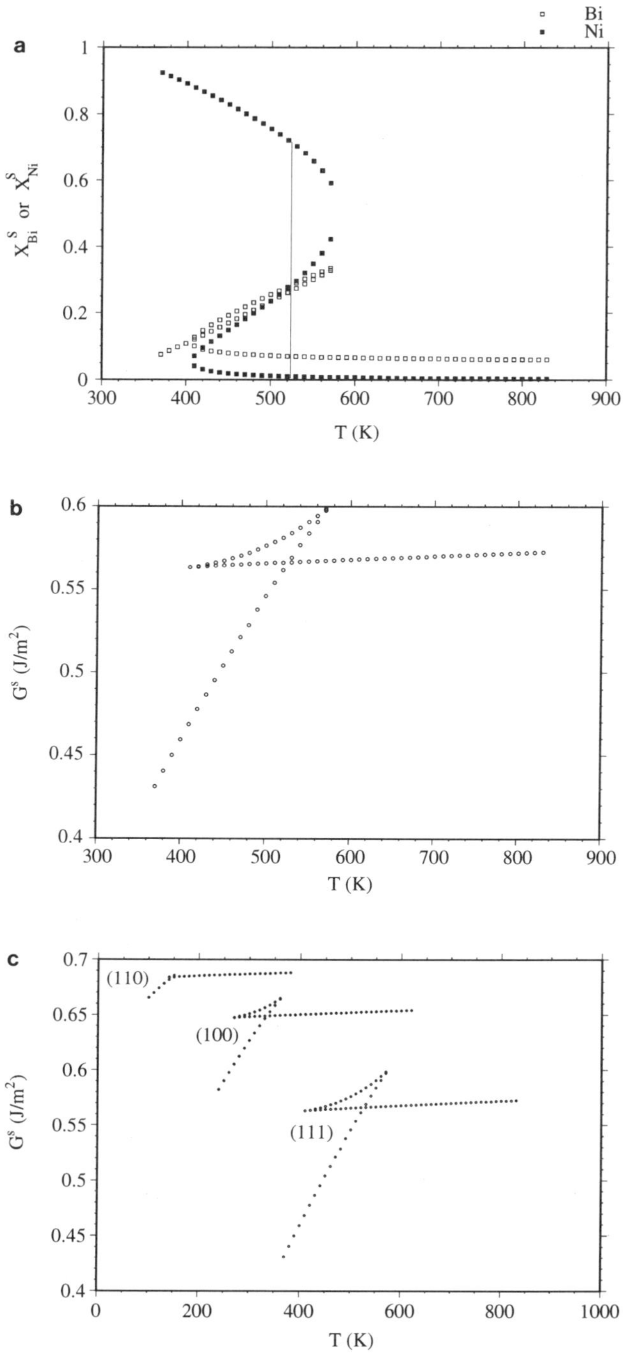

Figure 1. (a) Equilibrium surface atom fractions of Bi and Ni, on the {111} surface, for a Pb--5at%Bi-0.04at%Ni alloy, as a function of temperature, after ref. 14. The vertical line indicates the transition temperature. (b) Surface excess free energy of the (111) surface, vs. temperature, corresponding to (a), after ref. 14. (c) Same as (a), comparing the surface excess free energies of {111}, {100} and {110}, after ref. 17.

is normal to that particular direction. The equilibrium form of crystals may be obtained from γ-plots by means of the Wulff construction, which entails the construction of normal planes at the tips of the radii of the γ-plot. The inner envelope of these normal planes generates the equilibrium form. The existence of a cusp on the γ-plot is a necessary condition for the development of a facet on the equilibrium form.

At a temperature of absolute zero, the surface of a crystal is expected to be entirely bounded by flat facets of one or more distinct crystallographic orientations. As temperature is increased, thermal disordering is initiated on all facets by the formation of vacancies and adatoms. If the crystal is bounded by facets of several different orientations, the degree of thermal disorder will generally be different on the different facets. As the roughening transition temperature is approached, from a microscopic perspective, the energy to create atomic height steps at the surface will tend to vanish. Thus, while the surface atoms will continue to occupy sites of the crystal lattice, the width of the surface will diverge on an atomic scale. From a more macroscopic perspective, the cusp in the γ-plot associated with the facet which is undergoing roughening will lose its singular sharpness, and the facet will disappear from the equilibrium form of the crystal. With the blunting or disappearance of the cusp, the facet to be replaced by a curved surface covering a certain range of orientations lying between it and adjacent facets (some of which may not yet be roughened). If facets of different types are present at the surface, each facet type will disappear at its own roughening temperature. Figure 2 displays photomicrographs of crystallites of a Pb-5at%Bi-0.04at%Ni alloy which have undergone different thermal histories, as described in more detail below. Fig. 2a has been equilibrated at 260°C and shows rounded surfaces together with small {111} and {100} facets. At this temperature, therefore, any other facets which may have been present at lower temperatures have disappeared, and must therefore lie above their roughening transition temperatures.

EXPERIMENTS ON Pb-Bi-Ni ALLOYS

Observation of the Compositional Surface Phase Transition

The experiments summarized here were conducted on isolated microscopic single crystals of a Pb-5at%Bi-0.04at%Ni alloy (see Fig. 2), ranging in size from 10 to 15 μm, and prepared as described in ref 17. The composition of any region of the surface was determined by Auger electron spectroscopy in a scanning Auger microprobe, using an electron beam diameter of about 100nm. With such a small beam size, it is possible to perform analyses on individual facets of the crystallites. As an example, the composition of the {111} facets as a function of temperature is displayed in Fig. 3. It can be seen that the surface atom fractions of Bi and Ni undergo a sharp jump in composition at a temperature lying between 250 and 260°C, illustrating the occurrence of the compositional surface phase transition, as expected from the model for an alloy of this composition.

Observation of the Roughening-Faceting Transitions

Based on the description given above for the roughening-faceting transition, one possible approach for studying the phenomenon is through the study of the appearance and disappearance of facets from the equilibrium form of crystals. As mentioned above, Fig. 2 displays the equilibrium form of a given crystal of the Pb-Bi-Ni alloy, after equilibration at a series of different temperatures. Figure 2a shows a crystal equilibrated at 260°C. Auger analysis of the surface indicates that the rounded surfaces as well as the {111} and {100} facets all have approximately the same surface composition, namely: 10at%Bi and Ni at the noise level. Comparing with Fig. 3, it is clear that this state corresponds to the high

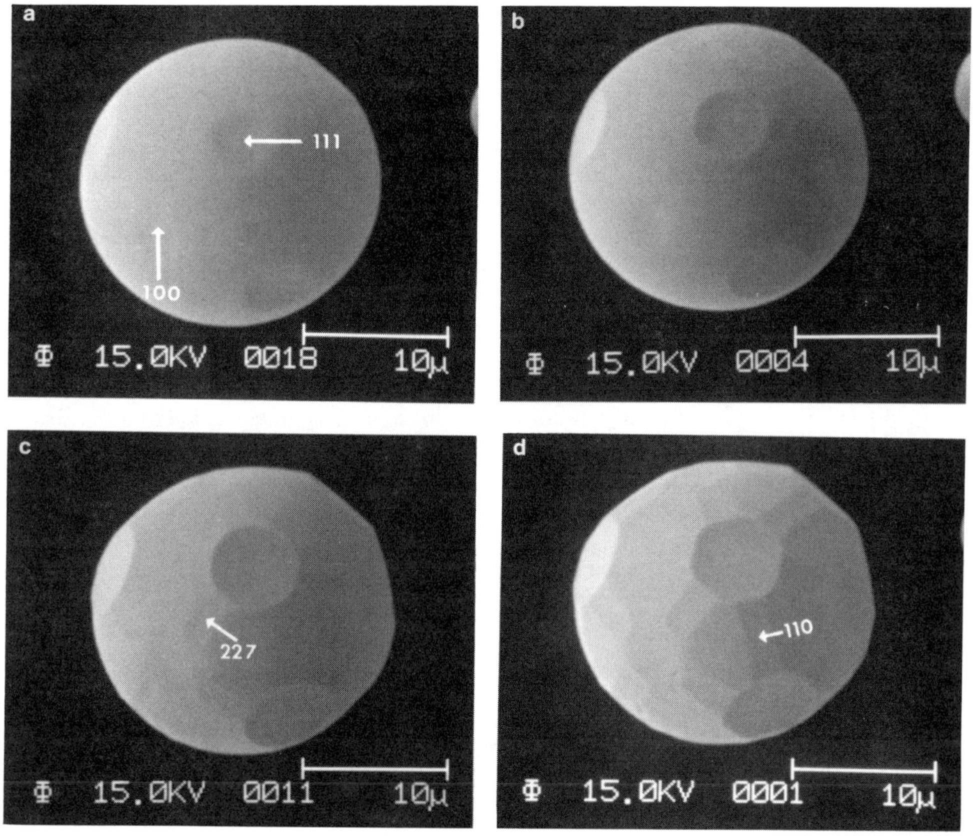

Figure 2. Photomicrographs of a Pb-5at%Bi-0.04at%Ni crystallite after equilibration at (a) 260°C, (b) 250°C, (c) 235°C, and (d) 225°C, after refs. 17 and 18.

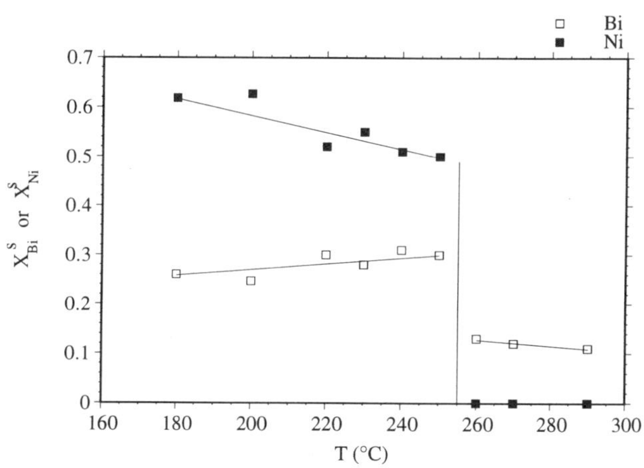

Figure 3. Experimentally determined surface atom fractions of Bi and Ni on the {111} facets of the alloy, vs. temperature.

temperature side of the chemical surface phase transition. Upon cooling to 250°C and holding for several hours to re-equilibrate, one observes the form shown in Fig. 2b. Surface analysis of this crystallite indicates that the surface concentrations of Bi and Ni of the {111} facets have now increased to about 30at% and 50at%, respectively, while the composition of the remaining parts of the surface are in the same state as they were in Fig. 2a. Thus, we can conclude that the {111} facets have undergone the chemical surface phase transition somewhere between 260 and 250 °C. Comparing the forms shown in Figs. 2a and 2b, we note that whereas the edges of {111} facets in Fig 2a are connected smoothly to the rounded parts of the equilibrium form, these facets have a sharp edge in Fig. 2b. Thus, the occurrence of the chemical surface phase transition on the {111} facets is accompanied by a smooth-to-sharp edge transition on those facets, indicating the disappearance of certain surface orientations from the equilibrium form of the alloy[18]. If the crystallite is now cooled further to 235°C and equilibrated, new {227} facets appears on the equilibrium form, as shown in Fig. 2c. Analysis of these facets reveals that the Bi and Ni surface concentrations have increased sharply to 25at% and 11at%, respectively. At this stage, all other surfaces (except the {111} facets} remain at relatively low concentrations of solutes. This indicates that the {227} facets have formed in conjunction with compositional surface phase transition at those orientations.

Figure 2d shows the crystal after further cooling to 225°C. In addition to {111}, {100} and {227} facets, new facets have appeared at {110} orientations, and the formation of those facets is also accompanied by higher Bi and Ni surface concentrations at those regions (about 30 at%Bi and 20 at%Ni). At this temperature, the {100} facets are also found to have undergone the compositional surface phase transition, and display surface compositions similar to those of the {110} surfaces.

These changes observed on cooling are reversible. Upon heating, the facets disappear at about the same temperatures as those at which they appeared, and the surface compositions in the region of the facets revert back to relatively low values of surface concentration of the two solutes as soon as the facets have disappeared. These results therefore demonstrate a coupling of compositional phase transitions with roughening-faceting transitions.

Interpretation of the Results in Terms of γ-plots

To begin with, we address the implications of the strong temperature dependence of the surface energy in the highly segregated state, in relation to the apparent coupling between the appearance of new facets on the equilibrium form of the alloy and the chemical surface phase transition. Facets are expected to appear on the equilibrium form of crystals only if cusps are present at those orientations in the γ-plot[19]. Heyraud and Metois[20] have determined the γ-plot from the equilibrium form of pure Pb, along the <110> crystallographic zone which passes through all of the new facets reported above, and have shown that there only exist cusps at {111} and {100} orientations in the temperature regime studied for the alloy. Also, the equilibrium form of the alloy at 260°C (Fig. 2a) is very similar to the equilibrium form of pure Pb. Thus, it is possible to conclude that the γ-plot of the alloy at 260°C is like that for pure Pb, and only possesses cusps at {111} and {100} orientations.

New {227} facets first appear on the stable equilibrium form of the alloy at 235°C. Once these facets appear, the surface of the facets show high concentrations of solute, characteristic of the compositional surface phase transition. Thus, between a temperature of 260°C where no cusp exists at {227} orientations, and 235°C where the facets appear, a cusp must form at those orientations. As noted earlier in connection with Figs. 1b and 1c, the change in surface energy with decreasing temperature is weak for all orientations at high temperatures in the low surface segregated state. At the compositional surface transition temperature, the two surface states (with low and high surface solute concentrations, respectively) will have identical surface energy; thus, if no cusp exists in the γ-plot before the compositional surface phase transition has occurred upon cooling, then no cusp will exist there at the transition temperature. However, as the temperature continues to decrease below the transition temperature, a rapid decrease in surface energy occurs at the orientation where the transition has taken place, thus producing a cusp at that orientation. As the cusp becomes sufficiently deep, it will eventually penetrate the envelope of tangent planes which define the equilibrium form of the crystal, and a new facet will appear at that orientation.

Our conclusions based on the above argument is that the appearance of new facets does not occur simultaneously with the sharp increase in surface solute concentration, but rather a few degrees below the transition. At the transition temperature, the composition of the "geometric points" which represent the orientations at which the compositional change has occurred on the rounded parts of the equilibrium form are indeed enriched in solute. However, the areas associated with the compositional change are too small to be detected by means of the 100 nm analytic probe of the SAM used in this study. The compositional changes can only be detected once the new facet has appeared, and has reached dimensions of the order of the probe size, or larger.

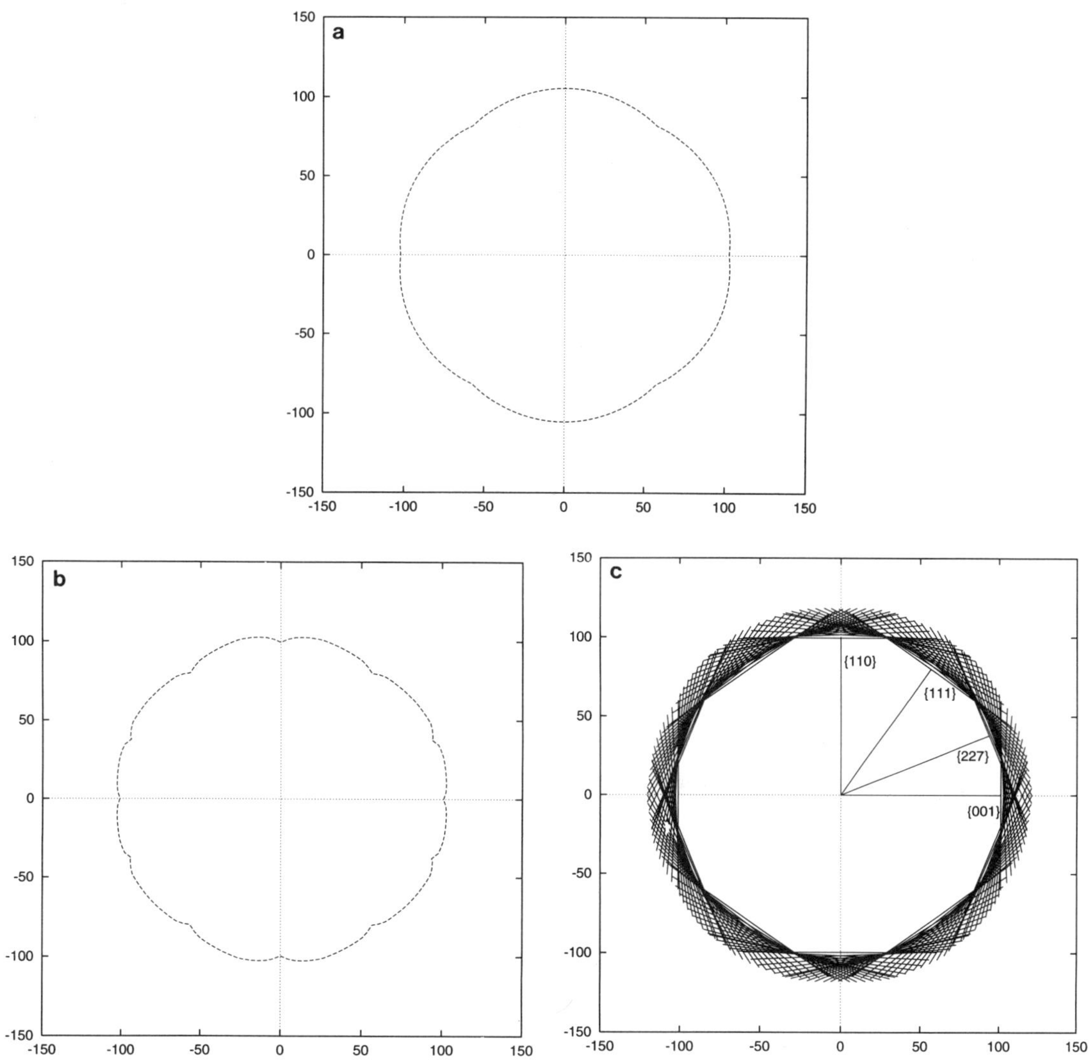

Figure 4. (a) γ-plot for pure Pb from data of ref. 20. (b) g-plot adjusted for equilibrium form observed in Fig. 2d. (c) Wulff plot corresponding to (b).

The arguments advanced here apply equally well to {110} facets. This conclusion is illustrated schematically in Fig. 4. Figure 4a shows the γ-plot obtained by Heyraud and Metois for pure Pb[20]. Figure 4b is a modification of Fig. 4a, in which the cusps at {111} and {100} have been deepened, and new cusps have been created at {227} and {110} orientations. The depth of the cusps has been adjusted to produce an equilibrium form by means of the Wulff construction (Fig. 4c) that matches the experimental equilibrium form of Fig. 2d. The deepening of the cusps needed to produce the desired shape is of the order of a few percent, and falls within the general range of anisotropy measured previously in the case of pure Pb.

CONCLUSIONS

This paper has summarized results relating to chemical surface phase transitions and roughening faceting transitions which have been observed in Pb-Bi-Ni alloys. The Bi and Ni solutes present in these alloys co-segregate to the surface, and their surface concentrations undergo an abrupt increase on cooling below some transition temperature. This transition temperature depends on the orientation of the surface; it is highest on surfaces of {111} orientation, and is progressively lower on the {227}, {110}, and {100} surfaces, respectively. The temperature dependence of the surface energies of the segregated surfaces is small in the high temperature weakly segregated state, but quite large in the low temperature strongly segregated state. As a consequence, once the transition has occurred at some particular surface orientation, a cusp forms in the γ-plot at that orientation, if one is not already present there. This process leads to the formation of new facets in the alloy which are not seen on the equilibrium form of pure Pb in the same temperature range.

Acknowledgments. Much of the material presented in this paper has been drawn from a series of papers co-authored with Dr. W-C. Cheng. The author also wishes to acknowledge with thanks support of this research by the National Science Foundation under grant DMR 9530469.

References

1. K. Muller, Ber. Bunsenges. Phys. Chem. 90:184 (1986).
2. R.J. Brehm, G. Ertl, V. Penka and R. Schwanker, *The Structure of Surfaces*, Springer, Berlin (1985).
3. M. Wortis, in: *Chemistry and Physics of Solid Surfaces VII*, R. Vanselow and R.F. Howe eds., Springer-Verlag, Berlin (1988).
4. J.G. Dash, Contemp. Phys. 30:89 (1989).
5. T.M. Buck, G.H. Wheathley, and L. Marchut, Phys. Rev. Lett. 51:43 (1983).
6. J.C. Hamilton and J.M. Blakely, Surf. Sci. 91:199 (1980).
7. J. Eugene, B. Aufray and F. Cabane, Surf. Sci. 241:1 (1991).
8. Y. Liu and P. Wynblatt, Surf. Sci. Lett. 241:L21 (1991).
9. J. W. Cahn, J. Chem Phys. 66:3667 (1977).
10. M. Schick, in: *Liquids at Interfaces*, J. Charvolin, J. F. Joanny and J. Zinn-Justin eds, Elsevier, Amsterdam (1990).
11. W-C. Cheng and P. Wynblatt, Surf. Sci. 302:185 (1994).
12. W-C. Cheng and P. Wynblatt, Surf. Sci. 302:179 (1994).
13. P. Wynblatt and Y. Liu, J. Vac. Sci. Technol. A10:2709 (1992).
14. W-C. Cheng and P. Wynblatt, Surf. Sci. 364:409 (1996).
15. J.C. Heyraud and J.J. Metois, J. Cryst. Growth 82:269 (1987).
16. H.N. Yang, T-M. Lu and G-C. Wang, Phys. Rev. Lett. 63:1621 (1989), and Phys. Rev. B43:4714 (1991).
17. W-C. Cheng and P. Wynblatt, J. Cryst. Growth, to be published
18. W-C. Cheng and P. Wynblatt, Surf. Sci. 364:417 (1996).
19. C. Herring, Phys. Rev. 82:87 (1951).
20. J.C. Heyraud and J.J. Metois, Surf. Sci. 177:213 (1986).

Dr. M. Alava
Michigan State University
Physics and Astronomy Department
E. Lansing, MI 48824

Dr. Jacques Amar
Emory University
Department of Physics
1510 Clifton Road
Atlanta, GA 30322
phsjga@emory.edu

Prof. A.-L. Barabasi
Department of Physics
University of Notre Dame
Notre Dame, IN 46556
alb@nd.edu

Dr. N.C. Bartelt
Senior Member of the Technical Staff
Sandia National Laboratories
Livermore, CA 94551

Dr. Aniket Bhattacharya
Michigan State University
Department of Physics and Astronomy
202A Physics - Astronomy Bldg.
E. Lansing, MI 48824-1116
aniket@rigid1.pa.msu.edu

Branislav Blagojevic
Michigan State University
Department of Physics and Astronomy
350 Giltner Hall
E. Lansing, MI 48824
blagojevic@msupa.pa.msu.edu

Dr. John Blendell
NIST
Building 223, Ceramics Department
Mail Room A215
Gaithersberg, MD 20899

Professor H.P. Bonzel
Forschungszentrum Jülich
IGV
D-52425 Jülich
GERMANY

Lothar Brendel
HLRZ
Forschungzentrum Julich/RFA
Julich, 52425
GERMANY
l.brendel@kfa-juelich.de

Dr. David Cahill
Materials Science Department
University of Illinois
1101 W. Springfield
Urbana, IL 61801
d-cahill@uiuc.edu

Dr. Dominique Chatain
CRMC2-CNRS
Campus de Luminy
Case 1913
13288 Marseille Cedex 09,
FRANCE
chatain@crmc2.univ-cnrs.fr

Professor Mei-Yin Chou
School of Physics
Georgia Institute of Technology
Atlanta, GA 30332
ph279mc@prism.gatech.edu

Professor Philip Clapp
University of Connecticut
Institute of Materials Science
Edward V. Gant Building
Storrs, CT 06269-3136

Professor Roy Clarke
Department of Physics
Randall Lab
University of Michigan
Ann Arbor, MI 48109
clarke@umiphys

Dr. Jim De Yoreo
Lawrence Livermore National Laboratory
University of California
P.O. Box 808
Livermore, CA 94551-9900

Diana C. Dismus-Campbell
Rutgers University
Department of Mathematics
95 Hollywood Ave.
Somerset, NJ 08873
dismus@math.rutgers.edu

Professor Phillip Duxbury
Michigan State University
Department of Physics and Astronomy
Physics - Astronomy Bldg., Rm 205
E. Lansing, MI 48824-1116
duxbury@pa.msu.edu

Professor Charles Ebner
Ohio State University
Department of Physics
174 West Eighteenth Avenue
Columbus, OH 43210
ebner@mps.ohio-state.edu

Professor T. L. Einstein
University of Maryland
Department of Physics
College Park, MD 20742
einstein@surface.umd.edu

Jonah Elebacher
Harvard University
Applied Physics
Lyman Lab 324A
Cambridge, MA 02138

Professor J. Frenken
FOM-Institute for Atomic & Molecular Physics
Surface Crystallography
Kruislaan 407
Amsterdam, NL-1098 SJ
NETHERLANDS
frenken@fkserv.amolf.nl

Professor Eric Ganz
Department of Physics
University of Minnesota
116 Church Street SE
Minneapolis, MN 55455
ganz@physics.spa.umn.edu

Dr. M. Giesen
Forschungszentrum Jülich GmbH
Institut fur Grenzflachenforschung
und Vakuumphysik
D-52425 Jülich,
GERMANY

Dr. George Gilmer
Lucent Technologies
600 Mountain Ave
Murray Hill, NJ 07974
ghg@allwise.att.com

Professor Gunter Gottstein
Inst. fur Metallkunde und Metallphysik
RWTH Aachen
52056 Aachen
GERMANY
gg@hpl.imm.rwth-Aachen.de

Pekka Heino
Tampere University of Technology
Deptartment of Electrical Engineering
P.O. Box 692
Tampere, 33101
FINLAND
pah@ele.tut.fi

Dr. Elizabeth Holm
Department of Material Science
Sandia National Laboratory
P.O. Box 5800, MS 1411
Alburquerque, 87185-1411
NEW MEXICO
eaholm@sandia.gov

Professor Harald Ibach
Forschungszentrum Jülich GmbH
Institut fur Grenzflachenforschung
und Vakuumphysik
D-52425 Jülich,
GERMANY
ibach@igv016.igv.kfa-juelich.de

Dr. D.E. Jesson
Oak Ridge National Lab
Solid State Division, Martin Marietta Energy Sys.
P.O. Box 2008
Oak Ridge, TN 37831-6033
ejp@ornl.gov

Professor D. Kandel
c/o E. Kaxiras
Harvard University
Dept. of Physics, Lyman Lab 324A
Cambridge, MA 02138

Professor Efthimios Kaxiras
Department of Physics
Harvard University
Lyman Lab 324A
Cambridge, MA 02138

Dr. M. Jeannette Kelley
Mathematics Department, MS 3368
Texas A & M University
College Station, TX 77843-3368
mkelley@math.rutgers.edu

Dr. Sergei E. Kiriukhin
Georgia Institute of Technology
Department of Physics
Atlanta, GA 30332
gt0754a@priom.gatech.edu

Professor Mohan Krishnamurthy
Michigan Tech
Department of Metallurgical and Materials Engineer
1400 Townsend Drive
Houghton, MI 49931
mohan@mtu.edu

Professor Jong K. Lee
Michigan Technological University
Dept. of Metallurgical and Materials Engineering
1400 Townsend Drive
Houghton, MI 49931-1295
JKL103@MTU.EDU

Juitao Li
Institut de Physique Experimental
Universite de Lausame
Lausanne, 1015
SWITZERLAND
jli@eliot.unil.ch

Xingquan Li
Randall Lab
Physics Department
University of Michigan
Ann Arbor, MI 48109-1120
xli@umich.edu

Maxim Makeev
University of Notre Dame
Department of Physics
Notre Dame, IN 46556

Professor Simon Mochrie
Room 13-2010
Massachusetts Institute of Technology
77 Massachusetts Avenue
Cambridge, MA 02139
simon@lindy.mit.edu

Professor Frank Morgan
Department of Mathematics
Williams College
Williamstown, MA 01267
frank.morgan@williams.edu

Dr. Frieder Mugele
Universitat Konstanz
Fak fur Physik/Ls Leiderer
Konstanz, 78434
GERMANY
frieder.mugele@uni-konstanz.de

Professor W.W. Mullins
Materials Science and Engineering Department
Carnegie-Mellon University
Pittsburgh, PA 15213
ww06+@andrew.cmu.edu

Dr. M.V. Ramana Murty
Cornell University
Department of Physics
Ithaca, NY 14853
murty@msc.cornell.edu

Dr. Cheng-Hsun Nien
Rutgers University
Physics Department
P.O. Box 849
Piscataway, NJ 08855-0849
nien@physics.rutgers.edu

Professor Bradford Orr
University of Michigan
Department of Physics
Ann Arbor, MI 48109-1120

Kalman Pelhos
Rutgers University
Department of Physics and Astronomy
P.O. Box 849
Piscataway, NJ 08855-0849
pelhos@physics.rutgers.edu

Professor John Pelz
Department of Physics
Ohio State University
174 W. 18th Avenue
Columbus, OH 43210
jpelz@mps.ohio-state.edu

Professor Thomas Pence
Michigan State University
Materials Science and Mechanics
E. Lansing, MI 48824

Professor Janet Rankin
Department of Engineering
Brown University
Box D
Providence, RI 02912
ellie@en3nl.engin.brown.edu

Professor Chris Roland
North Carolina State University
Department of Physics
Box 8202
Raleigh, NC 27695
roland@ajax.physics.ncsu.edu

Professor Leonard Sander
University of Michigan
Department of Physics
Ann Arbor, MI 48109-1120

Michael Schroeder
Theoretische Physik
GM-Universitat Duisburg
Duisburg, 47048
GERMANY
schroed@comphys.uni-duisourg.de

Professor W. Selke
Institut fur Theoretische physik
Technische Hochschule
D-52056 Aachen,
GERMANY

Professor Roni Shneck
University of Virginia
Department of Materials Science and Engineering
Charlottesville, VA 22903
rs2p@virginia.edu

Ellak Somfai
University of Michigan
Department of Phyiscs
Ann Arbor, MI 48109-4420
ellak@umich.edu

Dr. Bruce Taggart
National Science Foundation
4201 Wilson Boulevard
Arlington, VA 22230

Dr. L.H. Tang
Blackett Lab., Condensed Matter Theory Group
Imperial College
Prince Consort Road
London, SW7 2BZ
UNITED KINGDOM
l.h.Tang@ic.ac.uk

Professor Jean Taylor
83 Riverside Drive
Princeton, NJ 08540
taylor@math.rutgers.edu

Dr. Jerry Tersoff
IBM T.J. Watson Research Center
P.O. Box 218
Yorktown Heights, NY 10598
tersoff@watson.ibm.com

Professor L. Truskinovsky
University of Minnesota
107 Akerman Hall
110 Union St. Southeast
Minneapolis, MN 55455
trusk@aem.umn.edu

Professor Ignatius Tsong
Department of Physics
Arizona State University
Tempe, AZ 85287
tsong@phyast.la.asu.edu

Dr. Christopher Umbach
Cornell University
Department of Materials Science and Engineering
359 Band Hall
Ithaca, NY 14853
umbach@msc.cornell.edu

Professor H. van Beijeren
Utrecht University
Institute of Theoritical Physics
Postbus 80 006
NL-3508 TA Utrecht,
NETHERLANDS
h.van.beijeren@fys.tuu.nl

Professor Peter Voorhees
Department of Material Science
Northwestern University
Evanston, IL 60208
p-voorhees@nwu.edu

Professor John Weeks
IPST
University of Maryland
College Park, MD 20742
jdw@ella.umd.edu

Dr. Walter L. Winterbottom
30106 Pipers Lane Court
Farmington Hills, MI 48334

Frank Worthington
Michigan State University
Materials Science and Mechanics
E. Lansing, MI 48824

Professor Paul Wynblatt
Materials Science and Engineering Department
Carnegie-Mellon University
Pittsburgh, PA 15213

Professor Steven M. Yalisove
University of Michigan
Dept. of Mat. Sci. and Eng.
2300 Hayward Street
Ann Arbor, MI 48103
smy@umich.edu

Professor Andrew Zangwill
School of Physics
Georgia Institute of Technology
Atlanta, GA 30332-0430

Dr. S.B. Zhang
National Renewable Energy Lab
1617 Cole Boulevard
Golden, CO 80401
zhangs@sst.nrel.gov

Zili Zhang
2071 Randall lab
500 E. University
Ann Arbor, MI 48109-1120
zili@umich.edu

INDEX